Geographically Weighted Regression

Geographically Weighted Regression

the analysis of spatially varying relationships

A. Stewart Fotheringham
Chris Brunsdon
Martin Charlton
University of Newcastle, UK

JOHN WILEY & SONS, LTD

The Atrium, Southern Gate, Chichester,
West Sussex PO19 8SQ, England

Telephone (+44) 1243 779777
Email (for orders and customer service enquiries):
cs-books@wiley.co.uk
Visit our Home Page on www.wileyeurope.com
or www.wiley.com

Other Wiley Editorial Offices

John Wiley & Sons Inc., 111 River Street,
Hoboken, NJ 07030, USA

Jossey-Bass, 989 Market Street,
San Francisco, CA 94103-1741, USA

Wiley-VCH Verlag GmbH, Boschstr. 12,
D-69469 Weinheim, Germany

John Wiley & Sons Australia Ltd, 33 Park Road, Milton,
Queensland 4064, Australia

John Wiley & Sons (Asia) Pte Ltd, 2 Clementi Loop #02-01,
Jin Xing Distripark, Singapore 129809

John Wiley & Sons Canada Ltd, 22 Worcester Road,
Etobicoke, Ontario, Canada M9W 1L1

British Library Cataloguing in Publication Data

A catalogue record for this book is available from the British Library

ISBN 0-471-49616-2

Typeset in 10/12pt Times by Kolam Information Services Pvt. Ltd, Pondicherry, India
Printed and bound in Great Britain by Antony Rowe Ltd, Chippenham, Wiltshire
This book is printed on acid-free paper responsibly manufactured from sustainable forestry in which at least two trees are planted for each one used for paper production.

ASF: *To Barbara, Iain and Neill*
CB: *To Francis and Anne*
MEC: *To Ted and Avril*

Contents

Acknowledgements

The UK maps in this book are based on copyright digital map data owned and supplied by Bartholomew Ltd and are used with permission. Some of the maps are also based on census data provided with the support of the ESRC and JISC and use boundary material which is the copyright of the Crown and the ED-LINE Consortium. The US census data and boundaries were obtained from CensusCD+Maps, a product of Geolytics Inc. The authors are grateful for the enlightened attitude of the US Government in making spatial data relatively freely available.

Throughout the book we make extensive use of house price data that has been supplied by the Nationwide Building Society to the University of Newcastle upon Tyne and we are extremely grateful for their generosity.

Dr Robin Flowerdew of the Department of Geography at the University of St Andrews supplied the school performance data used in Chapter 5 as part of a conference on local modelling with spatial data.

A number of people also deserve credit for assisting with various aspects of this book. Ann Rooke applied her usual superlative cartographic skills to some of the figures. Stamatis Kalogirou wrote the Visual Basic front end to the GWR software and Barbara Fotheringham did a very professional job of helping to proofread the manuscript.

We would also like to thank Sally Wilkinson, Lyn Roberts and Keily Larkins at John Wiley & Sons, Ltd for their encouragement, patience, assistance and good nature during the various evolutionary stages of this book. Further, we acknowledge a great debt to the reviewers of both the initial book proposal and an earlier version of the finished product for their strong support and useful insights. The book is far better for their comments.

Finally, we make the usual disclaimer that any errors remaining in the book are the sole responsibility of the authors – apologies for not catching them all!

This publication contains maps based on copyright digital map data owned and supplied by Bartholomew Ltd and is used with permission.

This applies to: Figures 2.1, 2.2, 2.4, 2.9, 2.12, 2.14–2.19, 5.1, 6.2–6.7, 7.1–7.6, 7.8–7.12, 8.1–8.3. Data taken from Bartholomews.

Figures 2.1, 2.2, 2.9, 2.12, 2.13–2.19, 5.1. Data taken from Bartholomews and UKBorders.

Figures 2.2, 5.1. Data taken from Bartholomews, UKBorders and Nationwide Building Society.

Maps are based on data provided with the support of the ESRC and JISC and use boundary material which is copyright of the Crown and the EDLINE consortium.

This applies to: Figures 2.1–2.3, 2.5–2.7, 2.9, 2.12, 2.14–2.19, 3.1, 3.3–3.5, 3.8, 3.9, 3.11, 3.12, 5.1–5.4, 5.6–5.10, 6.8–6.13.

Table 2.1 is calculated from data supplied by the Nationwide Building Society to the University of Newcastle upon Tyne.

1

Local Statistics and Local Models for Spatial Data

1.1 Introduction

Imagine reading a book on the climate of the United States which contained only data averaged across the whole country, such as mean annual rainfall, mean annual number of hours of sunshine, and so forth. Many would feel rather short-changed with such a lack of detail. We would suspect, quite rightly, that there is a great richness in the underlying data on which these averages have been calculated; we would probably want to see these data, preferably drawn on maps, in order to appreciate the spatial variations in climate that are hidden in the reported averages. Indeed, the averages we have been presented with may be practically useless in telling us anything about climate in any particular part of the United States. It is known, for instance, that parts of the north-western United States receive a great deal more precipitation than parts of the Southwest and that Florida receives more hours of sunshine in a year than New York. In fact, it might be the case that not a single weather station in the country has the characteristics depicted by the mean climatic statistics.

The average values in this scenario can be termed global observations: in the absence of any other information, they are assumed to represent the situation in every part of the study region. The individual data on which the averages are calculated can be termed local observations: they describe the situation at the local level.[1]

[1] There is at least one other slightly different definition of 'local' and 'global' in the literature. Thioulouse *et al.* (1995) define a local statistic as one which is calculated on pairs of points or areas which are adjacent and a global statistic as one calculated over all possible pairs of points or areas. Their use of the term 'local', however, is not the same as used throughout this book because it still produces a global model; it merely separates the model applications into different spatial regimes.

Only if there is little or no variation in the local observations do the global observations provide any reliable information on the local areas within the study area. As the spatial variation of the local observations increases, the reliability of the global observation as representative of local conditions decreases.

While the above scenario might appear rather ludicrous (surely no one would publish a book containing average climatic data without describing at least some of the local data?), consider a second scenario which is much more plausible and indeed describes a methodology which is exceedingly common in spatial analysis. Suppose we had data on house prices and their determinants across the whole of England and that we wanted to model house price as a function of these determinants (such models are often referred to as hedonic price models and an example of the calibration of these models is provided in Chapter 2). Typically, we might run a regression of house prices on a set of structural attributes of each house, such as the age and floor area of the house; a set of neighbourhood attributes, such as crime rate or unemployment rate; and a set of locational attributes, such as distance to a major road or to a certain school. The output from this regression would be a set of parameter estimates, each estimate reflecting the relationship between house price and a particular attribute of the house. It would be quite usual to publish the results of such an analysis in the form of a table describing the parameter estimates for each attribute and commenting on their sign and magnitude, possibly in relation to some *a priori* set of hypotheses. In fact this is the standard approach of the vast majority of empirical analyses of spatial data.

However, the parameter estimates in this second scenario are *global statistics* and are possibly just as inadequate at representing local conditions as are the average climatic data described above. Each parameter estimate describes the *average* relationship between house price and a particular attribute of the house across the study region (in this case, the whole of England). This average relationship might not be representative of the situation in any particular part of England and may hide some very interesting and important local differences in the determinants of house prices. For example, suppose one of the determinants of house prices in our model is the age of the house and the global parameter estimate is close to zero. Superficially this would be interpreted as indicating that house prices are relatively independent of the age of the property. However, it might well be that there are contrasting relationships in different parts of the study area which tend to cancel each other out in the calculation of the global parameter estimate. For example, in rural parts of England, old houses might have character and appeal, thus generating higher prices than newer houses, *ceteris paribus*, whereas in urban areas, older houses, built to low standards to house workers in rapidly expanding cities at the middle of the nineteenth century, might be in poor condition and have substantially lower prices than newer houses. This local variation in the relationship between house price and age of the house would be completely lost if all that is reported is the global parameter estimate. It would be far more informative to produce a set of *local statistics*, in this case local parameter estimates, and to map these than simply to rely on the assumption that a single global estimate will be an accurate representation of all parts of the study area.

The only difference between the examples of the US climate and English house prices presented above is that the first describes the representation of spatial data, whereas the second describes the representation of spatial relationships. It would seem that while we generally find it unhelpful to report solely global observations on spatial data, we are quite happy to accept global statements of spatial relationships. Indeed, as hinted at above, journals and textbooks in a variety of disciplines dealing with spatial data are filled with examples of global forms of spatial analysis. Local forms of spatial analysis or spatial models are very rare exceptions to the overwhelming tide of global forms of analysis that dominates the literature.

In this book, through a series of examples and discussions, we hope to convince the reader of the value of local forms of spatial analysis and spatial modelling, and in particular, the value of one form of local modelling which we term *Geographically Weighted Regression (GWR)*. We hope to show that in many instances undertaking a global spatial analysis or calibrating a global spatial model can be as misleading as describing precipitation rates across the USA with a single value.

1.2 Local Aspatial Statistical Methods

Spatial data contain both attribute and locational information: aspatial data contain only attribute information. For instance, data on the manufacturing output of firms graphed against the number of their employees are aspatial, whereas the numbers of people suffering from a certain type of disease in different parts of a country are spatial. Unemployment rates measured for one location over different time periods are aspatial but unemployment rates at different locations are spatial and the spatial component of the data might be very useful in understanding why the rates vary. The difference between aspatial and spatial data is important because many statistical techniques developed for aspatial data are not valid for spatial data. The latter have unique properties and problems that necessitate a different set of statistical techniques and modelling approaches (for more on this, see Fotheringham *et al.* 2000, particularly Chapter 2). This is also true in local analysis.

There is a growing literature and an expanding array of techniques for examining local relationships in aspatial data. For example, there are techniques such as the use of spline functions (Wahba 1990; Friedman 1991; Green and Silverman 1994); LOWESS regression (Cleveland 1979); kernel regression (Cleveland and Devlin 1988; Wand and Jones 1995; Fan and Gijbels 1996; Thorsnes and McMillen 1998); and variable parameter models in the econometric literature (Maddala 1977; Johnson and Kau 1980; Raj and Ullah 1981; Kmenta 1986; Casetti 1997) that are applicable to the local analysis of aspatial data. Good general discussions of local regression techniques for aspatial data are given by Hardle (1990), Barnett *et al.* (1991), Loader (1999) and Fox (2000a; 2000b).

The basic problem that local statistics attempt to solve is shown in Figure 1.1. Here there is a relationship between two aspatial variables, Y and X, which needs to be determined from the observed data. A global linear regression model, for example, would produce a relationship such as that depicted by line A; although the model gives a reasonable fit to the data, it clearly misses some important local variations in the relationship between Y and X. Here, notice, 'local' means in terms of attribute space, in this case that of the X variable, rather than geographical

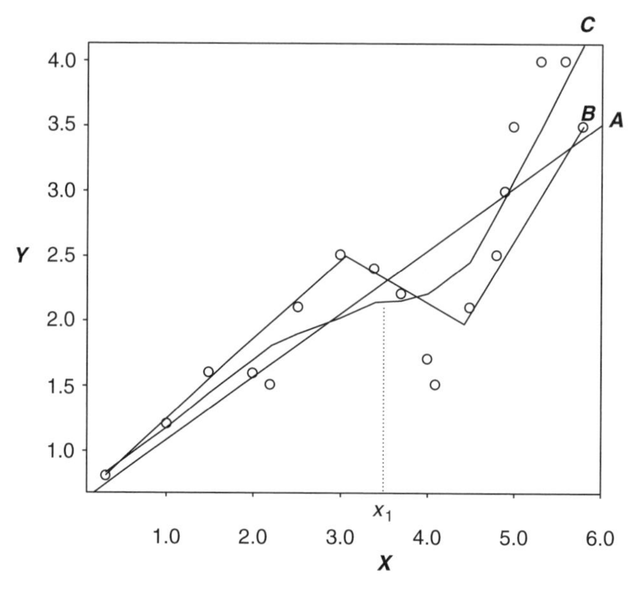

Figure 1.1 *Global and local aspatial relationships*

space.[2] A local technique, such as a linear spline function, depicted by line B, would give a more accurate picture of the relationship between Y and X. This would be obtained by essentially running four separate regressions over different ranges of the X variable with the constraint that the end points of the local regression lines meet at what are known as 'knots'.[3] Finally, a very localised technique such as LOWESS regression would yield line C where the relationship between Y and X is evaluated at a large number of points along the X axis and the data points are weighted according to their 'distance' from each of these regression points.[4] For example, suppose the regression point were at x_1. Then the data points for the regression of Y on X would be weighted according to their distance from the point x_1 with points closer to x_1 being weighted more heavily than points further away. This weighted regression yields a local estimate of the slope parameter for the relationship between Y and X. The regression point is moved along the X axis in small intervals until a line such as that in C can be constructed from the set of local parameter estimates.

[2] For something of a hybrid application of local modelling the reader is referred to McMillan (1996) in which land values in Chicago are regressed on distance to various features within the city. Although this is essentially an aspatial model because the local regressions are calibrated only in attribute space and not in geographical space, the use of distance as an independent variable does allow a spatial interpretation of the results to be made. As such, McMillan's application can be thought of as 'semi-spatial'.

[3] Although a linear spline function is depicted in this example, cubic spline functions are often used in curve fitting exercises. The linear spline is shown here to distinguish it from the LOWESS fit.

[4] The terms LOWESS and LOESS are used interchangeably in the literature; use is based on personal preference.

The difference between applying local techniques to aspatial data and to spatial data is that the relationship between Y and X, as shown in Figure 1.1 might vary depending on the location at which the regression is undertaken. That is, instead of simply having the problem of fitting a non-linear function to a set of data, this non-linear function itself might vary over space as shown for two locations in Figure 1.2. Consequently, local statistical analyses for spatial data have to cope

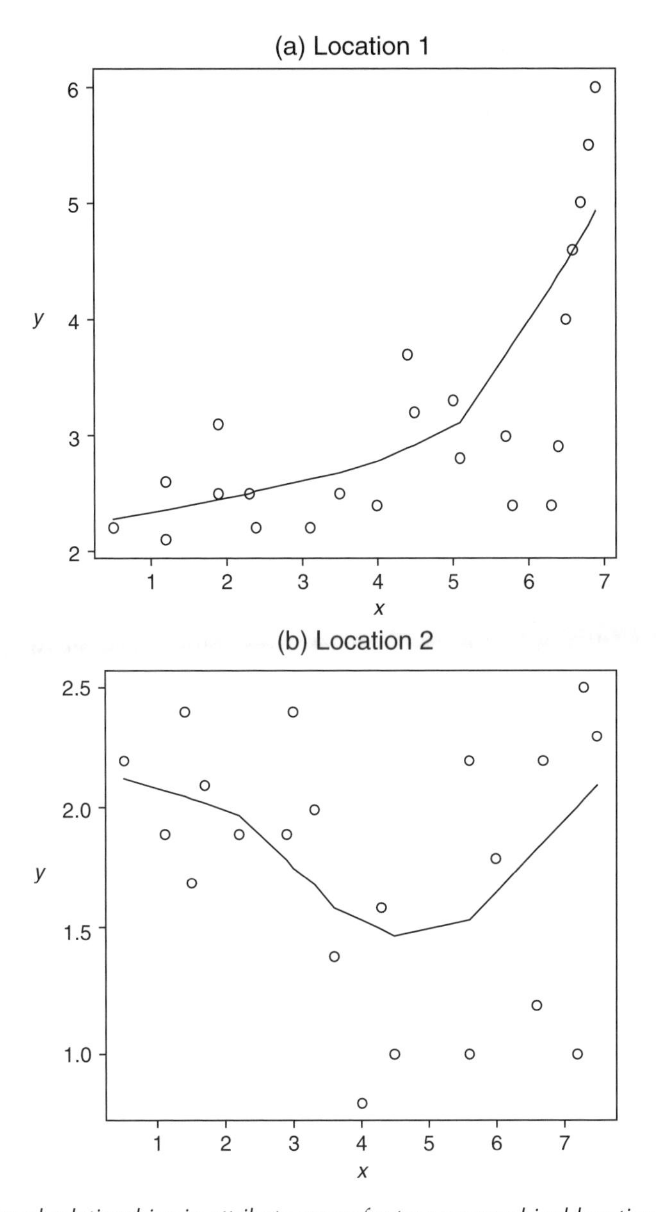

Figure 1.2 *Local relationships in attribute space for two geographical locations*

with two potential types of local variation: the local relationship being measured in attribute space and the local relationship being measured in geographical space. Compounding the problem of measuring spatial variations in relationships is the fact that the relationships in geographical space can vary in two dimensions rather than just in one. That is, local variations in attribute space, such as those shown in Figure 1.1, take place along a line and the dependency between relationships is easier to establish than in the two-dimensional equivalent of geographical space.

Because local statistical techniques for aspatial data are already fairly well established and because such techniques do not always translate easily to spatial data, the remainder of this book concentrates almost exclusively on the local analysis of spatial data. Henceforth, any discussion of local analysis is assumed to refer to spatial data unless otherwise stated.

1.3 Local versus Global Spatial Statistics

Local statistics are treated here as spatial disaggregations of global statistics. For instance, the mean rainfall across the USA is a global statistic; the measured rainfall in each of the recording stations, i.e. the data from which the mean is calculated, represent the local statistics. A model calibrated with data equally weighted from across a study region is a global model that yields global parameter estimates. A model calibrated with spatially limited sets of data is a local model that yields local parameter estimates. Local and global statistics differ in several respects as shown in Table 1.1.

Global statistics are typically single-valued: examples include a mean value, a standard deviation and a measure of the spatial autocorrelation in a data set. Local statistics are multi-valued: different values of the statistic can occur in different locations within the study region. Each local statistic is a measure of the attribute or the relationship being examined *in the vicinity of* a location within the study

Table 1.1 *Differences between local and global statistics*

Global	Local
Summarise data for whole region	Local disaggregations of global statistics
Single-valued statistic	Multi-valued statistic
Non-mappable	Mappable
GIS – unfriendly	GIS – friendly
Aspatial or spatially limited	Spatial
Emphasise similarities across space	Emphasise differences across space
Search for regularities or 'laws'	Search for exceptions or local 'hot-spots'
Example:	*Example:*
Classic Regression	*Geographically Weighted Regression (GWR)*

region: as this location changes, the local statistic can take on different values.[5] Consequently, global statistics are unmappable or 'GIS-unfriendly', meaning they are not conducive to being analysed within a Geographic Information System (GIS) because they consist of a single value. Local statistics, on the other hand, can be mapped and further examined within a GIS. For instance, it is possible to produce a map of local parameter estimates showing how a relationship varies over space and then to investigate the spatial pattern of the local estimates to establish some understanding of possible causes of this pattern. Indeed, given that very large numbers of local parameter estimates can be produced, it is almost essential to map them in order to make some sense of the pattern they display. Local statistics are therefore spatial statistics whereas global statistics are aspatial or spatially limited.

By their nature, local statistics emphasise differences across space whereas global statistics emphasise similarities across space.[6] Global statistics lead one into thinking that all parts of the study region can be accurately represented by a single value whereas local statistics can show the falsity of this assumption by depicting what is actually happening in different parts of the region. Consequently, local statistics are useful in searching for exceptions or what are known as local 'hot spots' in the data. This use places them in the realm of exploratory spatial data analysis where the emphasis is on developing hypotheses from the data, as opposed to the more traditional confirmatory types of analysis in which the data are used to test *a priori* hypotheses (Unwin and Unwin 1998; Fotheringham *et al.* 2000). It also suggests the techniques are not rooted fully in the positivist school of thought where the search for global models and 'laws' is important. However, this issue is not as clear-cut as it might seem because local statistics can also play an important role in confirmatory analyses as well as in building more accurate global models, a point expanded upon below.

The extent to which global estimates of relationships can present very misleading interpretations of local relationships is shown in Figure 1.3, a spatial example of Simpson's Paradox (Simpson 1951).[7] Simpson's paradox refers to the reversal of results when groups of data are analysed separately and then combined. In the spatial example presented in Figure 1.3, data are plotted showing the relationship between the price of a house and the population density of the area in which the house is located. In Figure 1.3(a) data from more than one location are aggregated

[5] This is the case even for statistics which measure the degree to which observations vary, such as a standard deviation, or the degree to which they covary, such as a covariance. A standard deviation presents a global average degree of variation in the data; it supplies no information on whether the degree of variation in the data varies spatially. For example, in some parts of the region, the data could be very stable, whereas in other parts, the data might vary wildly. A similar statement can be made for covariance. The traditional measure of covariance is a global statistic because it measures the degree of covariance between two variables averaged over a region. One could produce a local covariance measure that describes how the covariance between two variables differs across the region; in some areas, the two variables might exhibit considerable covariation, whereas in others the covariance might be negligible.

[6] Again, this statement is true even with statistics that measure the degree to which data vary over space. In such cases it is the degree of variation that is measured globally and we are led into thinking that this degree of variation is constant over space when in fact it might not be.

[7] For an example of Simpson's Paradox in aspatial data, see Appleton *et al.* (1996).

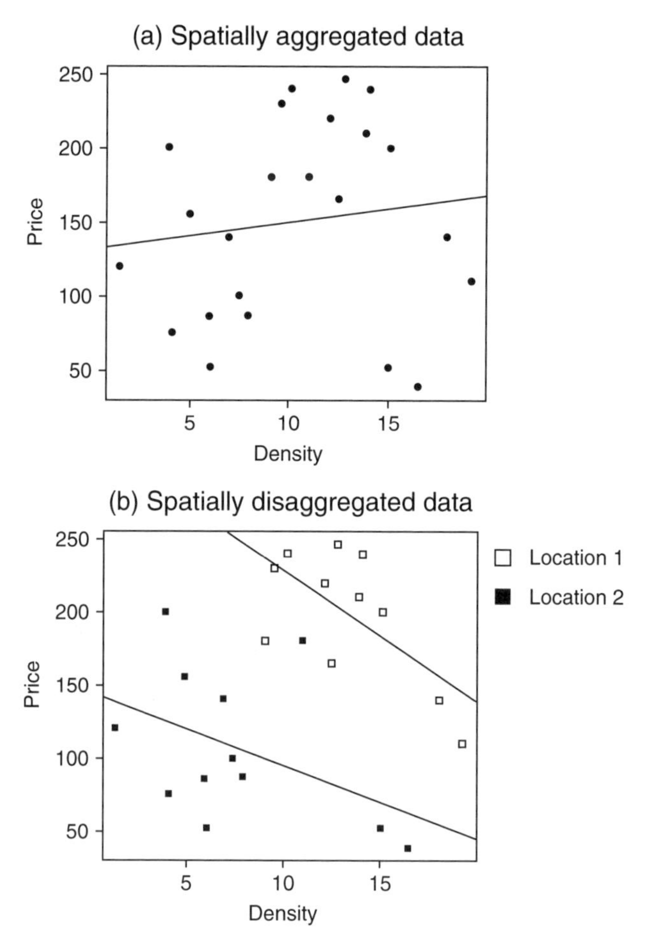

Figure 1.3 *A spatial example of Simpson's Paradox*

and the relationship, shown by the included linear regression line, is a positive one which suggests that house prices rise with increasing population density. However, in Figure 1.3(b) the data are separated by location and in both locations the relationship between house price and population density is a negative one. That is, for both individual locations, there is a negative relationship between house price and density but when the data from the two locations are aggregated, the relationship appears to be a positive one. Simpson's Paradox highlights the dangers of analysing aggregate data sets. Whilst it is normally demonstrated in aspatial data sets where the aggregation is over population subgroups, the paradox applies equally to spatial data where the aggregation is over locations.

1.4 Spatial Non-stationarity

Social scientists have long been faced with a difficult question and a potential dilemma: are there any 'laws' that govern social processes, and if there are not, does a quantitative approach have any validity? The problem is more clearly seen as two sub-problems. The first is that models in social sciences are not perfectly accurate. There is always some degree of error (sometimes quite large) indicating that a model has not captured fully the process it is being used to examine. We continually strive to produce more accurate models but the goal of a perfect model is elusive. The second is that the results derived from one system can rarely, if ever, be replicated exactly in another. An explanatory variable might be highly relevant in one application but seemingly irrelevant in another; parameters describing the same relationship might be negative in some applications but positive in others; and the same model might replicate data accurately in one system but not in another. These issues set social science apart from other sciences where the goal of attaining a global statement of relationships is a more realistic one. Physical processes tend to be *stationary* whereas social processes are often not. For instance, in physics, the famous relationship relating energy and mass, $E = mc^2$, is held to be the same no matter where the measurement takes place: there is not a separate relationship depending on which country or city you are in.[8] Social processes, on the other hand, appear to be *non-stationary*: the measurement of a relationship depends in part on where the measurement is taken. In the case of spatial processes, we refer to this as *spatial non-stationarity*. In essence, the process we are trying to investigate might not be constant over space. Clearly, any relationship that is not stationary over space will not be represented particularly well by a global statistic and, indeed, this global value may be very misleading locally. It is therefore useful to speculate on why relationships might vary over space; in the absence of a reason to suspect that they do vary, there is little or no need to develop local statistical methods.

There are several reasons why we might expect measurements of relationships to vary over space. An obvious one relates to sampling variation. Suppose we were to take spatial subsets of a data set and then calibrate a model separately with each of the subsets. We would not expect the parameter estimates obtained in such calibrations to be *exactly* the same: variations would exist because of the different samples of data used. This variation is relatively uninteresting in that it relates to a statistical artefact and not to any underlying spatial process, but it does need to be accounted for in order to identify more substantive causes of spatial non-stationarity.

A second possible cause of observed spatial non-stationarity in relationships is that, for whatever reasons, some relationships are intrinsically different across space. Perhaps, for example, there are spatial variations in people's attitudes or preferences or there are different administrative, political or other contextual issues

---

[8] Even with this equation there is a controversy over whether the speed of light is actually a constant everywhere. However, the argument is only about extreme conditions not met in any practical circumstances and the argument has far from universal acceptance.

that produce different responses to the same stimuli over space. Contextual effects appear to be well documented, for example, in studies of voting behaviour as evidenced by, *inter alia*, Cox (1969); Agnew (1996) and Pattie and Johnston (2000). The idea that human behaviour can vary intrinsically over space is consistent with post-modernist beliefs on the importance of place and locality as frames for understanding such behaviour (Thrift 1983). Within this framework the identification of local variations in relationships would be a useful precursor to more intensive studies that might highlight why such differences occur.

A third possible cause of observed spatial non-stationarity is that the model from which the relationships are estimated is a gross misspecification of reality and that one or more relevant variables are either omitted from the model or are represented by an incorrect functional form. This view, more in line with the positivist school of thought and very much in line with that in econometrics, runs counter to that discussed above: it assumes a global statement of behaviour can be made but that the structure of the model is not sufficiently well formed to allow this global statement to be made. Within this framework, mapping local statistics is useful in order to understand more clearly the nature of the model misspecification. The spatial pattern of the measured relationship can provide a good clue as to what attribute(s) might have been omitted from the model and what might therefore be added to the global model to improve its accuracy. For example, if the local parameter estimates for a particular relationship tend to have different signs for rural and urban areas, this would suggest the addition of some variable denoting the 'urban-ness' or the 'rural-ness' of an area. In this sense, local analysis can be seen as a model-building procedure in which the ultimate goal is to produce a global model that exhibits no significant spatial non-stationarity. In such instances, the role of local modelling is essentially that of a diagnostic tool which is used to indicate a problem with the global model; only when there is no significant spatial variation in measured relationships can the global model be accepted.

Alternatively, it might not be possible to reduce or remove the misspecification problem with the global model by the addition of one or more variables: for example, it might be impossible to collect data on such variables. In such a case, local modelling then serves the purpose of allowing these otherwise omitted effects to be included in the model through locally varying parameter estimates.

The above discussion on the possible causes of spatial non-stationarity raises an interesting and, as yet unsolved puzzle in spatial analysis. If we do observe spatial variations in relationships, are they due simply to model misspecification or are they due to intrinsically different local spatial behaviour? In a nutshell, can all contextual effects be removed by a better specification of our models (Hauser 1970; Casetti 1997)? Is the role of place simply a surrogate for individual-level effects which we cannot recognise or measure? If the nature of the misspecification could be identified and corrected, would the local variations in relationships disappear? We can only speculate on whether, if one were to achieve such a state, all significant spatial variations in local relationships would be eliminated (see also Jones and Hanham 1995 for a useful discussion on this and the role of local analysis in both realist and positivist schools of thought). We can never be completely confident that our models are correct specifications of reality because of our lack of theoretical understanding of the processes governing human spatial behaviour. In some ways, this is a chicken-and-egg dilemma. We can never completely test theories of spatial behaviour because of model misspecification, but model misspecification is the product of inadequate spatial theory.

However, the picture is not so bleak: in specific applications of any form of spatial model, we can ask whether the current form of the model we are using produces significant local variations in any of the relationships in which we are interested. If the answer is 'yes', then an examination of the nature of the spatial variation can suggest to us a more accurate model specification or the nature of some intrinsic variation in spatial behaviour. In either case, our knowledge of the system under investigation will be improved, in some cases dramatically.

Given the potential importance of local statistics and local models to the understanding of spatial processes, it is surprising that local forms of spatial analysis are not more frequently encountered. However, there have been some notable contributions to the literature on spatially varying parameter models that we now describe. These developments can be divided into three categories: those that are focussed on local statistics for univariate spatial data, including the analysis of point patterns; those that are focussed on more complex multivariate spatial data; and those that are focussed on spatial patterns of movement. We now describe some of the literature on local models and local statistics prior to a full description in Chapter 2 of one local modelling technique, Geographically Weighted Regression, that forms the focus of this book.

1.5 Examples of Local Univariate Methods for Spatial Data Analysis

Four types of local univariate analysis for spatial data can be identified. These are: local forms of point pattern analysis; local graphical analysis; local filters; and local measures of spatial dependency.

1.5.1 Local Forms of Point Pattern Analysis

Many data, such as the locations of various facilities, or the incidence of a particular disease, consist of a set of geocoded points that make up a spatial point pattern. The analysis of spatial point patterns has long been an important concern in geographical enquiry (*inter alia*, Getis and Boots 1978; Boots and Getis 1988). Traditionally, most methods of spatial point pattern analysis, such as quadrat analysis and neighbour statistics, have involved the calculation of a global statistic that describes aspects of the whole point pattern (*inter alia* Dacey 1960; King 1961; Tinkler 1971; Boots and Getis 1988). From this global analysis, a judgement would be reached as to whether the overall pattern of points was clustered, dispersed or random. Clearly, such analyses are potentially flawed because interesting spatial variations in the point pattern might be subsumed in the calculation of the average or global statistic. In many instances, particularly in the study of disease, such an approach would appear to be contrary to the purpose of the study, which is to identify any interesting local clusters of disease incidence (see, for example, Lin and Zeng 1999). Typically, we are not interested in some general statistic referring to the whole point pattern: it is more useful to be able to identify particular parts of the study region in which there is a raised incidence of the disease. Consequently, there has been a growing interest in developing local forms of point pattern analysis.

One of the first of these was the Geographical Analysis Machine (GAM) developed by Openshaw *et al.* (1987) and updated by Fotheringham and Zhan (1996). As Fotheringham and Brunsdon (1999) note, the basic components of a GAM are:

1. a method for defining sub-regions of the data;
2. a means of describing the point pattern within each of these sub-regions;
3. a procedure for assessing the statistical significance of the observed point pattern within each sub-region considered independently of the rest of the data;
4. a procedure for displaying the sub-regions in which there are significant patterns as defined in 3.

The basic idea outlined in Fotheringham and Zhan (1996) demonstrates the emphasis of this type of technique on identifying interesting local parts of the data set rather than simply providing a global average statistic. Within the study region containing a spatial point pattern, random selection is made initially of a location, and then of a radius of a circle to be centred at that location. Within this random circle, the number of points is counted and this observed value compared with an expected value based on an assumption about the process generating the point pattern (usually that it is random). The population-at-risk within each circle is then used as a basis for generating an expected number of points which is compared to the observed number. The circle can then be drawn on a map if it contains a statistically interesting count (that is, a much higher or lower observed count of points than expected). The process is repeated many times so that a map is produced which contains a set of circles centred on parts of the region where interesting clusters of points appear to be located. The GAM and similar statistics are a subset of a much broader class of statistics known as 'Scan Statistics' of which there are several notable spatial applications, particularly in the identification of disease clusters (*inter alia* Kulldorf and Nagarwalla 1995; Hjalmars *et al.* 1996; Kulldorf 1997; Kulldorf *et al.* 1997; Gangnon and Clayton 2001).[9]

1.5.2 Local Graphical Analysis

One of the by-products of the enormous increases in computer power that have taken place is the rise of techniques for visualising data (Fotheringham, 1999a; Fotheringham *et al.* 2000, Chapter 4). Within spatial data analysis, exploratory

[9] At the time of writing, software for calculating spatial, temporal and space-time scan statistics can be downloaded from http://dcp.nci.nih.gov/bb/SaTScan.html

graphical techniques which emphasise the local nature of relationships have become popular. For example, using software such as MANET (Unwin *et al.* 1996), or XLispstat (Tierney 1990; Brunsdon and Charlton 1996), it is possible to link maps of spatial data with other non-cartographical representations (such as scatterplots or dotplots). Selecting an object on one representation highlights the corresponding object on the other (an early example of this is Monmonier 1969). For example, if a scatterplot reveals a number of outlying observations, selecting these points will highlight their locations on a map. Similarly, selecting a set of points or zones on a map will highlight the corresponding points on a scatterplot. In this way, the spatial distribution of an attribute for a locally selected region can be compared to the distribution of the same attribute across the study area as a whole. Using techniques of this sort, combined with a degree of numerical pre-processing, it is possible to carry out a wide range of exploratory tasks on spatial data which are essentially local. For example, one can identify local clusters in data and investigate whether these are also associated with spatial clusters. Equally, one can also identify spatial outliers, cases that are *locally* unusual even if not atypical for the data set as a whole. More complex graphical techniques for depicting local relationships in univariate data sets include the spatially lagged scatterplot (Cressie 1984), the variogram cloud plot (Haslett *et al.* 1991) and the Moran scatterplot (Anselin 1996).

1.5.3 Local Filters

A number of techniques exist in image-processing that can be considered as 'local'. The data for an image is usually presented as a regular array of intensity values each value referring to a single cell of known area (or a pixel). In order to determine which pixels are likely to represent edges in the image, a high-pass filter can be applied; this acts to increase high-intensity values, and decrease low ones. To remove isolated high values, a low-pass filter can be employed; its action is to make the values in nearby pixels more similar. Other filters may be applied to enhance the values of linear objects in the image; these are known as directional filters. Such filters are usually a square array of weights, often 3×3 pixels. The output from a filter is a weighted mean value of the pixel at its centre and its immediate neighbours; the filter is applied to each pixel in the input image to produce an output image. The reader is referred to Lilliesand and Kiefer (1995) for further information on the use of filters in image processing.

These filtering techniques have also been applied to raster GIS data (i.e., data stored as a regular lattice). Tomlin (1990) proposed a wide variety of functions that can be applied to local neighbourhoods in such data. Examples of these include the 'focalmean', the 'focalmedian' and the 'focalvariety'. The focalmean function provides a weighted mean of the values in the raster which are immediate neighbours of the central one; in this way both high-pass and low-pass filters can be applied to raster GIS data. The focalmedian will return the median of the nine values in the surrounding 3×3 matrix (in some implementations the filter size can be varied). If the values in the raster are categorical (for example, they may represent land uses), then focalvariety will count the number of different values in the 3×3 matrix.

Some early examples of the use of filters for spatial analysis are contained in Schmid and MacCannell (1955) and Unwin (1981). More sophisticated examples are given by Cheng *et al.* (1996) who use variable window sizes and shapes for the local filtering of geochemical images. Rushton *et al.* (1995) apply a spatial filter to student enrolment projections.[10] A similar technique, popular in fields such as geodesy and meteorology, is that of optimal interpolation in which data weighted by spatial proximity are used to estimate unknown values (Liu and Gauthier 1990; Daley 1991, Reynolds and Smith 1995). The technique is also known as objective analysis (Cressman 1959).

1.5.4 Local Measures of Spatial Dependency

Spatial dependency is the extent to which the value of an attribute in one location depends on the values of the attribute in nearby locations. Although statistics for measuring the degree of spatial dependency in a data set have been formulated for almost three decades (*inter alia* Cliff and Ord 1972; Haining 1979), until very recently these statistics were only applied globally. Typically a single statistical measure is calculated which describes an overall degree of spatial dependency across the whole data set. Recently, however, local statistics for this purpose have been developed by Getis and Ord (1992), Ord and Getis (1995; 2001), Anselin (1995; 1998) and Rogerson (1999). Getis and Ord (1992), for example, develop a global measure of spatial association inherent within a data set that measures the way in which values of an attribute are clustered in space. A local variation of this global statistic is then formulated to depict trends in the data around each point in space. There are two variants of this localised value depending on whether or not the calculation includes the point i, around which the clustering is measured, although both are equivalent to spatially moving averages (Ord and Getis 2001). The local spatial association statistic allows that different trends in the distribution of one variable might exist over space. In some parts of the study area, for example, high values might be clustered; in other parts there might be a mix of high and low values. Such differences would not be apparent in the calculation of a single global statistic. In their empirical example, Getis and Ord (1992) find several significant local clusters of sudden infant death syndrome in North Carolina although the global statistic fails to identify any significant clustering.

Another local statistic for measuring spatial dependency is a local variant of the classic measure of spatial autocorrelation, Moran's I (Anselin 1995). When spatial data are distributed so that high values are generally located in close proximity to other high values and low values are generally located near to other low values, the data are said to exhibit positive spatial autocorrelation. When the data are

[10] At the time of writing, details of the application of spatial filters to health data, plus a downloadable copy of software for this purpose, DMAP, are provided by Rushton and his colleagues at http://www.uiowa.edu/%7Egeog/health/index11.html

distributed such that high and low values are generally located near each other, the data are said to exhibit negative spatial autocorrelation. However, it is possible that within the same data set, different degrees of spatial autocorrelation could be present; both positive and negative spatial autocorrelation could even exist within the same data set. Global measures of spatial autocorrelation would fail to pick up these different degrees of spatial dependency within the data. A global statistic might therefore misleadingly indicate that there is no spatial autocorrelation in a data set, when in fact there is strong positive autocorrelation in one part of the region and strong negative autocorrelation in another. The development of a localised version of spatial autocorrelation allows spatial variations in the spatial arrangement of data to be examined. Anselin (1995) presents an application of the localised Moran's I statistic to the spatial distribution of conflict in Africa and Sokal *et al.* (1998) demonstrate its use on a set of simulated data sets. Other studies of local Moran's I include those of Bao and Henry (1996), Tiefelsdorf and Boots (1997), and Tiefelsdorf (1998). Rosenberg (2000) provides a partially local measure of spatial autocorrelation through a directionally varying Moran's I coefficient and Brunsdon *et al.* (1998) describe a different method of estimating local spatial autocorrelation through Geographically Weighted Regression.

Finally, Rogerson (1999) derives a local version of the chi-square goodness-of-fit test and applies this to the problem of identifying relevant spatial clustering. This local statistic is related to Oden's (1995) modification of Moran's I that accounts for spatial variations in population density and is a special case of a test suggested by Tango (1995). The local statistic incorporates a spatially weighted measure of the degree of dissimilarity across regions.

1.6 Examples of Local Multivariate Methods for Spatial Data Analysis

The local univariate statistical methods described above are of limited use in the large and complex spatial data sets that are increasingly available. There is a need to understand local variations in more complex multivariate relationships (see, for example, the attempts by Ver Hoef and Cressie, 1993 and Majure and Cressie, 1997 to extend some of the local visual techniques described above to the multivariate case). Consequently, several attempts have been made to produce localised versions of traditionally global multivariate techniques. Perhaps the greatest challenge, given its widespread use, has been to produce local versions of regression analysis. The subject matter of this book, Geographically Weighted Regression, is one response to this challenge but there have been others. Here we describe five of these: the spatial expansion method; spatially adaptive filtering; multilevel modelling; random coefficient models; and spatial regression models. We leave the description of GWR to Chapter 2. Each of the five techniques described below has limited application to the analysis of spatially non-stationary multivariate relationships for reasons we now explain.

1.6.1 The Spatial Expansion Method

The Expansion Method (Casetti 1972; 1997; Jones and Casetti 1992) recognises explicitly that the parameters in a regression model can be functions of the context in which the regression model is calibrated. It allows the parameter estimates to vary locally by making the parameters functions of other attributes. If the parameters are functions of location (that is, if the relationships depicted by the parameter estimates are assumed to vary over space), a *spatial expansion model* results in which *trends* in parameter estimates over space can be measured (Brown and Jones 1985; Brown and Kodras 1987; Brown and Goetz 1987; Fotheringham and Pitts 1995; Eldridge and Jones 1991). Initially, suppose a global model is proposed such as:

$$y_i = \alpha + \beta x_{i1} + \ldots \tau x_{im} + \varepsilon_i \tag{1.1}$$

where y represents a dependent variable, the xs are independent variables, $\alpha, \beta, \ldots \tau$ represent parameters to be estimated, ε represents an error term and i represents a point in space at which observations on the ys and xs are recorded. This global model can be expanded by allowing each of the parameters to be functions of other variables. While most applications of the expansion method (see Jones and Casetti 1992) have undertaken aspatial expansions, Brown and Jones (1985), Eldridge and Jones (1991) and McMillen (1996) show that it is relatively straightforward to allow the parameters to vary over geographic space so that, for example:

$$\alpha_i = \alpha_0 + \alpha_1 u_i + \alpha_2 v_i \tag{1.2}$$

$$\beta_i = \beta_0 + \beta_1 u_i + \beta_2 v_i \tag{1.3}$$

and

$$\tau_i = \tau_0 + \tau_1 u_i + \tau_2 v_i \tag{1.4}$$

where u_i and v_i represent the spatial coordinates of location i. Equations (1.2)–(1.4) represent very simple linear expansions of the global parameters over space but more complex, non-linear, expansions can easily be accommodated.

Once a suitable form for the expansion has been chosen, the original parameters in the global model are replaced with their expansions. For instance, if it is assumed that parameter variation over space can be captured by the simple linear expansions in equations (1.2)–(1.4), the expanded model would be:

$$\begin{aligned} y_i = {} & \alpha_0 + \alpha_1 u_i + \alpha_2 v_i + \beta_0 x_{i1} + \beta_1 u_i x_{i1} + \beta_2 v_i x_{i1} \\ & + \ldots \tau_0 x_{im} + \tau_1 u_i x_{im} + \tau_2 v_i x_{im} + \varepsilon_i \end{aligned} \tag{1.5}$$

This model can then be calibrated by ordinary least squares regression to produce estimates of the parameters which are then fed back in to equations (1.2)–(1.4) to obtain spatially varying parameter estimates. These estimates, being specific to location i, can then be mapped to display spatial variations in the relationships represented by the parameters.

The expansion method has been very important in promoting awareness of spatial non-stationarity. However, it does have some limitations. One is that the technique is restricted to displaying trends in relationships over space with the complexity of the measured trends being dependent upon the complexity of the expansion equations. Consequently, the distributions of the spatially varying parameter estimates obtained through the expansion method might obscure important local variations to the broad trends represented by the expansion equations. A second limitation is that the form of the expansion equations needs to be assumed *a priori*, although more flexible functional forms than those shown above could be used. A third is that the expansion equations are assumed to be deterministic in order to remove problems of estimation in the terminal model.

1.6.2 Spatially Adaptive Filtering

Another approach to regression modelling that allows coefficients to vary locally is that of adaptive filtering (Widrow and Hoff 1960; Trigg and Leach 1968). When applied to multivariate time series data this method is used to compensate for drift of regression parameters over time. Essentially, this works on a 'predictor-corrector' basis. Suppose a model is assumed of the form

$$y_t = \sum_j x_{tj}\beta_{tj} + \varepsilon_t \tag{1.6}$$

where t is an index of discrete time points. When a new multivariate observation occurs at time t, the existing regression coefficients, $\hat{\boldsymbol{\beta}}_{t-1}$, are used to predict the dependent variable. However, if the prediction does not perform well, the values of the regression coefficient are 'adjusted' to improve the estimate. The adjusted coefficients are referred to as $\hat{\boldsymbol{\beta}}_t$. The degree of adjustment applied has to be 'damped' in some way to avoid problems of overcompensation. That is, in most cases a set of estimates of the β_j values could be found which gave a perfect prediction, but which also fluctuate wildly and do not give a good indication of the true values of $\boldsymbol{\beta}$ at time t. A typical damping approach is to use an update rule of the form

$$\hat{\beta}_{jt} = \hat{\beta}_{jt-1} + |\hat{\beta}_{jt-1}|\alpha_j(y_t - \hat{y}_t)/|\hat{y}_t| \tag{1.7}$$

where $\hat{\beta}_{jt}$ is the jth element of $\hat{\boldsymbol{\beta}}_t$, $\hat{y}_t$ is the predicted value of y_t based on $\hat{\boldsymbol{\beta}}_{t-1}$ and α_j is a damping factor controlling the extent to which the correction is applied for coefficient j.

Foster and Gorr (1986) and Gorr and Olligschlaeger (1994) suggest applying adaptive filtering ideas to spatial data to investigate the 'drift' of regression parameters. With spatial data the predictor-corrector approach then becomes iterative. With time series data, $\hat{\boldsymbol{\beta}}_{t-1}$ is simply updated in terms of its nearest *temporal* neighbour at time t; a given case has a unique neighbour and the flow of updating is only one way. However, when dealing with a spatial arrangement of data, zones (or points) typically do not have unique neighbours and the coefficient estimates have to be updated several times. In addition, the flow of updating is two-way between a pair of neighbouring zones which requires the process to iterate between coefficient estimates until some form of convergence is achieved. If convergence does occur, then the result should be a unique estimate of the regression coefficient vector $\boldsymbol{\beta}$ for each case. The fact that the casewise correction procedure is damped and based on incremental corrections applied between adjacent zones, suggests that some degree of spatial smoothing of the estimates of the individual elements of $\boldsymbol{\beta}$ must take place. Thus, the method tends to produce models in which regression parameters slowly 'drift' across geographical space. Local and regional effects may be investigated by mapping the coefficient estimates.

1.6.3 Multilevel Modelling

The typical spatial application of multilevel modelling attempts to separate the effects of personal characteristics and place characteristics (contextual effects) on behaviour (Goldstein 1987; Jones 1991a; 1991b; Duncan and Jones 2000). It is claimed that modelling spatial behaviour solely at the individual level is prone to what is known as the atomistic fallacy, missing the context in which individual behaviour occurs (Alker 1969). Equivalently, modelling behaviour solely at the aggregate or contextual level is prone to the ecological fallacy, that the results might not apply to individual behaviour (Robinson 1950). Multilevel modelling tries to avoid both these problems by combining an individual-level model representing disaggregate behaviour with a macro-level model representing contextual variations in behaviour. The resulting model has the form

$$y_{ij} = \alpha_j + \beta_j x_{ij} + \varepsilon_{ij} \tag{1.8}$$

where y_{ij} represents the behaviour of individual i living in place j; x_{ij} is the ith observation of attribute x at place j; and α_j and β_j are place-specific parameters where

$$\alpha_j = \alpha + \mu_j^{\alpha} \tag{1.9}$$

and

$$\beta_j = \beta + \mu_j^{\beta} \tag{1.10}$$

Each place-specific parameter is therefore viewed as consisting of an average value plus a random component. Substituting (1.9) and (1.10) into (1.8) yields the multi-level model,

$$y_{ij} = \alpha + \beta x_{ij} + (\varepsilon_{ij} + \mu_j^{\alpha} + \mu_j^{\beta} x_{ij}) \tag{1.11}$$

This model cannot be calibrated by OLS regression unless μ_j^α and μ_j^β are zero so that specialised software is needed such as HLM (Bryk *et al.* 1986), Mln (Rasbash and Woodhouse 1995) or MlwiN (Goldstein *et al.* 1998). Place-specific parameter estimates can be obtained by estimating separate variance effects and substituting these into equations (1.9) and (1.10).

Several refinements to the basic multilevel model described above are possible and are probably necessary for the accurate estimation of individual and contextual effects, making the modelling framework highly complex. These include adding place attributes in the specifications for α_j and β_j; extending the number of levels in the hierarchy beyond two (Jones *et al.* 1996); and the development of cross-classified multilevel models where each lower unit can nest into more than one higher order unit (Goldstein 1994). Although there are numerous examples of the application of multilevel modelling to spatial data, including those of Congdon (1995), Charnock (1996), Jones (1997), Verheij (1997), Duncan *et al.* (1996), Jones and Bullen (1993), Smit (1997), Duncan (1997), Reijneveld (1998) and Duncan and Jones (2000), Duncan and Jones (2000: 298) offer the following warning:

> Importantly, there are also other, more general, issues surrounding the use of multilevel models that require careful consideration. Until recently, researchers in many disciplines seem to have been carried away in an enthusiastic rush to use the new technique and such issues have tended to be ignored.

One particular problem with the application of multilevel modelling to spatial processes is that it relies on an *a priori* definition of a discrete set of spatial units at each level of the hierarchy. While this may not be an issue in many aspatial applications, such as the definition of what constitutes the sets of public and private transportation options, or what constitutes the sets of brands of decaffeinated and regular coffees, it can pose a serious problem in many spatial contexts. The definition of discrete spatial units in which spatial behaviour is modified by certain attributes of those units obviously depends critically on the units themselves being accurately identified. It also implies that the nature of whatever spatial process is being modelled is discontinuous. That is, it is assumed that the process is modified in exactly the same way throughout a particular spatial unit but that the process is modified in a different way as soon as the boundary of that spatial unit is reached. Most spatial processes do not operate in this way because the effects of space are continuous. Hence, imposing a discrete set of boundaries on most spatial processes is unrealistic. One exception would be where administrative boundaries enclose regions in which a policy that affects the behaviour of individuals is applied evenly throughout the region and where such policies vary from region to region. Duncan and Jones (2000) and de Leeuw and Kreft (1995) raise other issues with the application of multilevel models. Consequently, the application of the multilevel modelling framework to most spatial processes appears limited and the application to continuous processes awaits development (see Langford *et al.*, 1999, for a start in this direction).

1.6.4 Random Coefficient Models

While many local statistical techniques assume that the local relationships being measured vary smoothly, an alternative approach allows coefficients to vary *randomly* for each case (Rao 1965; Hildreth and Houck 1968; Raj and Ullah 1981; Swamy *et al.* 1988a; 1988b; 1989). While most applications of random coefficients models have been aspatial, Swamy (1971) presents an early example of parameter estimates that are allowed to vary cross-sectionally. To see how the random coefficients approach is applied, consider a study of house sales specifying a regression model where the dependent variable is house price, and the independent variables are characteristics of the houses. The classical linear regression approach would assume that the regression coefficient for a given variable would be the same for all cases – so that, say, the presence of a second bathroom had an identical effect on house price for any house in the study. The form this model would take, in matrix algebra, is the familiar

$$\boldsymbol{y} = \boldsymbol{X\beta} + \boldsymbol{\varepsilon} \tag{1.12}$$

where $\boldsymbol{X}$ is a matrix of predictor variables (in this case house characteristics), $\boldsymbol{\beta}$ is a vector of regression coefficients, $\boldsymbol{y}$ is a vector of response variables and $\boldsymbol{\varepsilon}$ is a vector of independent random error terms with distribution $N(0,\sigma^2)$. In terms of individual cases, the model can be written as

$$y_i = \sum_j x_{ij}\beta_j + \varepsilon_i \tag{1.13}$$

In random coefficients modelling, the parameters of this model are not assumed to be constant over space but are assumed to vary from case to case, and are drawn from some random distribution, typically the normal. The model can then be written as

$$y_i = \sum_j x_{ij}\beta_{ij} + \varepsilon_i \tag{1.14}$$

where β_{ij} is now a random variable. For each variable j there are i draws of the random regression coefficient from some distribution. Assuming this distribution to be normal,

$$\beta_{ij} \sim N(\beta_j,\sigma_j^2). \tag{1.15}$$

Calibrating a random coefficients model is then a task of estimating the parameters of the distributions from which casewise parameters are drawn – in this case $\{\beta_j,\sigma_j^2\}$ for all j and σ^2, the error term variance. Then, using Bayes' theorem, it is possible to estimate the value of the regression coefficient actually drawn for each case. A further, non-parametric, extension of the technique is to drop the assumption that the coefficients are drawn from a pre-specified distribution and to estimate the distribution itself from the data (Aitkin 1997).

An alternative approach is used in Besag (1986). Here it is assumed that local observations are governed by local parameters, but are independent of one another given these parameters. However, in a Bayesian framework it is also assumed that the parameters have a prior distribution which does exhibit spatial autocorrelation. Using a technique referred to as Iterated Conditional Modes (ICM), local parameter estimates are obtained. As with the Aitkin approach, coefficients are thought of as random, but here the interpretation is Bayesian: randomness is interpreted as beliefs about the coefficient values prior to data collection and analysis. The spatial autocorrelation in the prior distribution represents the belief that nearby coefficients are likely to have similar values.

The random coefficient modelling approach is not intrinsically spatial – local regression coefficients are drawn independently from some univariate distribution and no attention is paid to the *location* to which the parameters refer. Locations that are in close proximity to each other can have regression coefficients drawn from very different-looking distributions. However, once the local parameter estimates are obtained, they can be mapped and their spatial pattern explored. In this way, local variability of certain types of models can be considered. Brunsdon, Aitkin *et al.* (1999), for example, provide an empirical comparison of the application of GWR and the random coefficients model to a data set in which the spatial distribution of a health variable is related to the spatial variability of a set of socio-economic indicators.

1.6.5 Spatial Regression Models

Recognising that spatial data are not generally independent, so that statistical inference in ordinary regression models applied to spatial data is suspect, a number of attempts have been made to provide a regression framework in which spatial dependency is taken into account. These approaches may generally be described as spatial regression models. While such models are generally not thought of as local models, they do recognise the local nature of spatial data, for instance by relaxing the assumption that the error terms for each observation are independent. In particular, if each observation is associated with a location in space, it is assumed that the error terms for observations in close spatial proximity to one another are correlated. The vector of error terms, $\boldsymbol{\varepsilon}$, is assumed to have a multivariate Gaussian distribution with a zero mean and a variance–covariance matrix having non-zero terms away from the leading diagonal. This implies that although any given ε_i will have a *marginal* distribution centred on zero, its *conditional* distribution will depend on the values of the error terms for surrounding observations. For example, if nearby error terms tend to be positively correlated, then given a set of positive error terms one would expect the error term of another observation close to these to be positive also. That is, its conditional distribution would be centred on some positive quantity rather than zero. Although the output from such models still consists of a set of global parameter estimates, local relationships are incorporated into the modelling framework through the covariance structure of the error terms. In this sense, these models can be thought of as 'semi-local' rather than fully local.

There are a number of examples of models that can be classed as 'semi-local'. Perhaps the oldest such technique is that of Kriging (Krige 1966). Here, it is assumed that the spatial data are a set of measurements taken at n points. Suppose one of the data is a dependent variable and the others are predictor variables. Then one could fit an OLS regression model, but this would ignore the spatial arrangement of the locations at which the data are measured. An alternative is to assume that the covariance between any two error terms will be a function of the distance between them. That is, if $\boldsymbol{C}$ is the variance–covariance matrix for the n error terms, and $\boldsymbol{D}$ is the distance matrix for the sampling points, then

$$C_{ij} = f(d_{ij}) \tag{1.16}$$

where f is some distance-decay function and d_{ij} is an element of $\boldsymbol{D}$. There are a number of restrictions on the possible functional form of f, mainly due to the fact that $\boldsymbol{C}$ must be positive definite in order for the model to be well-defined. Typical functions might be the exponential

$$C_{ij} = \sigma^2 \exp(-3d_{ij}/k) \tag{1.17}$$

or the Gaussian

$$C_{ij} = \sigma^2 \exp(-3d_{ij}{}^2/k^2) \tag{1.18}$$

where the parameter σ^2 determines the level of variation of the error terms and k determines the spatial scale over which notable covariance between pairs of measurements occurs. Essentially, k controls the degree of locality in the model with small values of k suggesting that correlation only occurs between very close point pairs and large values of k suggesting that such effects exist on a larger spatial scale.

Calibrating such a model is typically treated as a two-stage problem. First, one has to estimate σ^2 and k, and once this has been done, the regression model itself is calibrated using the formula

$$\hat{\boldsymbol{\beta}} = (\boldsymbol{X}^{\mathrm{T}}\boldsymbol{C}\boldsymbol{X})^{-1}\boldsymbol{X}^{\mathrm{T}}\boldsymbol{C}\boldsymbol{y} \tag{1.19}$$

where $\hat{\boldsymbol{\beta}}$ is a vector of estimated regression coefficients, $\boldsymbol{X}$ is a matrix of independent variables and $\boldsymbol{X}^{\mathrm{T}}$ is its transpose, $\boldsymbol{y}$ is a vector of dependent variables. $\boldsymbol{C}$ is the covariance matrix from the error term estimated using the parameter estimates described above. However, this approach is not without its shortcomings. In particular, it is assumed here that $\boldsymbol{C}$ is known exactly, whereas in reality it is itself estimated from the data. It is also assumed that $\boldsymbol{C}$ is constant over space, which differentiates much of Kriging from GWR.[11] Further, the estimation of the semi-variogram, from which estimates of σ^2 and k are derived, is controversial: a good discussion of the estimation procedure is given in Bailey and Gatrell (1995) which reveals it to be something of a 'black art'.[12] Although useful results can be obtained from this approach, caution should be exercised when drawing formal statistical inferences.

Recently, these objections have been addressed to some extent by Diggle *et al.* (1998). Here, the analytically awkward form of the likelihood function for $\boldsymbol{\beta}$, σ^2 and k is dealt with in a Bayesian context. In particular, drawings from the posterior probability function for these unknown parameters are simulated using Monte Carlo Markov Chain (MCMC) techniques (Besag and Green 1993).

The last quarter century has also seen the growth of other kinds of spatial regression models, particularly those applied to zonal data such as states, counties or electoral wards. As with Kriging, one would expect that ordinary linear regression models applied to data aggregated in this way would fail to encapsulate any spatial interactions taking place because local relationships in the error terms are not represented in the simple non-spatial model. In spatial regression models, zonal proximities are taken as surrogates for local relationships and are typically measured by a *contiguity matrix* – an n by n matrix whose (i, j)th element is one if zones i and j are contiguous, and zero otherwise. Clearly this matrix is symmetrical and encapsulates the *relative* spatial arrangement of the zones. Note that this approach does not take into account the size or shape or absolute location of the zones; the information is solely topological. In most applications the contiguity matrix is standardised so that the rows sum to one and referred to as $\boldsymbol{W}$. A number of such spatial regression models exist: for example, the spatial autoregressive model of Ord (1975):

$$\boldsymbol{y} = \mu \mathbf{1} + \rho \boldsymbol{W}(\boldsymbol{y} - \mu \mathbf{1}) + \boldsymbol{\varepsilon} \tag{1.20}$$

where μ is an overall mean level of the random variate $\boldsymbol{y}$ multiplied by $\mathbf{1}$, a vector of ones; $\boldsymbol{\varepsilon}$ is a vector of independent normal error terms; and ρ is a coefficient determining the degree of spatial dependency of the model. The model can be extended so that the error vector $\boldsymbol{\varepsilon}$ also exhibits spatial autocorrelation. In this case, the coefficient ρ does not determine the distance decay rate of the spatial autocorrelation, but the degree to which the values at individual locations depend on their neighbours. A problem with this type of modelling is that neighbourhood influence is not calibrated in terms of the data but is generally prescribed by the specification

[11] Note that block Kriging is an attempt to overcome the problem of having to assume $\boldsymbol{C}$ is a constant over space by obtaining estimates of $\boldsymbol{C}$ separately for different spatial units. However, this is akin to running separate regressions in different spatial units and suffers from the same problems as this procedure, namely that the processes being modelled are likely to be continuous and not discrete; the processes are likely to vary within spatial units as well as between them and that there is usually no *a priori* justification for the selection of the spatial units in which the process is assumed to be stationary.

[12] Further discussion of the estimation procedure followed in Kriging, along with a worked example, can be found in Fotheringham *et al.* (2000: 171–8).

of W. Similar arguments may also be applied to pre-whitening procedures (Kendall and Ord 1973). With this approach, spatial filters to remove autocorrelation effects are applied to all variables (dependent and independent) prior to fitting a regression model. Again, however, the model being calibrated is a global one. Autoregressive models were also considered in a more general form by Besag (1974) for spatial data arranged on a regular lattice.

As mentioned above, spatial regression models are really mixed models in the sense that although they recognise the impact of local relationships between data, such relationships are almost always measured with a global autocorrelation statistic and the output of the model is a set of global parameter estimates. Brunsdon *et al.* (1998) provide an interesting example where GWR is applied to a spatially autoregressive model such as that in equation (1.20) so that the output from the model is a locally varying set of parameter estimates which includes a locally varying autocorrelation coefficient. GWR applied to spatially autoregressive models is therefore an alternative, and perhaps simpler, method of deriving local measures of spatial autocorrelation.

1.7 Examples of Local Methods for Spatial Flow Modelling

Although not directly applicable to what follows in this book because the disaggregations are for discrete points rather than for a continuous space, it is still worthwhile mentioning the literature on local models of spatial interaction because of its relatively long history. It was recognised quite early that global calibrations of spatial interaction models hid large amounts of spatial information on interaction behaviour and that localised parameters yielded much more useful information (Linneman 1966; Greenwood and Sweetland 1972; Gould 1975). This was very obvious when distance-decay parameters were estimated separately for each origin in a system instead of a single global estimate being provided (see Fotheringham 1981, for further discussion). The origin-specific parameter estimates could then be mapped to provide visual evidence of spatial variations and spatial patterns in their values.

Interestingly, it was the consistent, but counter-intuitive, spatial patterns of origin-specific distance-decay parameters in these local studies that led to the realisation that the global models were gross misspecifications of reality (Fotheringham 1981; 1984; 1986; Meyer and Eagle 1982; Fotheringham and O'Kelly 1989). This, in turn, led to the development of the competing destinations framework from principles of spatial information processing and a new set of spatial interaction models from which more accurate parameter estimates can be obtained (Fotheringham 1984; 1991). It is worth stressing that the global model misspecification only came to light through local parameter estimates being obtained and then mapped. The diagnostic spatial pattern of the local distance-decay parameter estimates would be completely missed in the calibration of a global model.

1.8 Summary

Interest in local forms of spatial analysis and spatial modelling is not new. The recognition that the calibration of global models produces parameter estimates which represent an 'average' type of behaviour, and are therefore of very limited use when behaviour does vary over space, dates back at least to Linneman's calibration of origin-specific models of international trade flows (Linneman 1966). Johnston (1973) also provides an early example of local analysis in the context of voting behaviour. However, as Fotheringham (1997) notes, the current high level of interest in the 'local' rather than the 'global' and the emergence of a battery of techniques for local modelling is notable for several reasons.[13]

Among these are that it refutes the criticism that those adopting a quantitative approach to investigate spatial processes are only concerned with the search for broad generalisations and have little interest in identifying local exceptions, an observation also made by Jones and Hanham (1995). Local forms of spatial analysis also provide a linkage between the outputs of spatial techniques and the powerful visual display capabilities of GIS and some statistical graphics packages. Perhaps most importantly though, they provide much more information on spatial relationships as an aid to both model development and the better understanding of spatial processes. Local statistics and local models provide us with the equivalent of a microscope or a telescope; they are tools with which we can see so much more detail. Without them, the picture presented by global statistics is one of uniformity and lack of variation over space; with them, we are able to see the spatial patterns of relationships that are masked by the global statistics.

To this point, we have discussed several types of local analytical techniques for spatial data. However, none of these is without problems. In the remainder of this book we turn our attention to a very general local modelling technique developed for spatial data termed Geographical Weighted Regression. In the next chapter we describe the development of GWR through an empirical example of house price determinants across London. We also describe another local modelling technique, moving-window regression, in Chapter 2 because this is shown to be a stepping stone towards GWR. Indeed, moving-window regression is a rudimentary form of GWR.

---

[13] As an example of the high level of interest in local techniques, see the two special issues of *Geographical and Environmental Modelling* devoted to this topic, details of which can be found in Fotheringham (1999b) and Flowerdew (2001).

2

Geographically Weighted Regression: The Basics

2.1 Introduction

In Chapter 1 we discussed the utility of local models and local statistics for spatial data. Section 1.6 contained a review of a range of techniques for the local analysis of multivariate data sets and we noted problems with each method. In this chapter, we develop an alternative method for the local analysis of relationships in multivariate data sets which we term Geographically Weighted Regression (GWR). One advantage of this alternative technique is that it is based on the traditional regression framework with which most readers will be familiar. Another advantage is that it incorporates local spatial relationships into the regression framework in an intuitive and explicit manner.

We now describe the development of GWR through an empirical example of the determinants of house prices in London. This is done in a series of steps that begin with relatively crude measures of examining local relationships but which eventually lead to GWR. Once the intuitive grounds for GWR are established in this manner, we then describe more formally the basic mathematical framework for GWR. Chapter 3 continues the same empirical example to demonstrate extensions to the basic model that might be useful in certain circumstances. While the use of GWR in Chapters 2 and 3 is primarily exploratory, we demonstrate in Chapter 4 that it is a statistical model that can be derived using the concept of local likelihood.

2.2 An Empirical Example

Rather than proceeding directly to the statistical description of GWR, we first provide the intuitive foundations for GWR through a set of empirical analyses using house price data for London.[1] It is interesting to note that at least two other local studies of house prices exist: McMillen and Thorsnes (2002) examine the effect on house prices of the location of toxic waste dumps and Pavlov (2000) uses essentially what is Geographically Weighted Regression to examine the determinants of house prices in Los Angeles County.

2.2.1 The Data

We obtained house price data for 12 493 properties sold within London during 1991.[2] The study area and the locations of the houses within the study area are depicted in Figures 2.1 and 2.2, respectively. The sample consists of all residential units sold with mortgages through the Nationwide Building Society that were not sales of former council houses. The latter are excluded from the sample because they were generally sold at below market value through a government scheme to increase home ownership and their inclusion would lead to bias in the results. A small proportion of other properties with miscoded data were also excluded.

To get a general indication of the spatial variation in our sample of house prices across London during 1991, average prices for the houses sold in each of the 33 London boroughs have been calculated and are displayed in Table 2.1 and in Figure 2.3. As one can see, there is a fair degree of variation even in the average prices across London (there will of course be even more variation in the observed prices themselves) with values ranging from a high of £135 063 in Kensington and Chelsea to a low of £61 826 in Newham. The general trend is for prices to be higher in the west than the east and in a swathe of boroughs from the centre towards the south-west.

2.2.2 A Global Regression Model

The prices reported in Table 2.1 and Figure 2.3 are unadjusted for housing mix so that we do not know at this point if the higher prices in some areas reflect higher property values per square metre or if the houses in those areas tend to be larger, or possess some other feature that makes them more expensive. This issue can be clarified by regressing prices on a series of structural and neighbourhood characteristics of each house. This produces what is known in the literature as a hedonic price model (*inter alia*, Goodman 1978; Cheshire and Sheppard 1995; Meen and

---

[1] We use the term 'house price' throughout even though the data include flatted properties which some would not call a 'house'. However, other terms such as 'property price' do not adequately reflect the fact that these are all residential properties and the term 'residential property price' seems overly pedantic.

[2] We are extremely grateful to the Nationwide Building Society for providing these data to the Department of Geography at the University of Newcastle. Although several years of house price data have generously been made available to us by the Nationwide Building Society, we have selected the 1991 period here because it matches the availability of data from the UK Census of Population held in April 1991.

Figure 2.1 *London boroughs*

Figure 2.2 *Location of sampled house prices*

Figure 2.3 *Average house price by London borough*

Andrew 1998).[3,4] In order to account for the variation in house prices in the data set, the following hedonic price model was constructed:

$$\begin{aligned} P_i = {} & \alpha_0 + \alpha_1 \mathrm{FLRAREA}_i + \alpha_2 \mathrm{BLDPWW1}_i + \alpha_3 \mathrm{BLDPOSTW}_i + \\ & \alpha_4 \mathrm{BLD60S}_i + \alpha_5 \mathrm{BLD70S}_i + \alpha_6 \mathrm{BLD80S}_i + \alpha_7 \mathrm{TYPDETCH}_i + \\ & \alpha_8 \mathrm{TYPTRRD}_i + \alpha_9 \mathrm{TYPBNGLW}_i + \alpha_{10} \mathrm{TYPFLAT}_i + \alpha_{11} \mathrm{GARAGE}_i + \\ & \alpha_{12} \mathrm{CENTHEAT}_i + \alpha_{13} \mathrm{BATH2}_i + \alpha_{14} \mathrm{PROF}_i + \alpha_{15} \mathrm{UNEMPLOY}_i + \\ & \alpha_{16} \mathrm{FLRDETCH}_i + \alpha_{17} \mathrm{FLRFLAT}_i + \alpha_{18} \mathrm{FLRBNGLW}_i + \\ & \alpha_{19} \mathrm{FLRTRRD}_i + \alpha_{20} \log_e (\mathrm{DISTCL}_i) + \varepsilon_i \end{aligned} \tag{2.1}$$

[3] Apart from providing an interesting example, there are other reasons why we have chosen a hedonic price model to demonstrate the application of GWR in this chapter. First, the technique is used frequently, primarily in economics and real estate studies, but very rarely is account taken of any spatial effects. Although the assumption of spatial stationarity has been strongly questioned (Meen and Andrew 1998), few studies have examined spatial non-stationarity in hedonic price models explicitly (see Pavlov 2000 for an example). Second, hedonic price models cannot hope to capture all the locational determinants of house prices and therefore spatial patterns in misspecification bias must almost always be present in the global calibration of such models.

[4] At the time of writing, the following web site contains an inventory of research papers on hedonic price modelling: http://www.sscnet.ucla.edu/ssc/labs/cameron/nrs98/hedoninv.htm

Table 2.1 *Average 1991 house prices for the 33 London boroughs*

Borough	Average house price (£)
Kensington and Chelsea	135 063
City	127 667
Westminster	114 456
Camden	110 654
Richmond	109 231
Hammersmith	93 112
Barnet	92 361
Bromley	91 436
Islington	91 109
Wandsworth	89 175
Harrow	85 140
Enfield	83 387
Kingston	82 828
Hounslow	81 368
Sutton	80 203
Lambeth	79 770
Ealing	79 200
Brent	78 596
Havering	78 314
Haringey	77 631
Merton	77 423
Hillingdon	76 900
Redbridge	75 955
Southwark	75 263
Greenwich	73 346
Tower Hamlets	73 115
Bexley	72 988
Croydon	72 386
Hackney	71 219
Lewisham	69 746
Waltham Forest	64 800
Barking	61 826
Newham	56 365

Note: Calculated from data supplied by the Nationwide Building Society to the University of Newcastle upon Tyne.

where P_i is the price in pounds sterling at which a house sold; FLRAREA is the floor area of the property in square metres; BLDxxx is a set of dummy or indicator variables that depict the age of the property as follows:

BLDPWW1 is 1 if the property was built prior to 1914, 0 otherwise;
BLDPOSTW is 1 if the property was built between 1940 and 1959, 0 otherwise;
BLD60S is 1 if the property was built between 1960 and 1969, 0 otherwise;
BLD70S is 1 if the property was built between 1970 and 1979, 0 otherwise;
BLD80S is 1 if the property was built between 1980 and 1989, 0 otherwise;

TYPxxx is a set of dummy variables that depict the type of house as follows:

TYPDETCH is 1 if the property is detached (i.e. it is a stand-alone house), 0 otherwise;
TYPTRRD is 1 if the property is in a terrace of similar houses (commonly referred to as a 'row house' in the USA), 0 otherwise;
TYPBNGLW is 1 if the property is a bungalow (i.e. it has only one floor), 0 otherwise;
TYPFLAT is 1 if the property is a flat (or 'apartment' in the USA), 0 otherwise.

GARAGE is 1 if the house has a garage, 0 otherwise; CENTHEAT is 1 if the house has central heating, 0 otherwise; BATH2 is 1 if the house has 2 or more bathrooms, 0 otherwise; PROF is the proportion of the workforce in professional or managerial occupations in the census ward in which the house is located;[5] UNEMPLOY is the rate of unemployment in the census ward in which the house is located; and FLRxxx is a set of interaction terms where:

FLRDETCH = FLRAREA * TYPDETCH
FLRFLAT = FLRAREA * TYPFLAT
FLRBNGLW = FLRAREA * TYPBNGLW
FLRTRRD = FLRAREA * TYPTRRD.

Finally, DISTCL is the straight-line distance from the property to the centre of London (taken here to be Nelson's column in Trafalgar Square) measured in kms; $\log_e$ denotes a natural logarithm; and α denotes a parameter to be estimated.

To avoid multicollinearity problems, the following variables are excluded from the regression model:

BLDINTW which is 1 if the property was built between 1914 and 1939 and 0 otherwise;
TYPSEMID is 1 if the property is semi-detached (i.e. it shares a common wall with one neighbour – often referred to as a 'duplex' in the USA), 0 otherwise;
FLRSEMID which is equal to FLRAREA * TYPSEMID

[5] Each census ward has an approximate population of 5 250.

Consequently, the model for a detached house built between 1960–69 with central heating, one bathroom but without a garage is:

$$P_i = \alpha_0 + \alpha_1 \text{FLRAREA}_i + \alpha_4 \text{BLD60S}_i + \alpha_7 \text{TYPDETCH}_i + \alpha_{11} \text{GARAGE}_i + \alpha_{14} \text{PROF}_i + \alpha_{15} \text{UNEMPLOY}_i + \alpha_{16} \text{FLRDETCH}_i + \alpha_{20} \log_e (\text{DISTCL}_i) + \varepsilon_i \quad (2.2)$$

where the gradient of the relationship between house price and floor area is given by adding the estimates of α_1 and α_{16} and the intercept is given by adding the estimates of α_0 and α_7.

Similarly, the model for a semi-detached house between 1914–39 with central heating, two or more bathrooms and no garage is:

$$P_i = \alpha_0 + \alpha_1 \text{FLRAREA}_i + \alpha_3 \text{BLDPOSTW}_i + \alpha_{12} \text{CENTHEAT}_i + \alpha_{13} \text{BATH2}_i + \alpha_{14} \text{PROF}_i + \alpha_{15} \text{UNEMPLOY}_i + \alpha_{20} \log_e (\text{DISTCL}_i) + \varepsilon_i \quad (2.3)$$

so that the gradient of the relationship between house price and floor area is given by the estimate of α_1 and the intercept is given by the estimate of α_0.

One problem in hedonic price modelling is that there is no single agreed functional form for hedonic models (Halvorsen and Pollakowski 1981). As Fleming (1999, 155) notes: 'Hedonic theory gives little guidance on the choice of the functional form for the hedonic specification.' However, the particular functional representation of each variable in equation (2.1) is justified on two grounds. The first is that for the continuous variables, partial regression plots revealed these forms to be reasonable representations of the relationship between house prices and each particular attribute. The second is that the additive nature of the model makes the interpretation of each parameter estimate highly intuitive. For instance, the addition of central heating to a house would be expected to yield an additional £α_{12} to its selling price and a detached house would be anticipated to sell for £α_7 more than a semi-detached house with similar characteristics. Interpretations of the parameter estimates in specific monetary terms are given below.

2.2.3 Global Regression Results

The model in equation (2.1) was calibrated using ordinary least squares regression to produce the parameter estimates reported in Table 2.2. The gradients of the relationship between house price and floor area, *ceteris paribus*, for the five types of housing are given in Table 2.3. It can be seen from the results in Table 2.2 that across the region and across all house types the average value of housing per square metre in London in 1991, *ceteris paribus*, was £697. When the parameter estimates for the interaction variables between house type and floor area are added, we obtain the house-type-specific relationships between price and floor area that are given in Table 2.3. These show that house prices ranged from £902/m^2 for detached houses to £574/m^2 for flats, *ceteris paribus*.

Table 2.2 Global regression parameter estimates

Variable	Parameter estimate	Standard error	T value
Intercept	58 900	2 524	23.3
FLRAREA	697	14	49.3
BLDPWW1	−2 340	601	−3.9
BLDPOSTW	−2 786	903	−3.1
BLD60S	−5 177	1 043	−5.0
BLD70S	−2 421	1 174	−2.1
BLD80S	6 315	920	6.9
TYPDETCH	−4 215	4 038	−1.0
TYPTRRD	3 465	2 120	1.6
TYPBNGLW	23 437	6 156	3.8
TYPFLAT	3 239	2 057	1.6
GARAGE	5 956	564	10.6
CENTHEAT	7 777	627	12.4
BATH2	22 297	1 166	19.1
PROF	72	24	3.0
UNEMPLOY	−211	38	−5.5
FLRDETCH	205	27	7.5
FLRFLAT	−123	22	−5.6
FLRBNGLW	−87	65	−1.4
FLRTRRD	−119	19	−6.2
$\log_e$(DISTCL)	−18 137	604	−30.1

Note: $R^2 = 0.60$

The parameter estimates in Table 2.2 relating house prices to the period in which the housing was built suggest that, *ceteris paribus*, newest housing (built during the 80s) had the highest value with a premium of £6 315 over the excluded time period variable, houses built between 1914 and 1939. From the age parameters, the order of relative contribution to the price of a house (from highest to lowest) is shown in Table 2.4. There is a big drop in value from newest housing to the next most valued, housing built between 1914–39, the inter-war period. Then there is another drop to housing built pre-WWI, in the 1970s and between 1940–59 which all have similar values. Finally, there is a big drop to the least valued housing stock, that built in the 1960s. In the United Kingdom, architectural style, planning decisions and housing quality during the period of the 1960s all have generally poor reputations; the latter is borne out in these results.

Table 2.3 *Price/m² of various house types estimated from the results in Table 2.2*

House type	Price/Sq. M. (£)	Derivation
Detached	902	$\alpha_1 + \alpha_{16}$
Semi-detached	697	α_1
Bungalow	610	$\alpha_1 + \alpha_{18}$
Terraced	578	$\alpha_1 + \alpha_{19}$
Flat	574	$\alpha_1 + \alpha_{17}$

Table 2.4 *Average value of houses by age of housing*

Period of housing	Difference (£) compared to equivalent houses built 1914–39
1980s	6 315
1914–39	0
Before 1914	−2 340
1970s	−2 421
1940–59	−2 786
1960s	−5 177

The global results in Table 2.2 suggest that the average value a garage adds to a house price is £5 956. The presence of central heating adds £7 777 to the value of a house and the presence of two or more bathrooms adds £22 297. Generally, houses with higher prices are associated with neighbourhoods with relatively large proportions of professionals and low unemployment rates. House prices decline rapidly as distance from the centre of London increases: for example, a property 1 km from the centre of London will be almost £42 000 more expensive than an identical property 10 km away. Of course, these results represent averages across London and it remains to be seen how representative they are and whether there are some interesting exceptions to these averages.

The R^2 value for the global regression is 0.60 indicating a reasonable explanatory performance but it still leaves 40% of the variance in house prices unexplained. Some of this unexplained variance probably results from assuming the relationships in the model to be constant over space – that is, we have assumed a stationary process to be operating when it might be non-stationary. Suppose, for instance, that there are intrinsic taste variations across London and that terraced houses are viewed differently in different parts of the city, or that the value of a garage is greater in some areas than in others. If such variations in relationships exist over space, then the hedonic price model in equation (2.1) will clearly be a misspecification of reality because it assumes these relationships to be constant.

The traditional method of investigating errors in a spatial model is to map the residuals from the model, as is done here in Figure 2.4. The distribution is clearly non-random with a belt of large positive residuals running north–south (the lighter shading on the map) and a parallel band of large negative residuals to the east (the

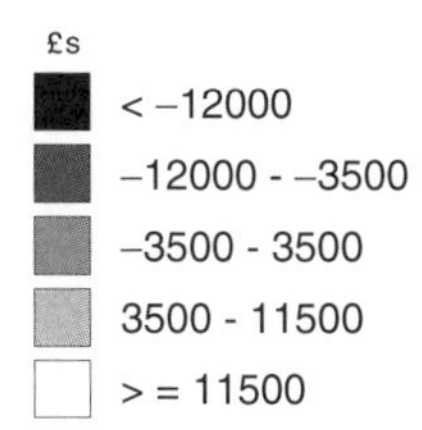

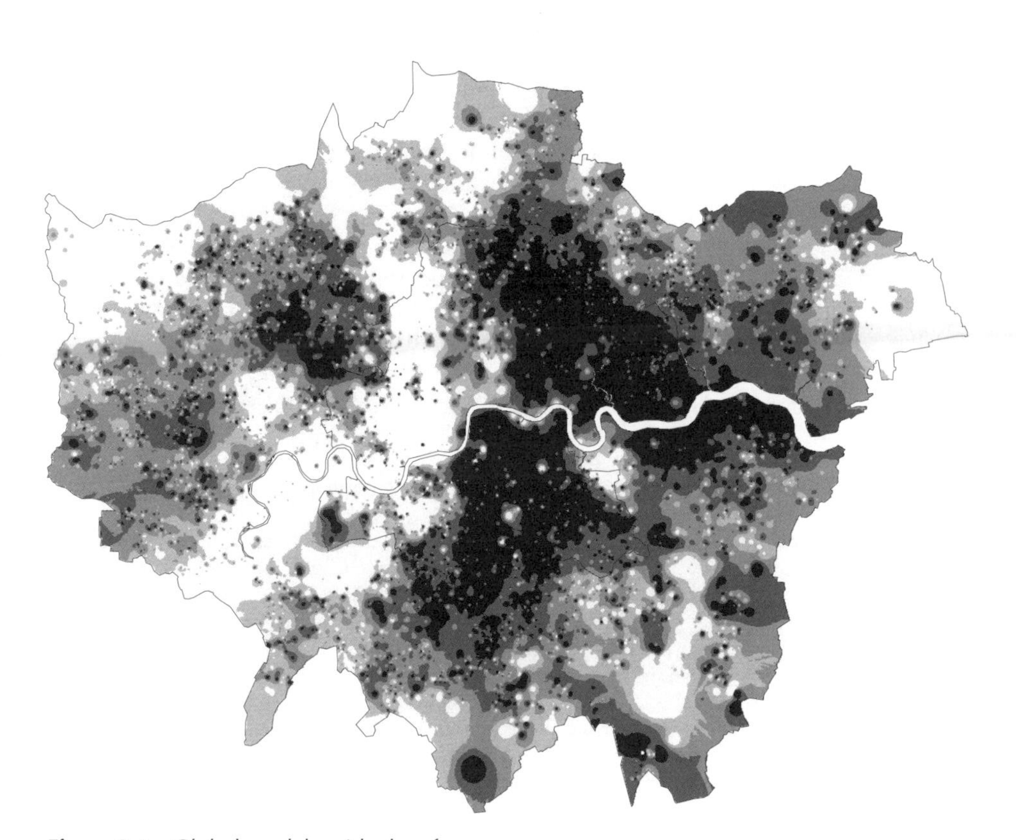

Figure 2.4 *Global model residual surface*

darker shading).[6] This conforms to a general view of the London housing market in which the west is held to be more attractive than the east, *ceteris paribus*. However, while the residual map indicates that there is a problem with the global model, it does not indicate which, if any, of the parameters in the model might exhibit spatial non-stationarity. For this, we need to examine maps of local parameter estimates as shown below.[7]

Having described a typical set of global regression results, and essentially ignored any potential variations in these results over space, we now describe several methods of examining whether spatial non-stationarity exists in any of the relationships depicted in equation (2.1). These methods range from the very crude, such as simply running separate global regressions for different spatial units, to the relatively sophisticated, such as Geographically Weighted Regression.

2.3 Borough-Specific Calibrations of the Global Model

One simple and obvious way of examining if the relationships being modelled in the global hedonic price equation are likely to be stationary over space would be to calibrate the global model separately for each of the 33 London boroughs. This would produce 33 sets of parameter estimates, the values of which could be mapped[8]. Two examples of such mapping are given in Figures 2.5 and 2.6 for the value per square metre of flats and terraced housing respectively.[9] For comparison, a third map, shown in Figure 2.7, describes the ratio of value of terraced property to that of flatted property for each borough. The values being mapped are listed in Table 2.5 which also gives the value of the R^2 statistic for each regression.

Figure 2.5 and Table 2.5 indicate that the value of flats varies substantially across London ranging from a high of £1 574 per m^2 in Kensington and Chelsea to a low of £104 per m^2 in Havering. In contrast, the global results in Table 2.3 suggest an average of £574 per m^2 throughout the city. In general, the value of flatted properties is higher close to the river. Similarly, the value of terraced property shows

---

[6] For a comparison with the residual map obtained from GWR, which exhibits practically zero spatial autocorrelation, see Figure 2.19.

[7] In Chapter 5 we revisit in more detail the issue of the spatial pattern of regression residuals. Amongst other things, we demonstrate how spatial patterns of residuals can be caused by applying a global model to processes that are non-stationary and how the residuals for GWR models typically do not exhibit any spatial pattern.

[8] One partial 'solution' to the problem of spatial non-stationarity in hedonic price models would be to add a series of dummy variables representing each spatial unit to the regression equation. This would allow a location-specific intercept to be estimated. The model proposed here goes further than this and allows location-specific versions of all the parameters in the model to be estimated. However, the model still has severe limitations. Somewhat more sophisticated treatment of the issue is found in Quandt (1958) in his discussion of 'switching regressions' which allow discrete changes in relationships over space to be modelled. Páez (2000) provides an example of switching regressions applied to spatial data.

[9] The value per square metre for flatted properties is obtained by adding the estimates of α_1 and α_{17} from equation (2.1) applied to each London borough separately. The value per square metre for terraced properties is obtained by adding the estimates of α_1 and α_{19}.

Figure 2.5 *Price/m² (£) of flats estimated from separate regressions for each London borough*

Figure 2.6 *Price/m² (£) of terraced properties estimated from separate regressions for each London borough*

Figure 2.7 *Rates of terraced to flatted property prices*

a magnitude of spatial variation from a high of £2019 per m^2 in Kensington and Chelsea to a low of only £80 per m^2 in Bexley. The single regression results suggest the value of terraced properties to be £578 per m^2 throughout the city. What is interesting is that the relative values of flatted properties and terraced properties, described in Figure 2.7, show a large variation over space. Whereas the global ratio of the relative values of terraced to flatted properties is almost exactly 1.0, this ratio varies between 5.34 (Havering) and 0.2 (Camden). In Havering, terraced properties are worth 5.34 times as much as flatted properties per m^2 while in Camden flatted properties are worth 5 times as much as terraced properties per m^2. Presumably flatted properties in Havering have a poor reputation and probably consist mainly of 1960s-style tower blocks. Conversely, flats in Camden are seen as relatively desirable and probably consist of either newer luxury flats or subdivided older and larger properties.

Although calibrating the global model separately for each of the 33 London boroughs produces information on possible spatial variations in the relationships being examined, this is a very naïve method that has several problems. One is a statistical issue in that some of the London boroughs contain relatively few houses of a certain type in our sample. There are no results reported, for example, for the City of London, which is a borough containing a large number of high rise office buildings and very few, if any, bungalows. Similarly, although results are reported for Tower Hamlets, the number of certain types of houses in the sample is very small and some of the individual parameter estimates have large standard errors.

Table 2.5 *Price/m²(£) of flats and terraced housing in each London borough from separate calibrations of the global hedonic model*

Borough	Price/m² (£) Flat	Terraced	Ratio Terrace/Flat	R^2
Barking	310	609	1.96	0.70
Barnet	528	579	1.10	0.75
Bexley	106	80	0.75	0.86
Brent	263	310	1.18	0.73
Bromley	399	427	1.07	0.83
Camden	897	179	0.20	0.69
City	***	***	***	***
Croydon	329	216	0.66	0.83
Ealing	464	350	0.75	0.63
Enfield	326	615	1.89	0.85
Greenwich	629	611	0.98	0.53
Hackney	432	612	1.42	0.71
Hammersmith	524	1 272	2.43	0.82
Haringey	543	623	1.15	0.73
Harrow	233	444	1.91	0.47
Havering	104	555	5.34	0.67
Hillingdon	265	270	1.02	0.71
Hounslow	513	733	1.43	0.65
Islington	595	889	1.49	0.80
Kensington	1 574	2 019	1.28	0.75
Kingston	141	605	4.29	0.81
Lambeth	350	606	1.73	0.72
Lewisham	268	513	1.91	0.76
Merton	517	554	1.07	0.64
Newham	267	249	0.93	0.56
Redbridge	420	518	1.23	0.77
Richmond	866	713	0.82	0.75
Southwark	667	498	0.75	0.72
Sutton	311	572	1.84	0.82
Tower Hamlets	628	381	0.61	0.79
Waltham Forest	257	320	1.25	0.80
Wandsworth	563	780	1.39	0.68
Westminster	626	1 672	2.67	0.64

*** Insufficient data to calibrate model

Another problem is akin to that of the global model in that the process being modelled is assumed to be constant within each borough. If there is spatial variation in at least some of the relationships being examined across London, then it is reasonable to assume that such spatial variation might also be present within each borough. Finally, simply calibrating the model separately for each London borough assumes that the process being modelled is discrete and that discontinuities occur exactly at the locations of the administrative boundaries of the boroughs. However, most spatial processes are continuous and many will be unrelated to the location of administrative boundaries. For these reasons, a better technique of local regression modelling is needed.

2.4 Moving Window Regression

Moving window regression overcomes the problem that the limits of the London boroughs are not necessarily the boundaries of the spatial processes determining house prices within London. In moving window regression, a grid of regression points is first constructed over the study area (see Figure 2.8). A set of regions is then defined around each regression point: these regions are usually square or circular although in theory any shape of region could be defined. In Figure 2.8, for example, one might define a region around each regression point as the four cells on which the regression point is centred. The regression model is then calibrated on all data that lie within the region described around a regression point and the process is repeated for all regression points. The resulting local parameter estimates can then be mapped at the locations of the regression points to view possible non-stationarity in the relationships being examined.

Examples of the use of moving window techniques are quite common in a variety of disciplines including geography (Hagerstrand 1965; Martin 1989; Fotheringham

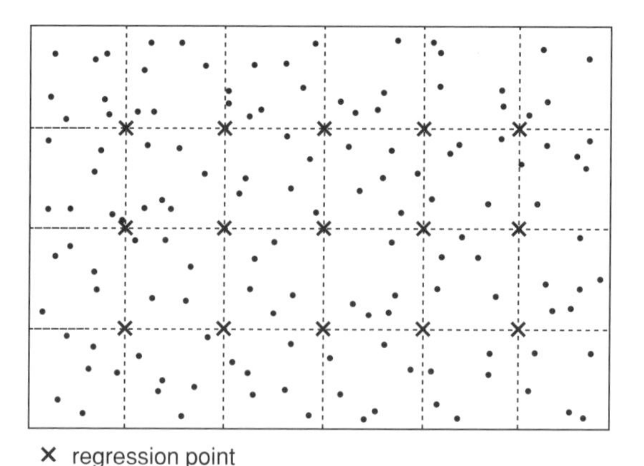

Figure 2.8 An example of moving window regression

et al. 1996); geosciences (Murray and Baker 1991); seismology (Ratdomopurbo and Poupinet 1995; Chandler and Malek 1991); spectral analysis (Mommersteeg *et al.* 1995) and signal processing (Gandhi and Khassam 1991). Moving window regression is often used when data are measured on regular grids or when dealing with time series data. An example of its use here is shown in Figure 2.9 which describes the value of flatted property across London using a circular moving window of radius 10 km applied to a regular grid of regression points. The map indicates a large degree of spatial variation in the value of flatted property (the values are comparable to those of Figure 2.5) although the spatial pattern is suspiciously simple with the values tending to decrease uniformly away from a high point which is located near the centre of the city.

Moving window regression, although producing a smoother surface of parameter estimates than the separate regressions for each borough, still represents a discontinuous technique. The data points within each local region are given a weight of 1, and data points outside the region are given a weight of 0. As discussed above, most spatial processes are continuous, so this weighting seems rather arbitrary. Also, the results will clearly be dependent on the size of the window or region chosen: larger windows will produce smoother surfaces than smaller windows. Finally, the technique will suffer from edge effects unless various steps are taken to prevent these. Typically, windows towards the edge of the map will contain fewer

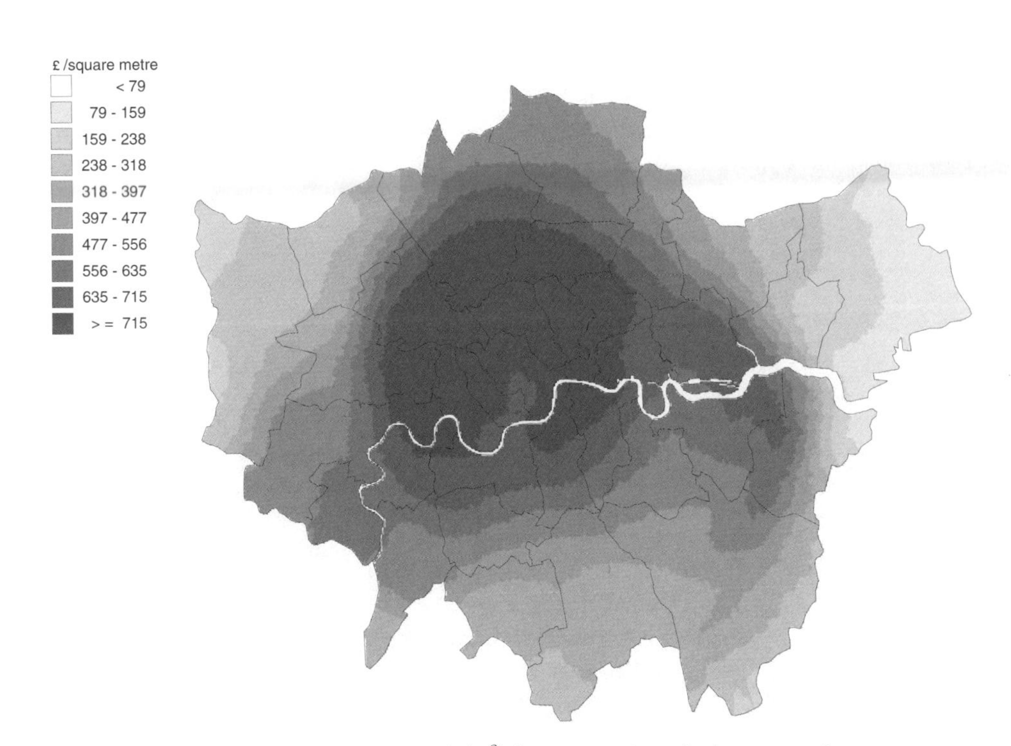

Figure 2.9 *Value of flatted properties (£/m²) from a moving window regression*

regression points than those towards the centre so that the parameter estimates obtained from such windows will have higher standard errors.

2.5 Geographically Weighted Regression with Fixed Spatial Kernels

To solve the problem of representing a continuous spatial process with a discrete weighting system, we now turn to a simple description of the essence of GWR. A mathematical description is left until Section 2.7 and more advanced features of GWR are described in subsequent chapters.

In the moving window example presented above, a region was described around a regression point and all the data points within this region or window were then used to calibrate a model. This process was repeated for all regression points. GWR works in the same way except that each data point is weighted by its distance from the regression point; hence, data points closer to the regression point are weighted more heavily in the local regression than are data points farther away. Graphically, the method is that of fitting a spatial kernel to the data as described in Figures 2.10 and 2.11. For a given regression point, the weight of a data point is at a maximum when it shares the same location as the regression point. This weight decreases continuously as the distance between the two points increases. In this way, a regression model is calibrated locally simply by moving the regression point across the region. For each location, the data will be weighted differently so that the results of any one calibration are unique to a particular location. By plotting the results of these local calibrations on a map, surfaces of parameter estimates, or any other display which is appropriate, can be generated. In practice, the results of GWR are relatively insensitive to the choice of weighting function but they are sensitive to the bandwidth of the particular weighting function chosen so that the

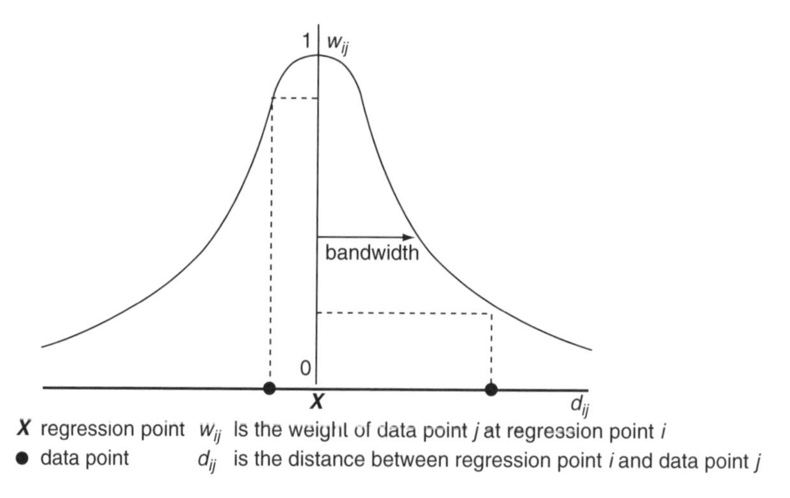

Figure 2.10 A spatial kernel

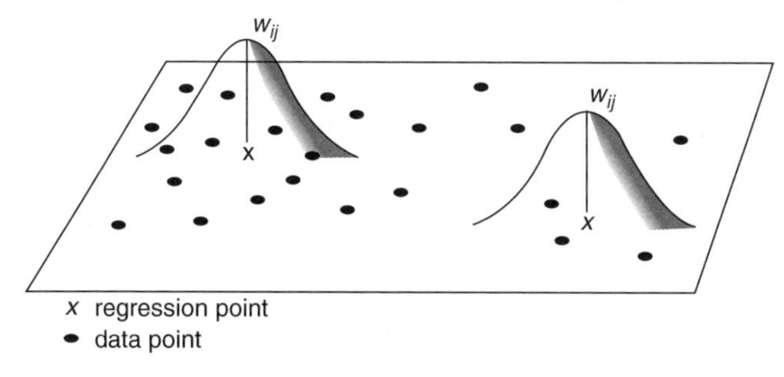

Figure 2.11 *GWR with fixed spatial kernels*

determination of the optimal value of the bandwidth is necessary as part of GWR.[10] More details on this are provided in Section 2.7 and in Chapter 9.

The results of calibrating a GWR with a fixed Gaussian spatial kernel to the London house price data provide a map of the spatial variation of each parameter estimate in the hedonic price model given in equation (2.1). As an example, the spatial variation in the value of flatted properties is displayed in Figure 2.12 obtained with a bandwidth of 2.5 km. Flatted property values are higher along both sides of the river from Richmond in the west to the western half of Greenwich in the east. The values reach a peak in the boroughs of Kensington and Chelsea, Westminster and Camden. Lower values tend to be found on the periphery of London but the map shows interesting ridges and hollows within this general pattern. For instance, there are local areas of very low values in Kingston and in Brent adjacent to very high values. The lowest values for flatted properties are in Bexley, Havering, Harrow and Hillingdon.

Although it might be tempting to compare the complexities of the two surfaces in Figures 2.9 and 2.12 as a means of inferring the superiority of one method over the other, such a comparison would not be valid because of the different scales of the kernels. The moving window results are in effect generated by a 'box-shaped' kernel which is broader than the Gaussian one used to generate the GWR surface. However, there is evidence to suggest that box-shaped kernels *do* generate smoother surfaces than Gaussian kernels (Siegmund and Worsley 1995). Intuitively, one would expect that moving windows would provide poorer results than a continuous kernel because of the different nature of the weighting schemes. At comparable scales, distant points carry more weight in moving window regressions than in GWR with Gaussian kernels.

[10] As shown in Figure 2.10, the bandwidth is a measure of the distance-decay in the weighting function and indicates the extent to which the resulting local calibration results are smoothed. Spatial kernels with a small bandwidth have a steeper distance-decay weighting function and produce rougher surfaces than spatial kernels with a large bandwidth. In a Gaussian kernel, for example, the weighting function is given by:

$$w_{ij} = \exp[-{}^1/_2(d_{ij}/b)^2]$$

where b is the bandwidth.

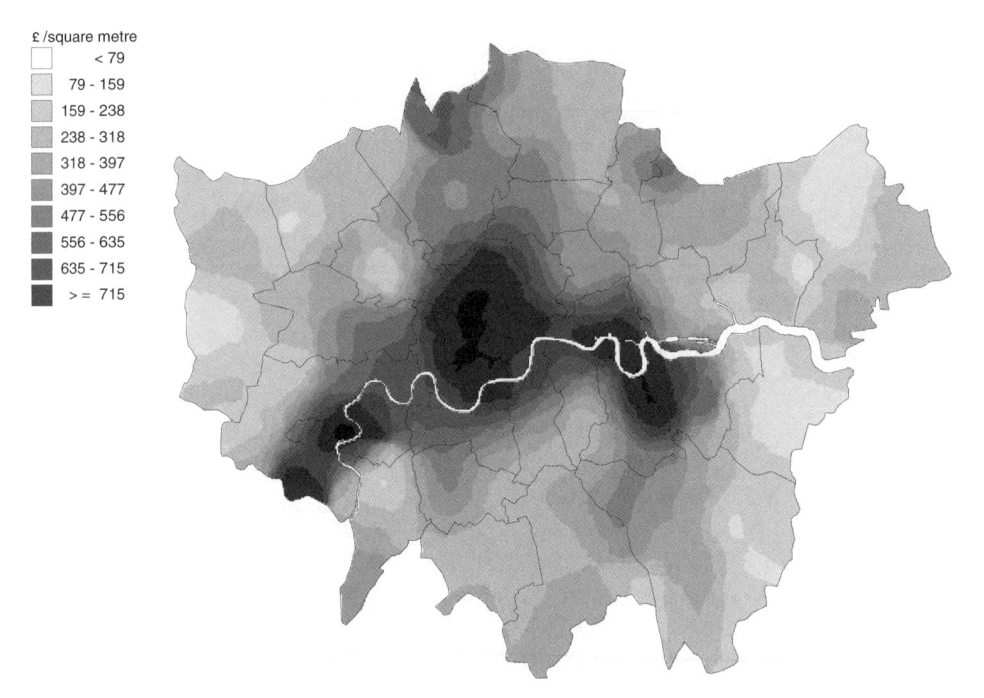

Figure 2.12 *Value of flatted properties (£/m^2) from GWR with a fixed spatial kernel*

2.6 Geographically Weighted Regression with Adaptive Spatial Kernels

A potential problem that might arise in the application of GWR with fixed spatial kernels is that for some regression points, where data are sparse, the local models might be calibrated on very few data points, giving rise to parameter estimates with large standard errors and resulting surfaces which are 'undersmoothed'. In extreme cases, the estimation of some parameters might be impossible due to insufficient variation in small samples. Accordingly, to reduce these problems, the spatial kernels in GWR can be made to adapt themselves in size to variations in the density of the data so that the kernels have larger bandwidths where the data are sparse and have smaller bandwidths where the data are plentiful. There are various ways in which such adaptive bandwidths can be achieved and some of these are described in the following section. Graphically, the application of GWR with adaptive spatial kernels is described in Figure 2.13.

The results of applying GWR with an adaptive spatial kernel to the house price data are shown for the estimated value of flatted properties in Figure 2.14.[11] As can be seen, the spatial pattern of these values is very similar to those derived from

[11] In the calibration of the adaptive spatial kernel a bi-square weighting function based on nearest neighbours was used (see Section 2.7, equation 2.20) and the optimal number of nearest neighbours was found to be 931.

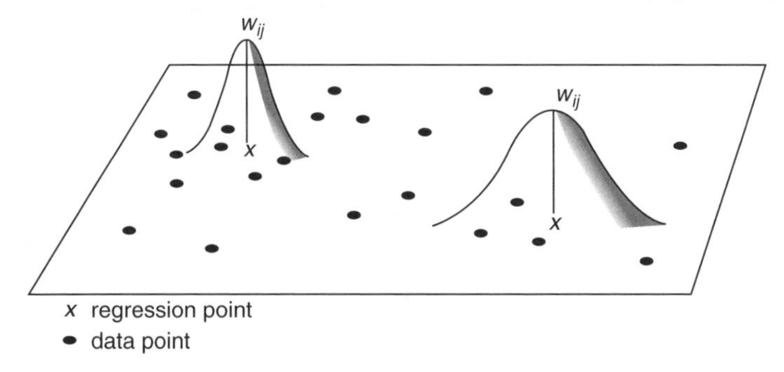

Figure 2.13 *GWR with adaptive spatial kernels*

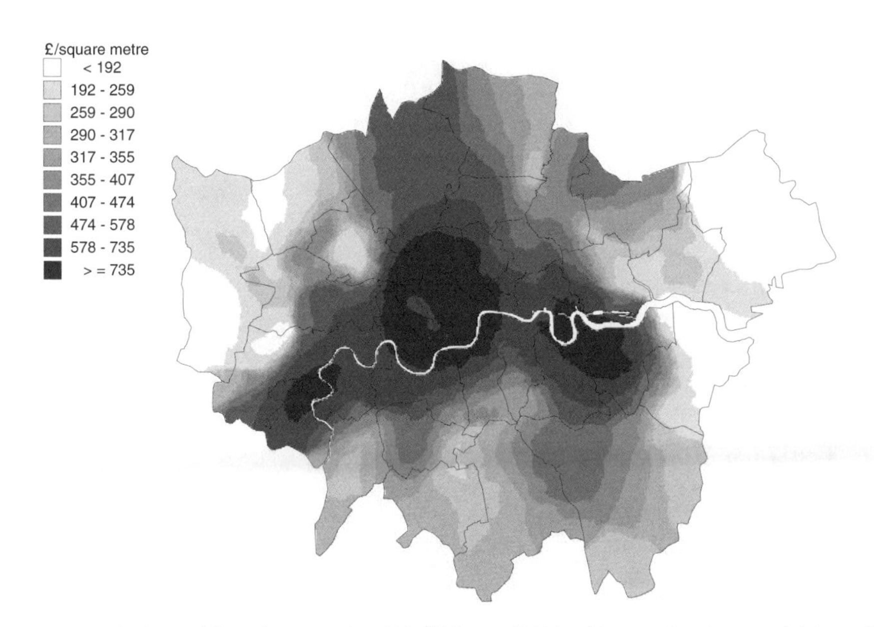

Figure 2.14 *Value of flatted properties (£/m²) from GWR with an adaptive spatial kernel*

the fixed spatial kernel method except that the map is slightly smoother. This is to be expected because with the fixed method there is a greater likelihood that some local calibrations will be based on only a few data points and so the resulting distribution of local estimates will exhibit greater variation.

Of course, maps for each parameter estimated in equation (2.1) can be produced in this way through GWR. Examples of these are provided in Figures 2.15 to 2.18. In Figure 2.15 the value of terraced property per square metre is mapped (by summing the local estimates of the α_1 and α_{19}). The values peak in the boroughs of Kensington and Chelsea and Westminster where terraces are typically Georgian or Regency mainly built for the more wealthy. Terraced properties have relatively

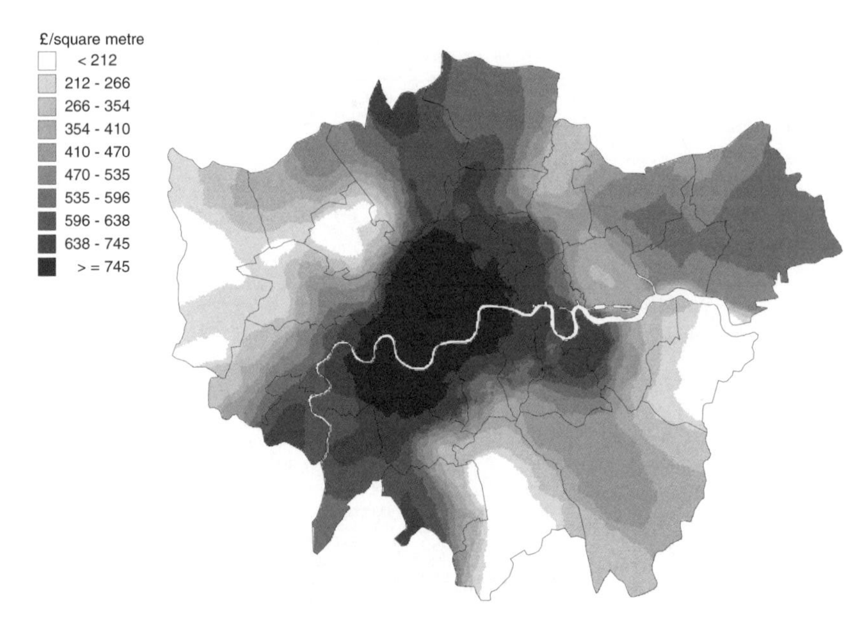

Figure 2.15 *Value of terraced property (£/m²) from GWR*

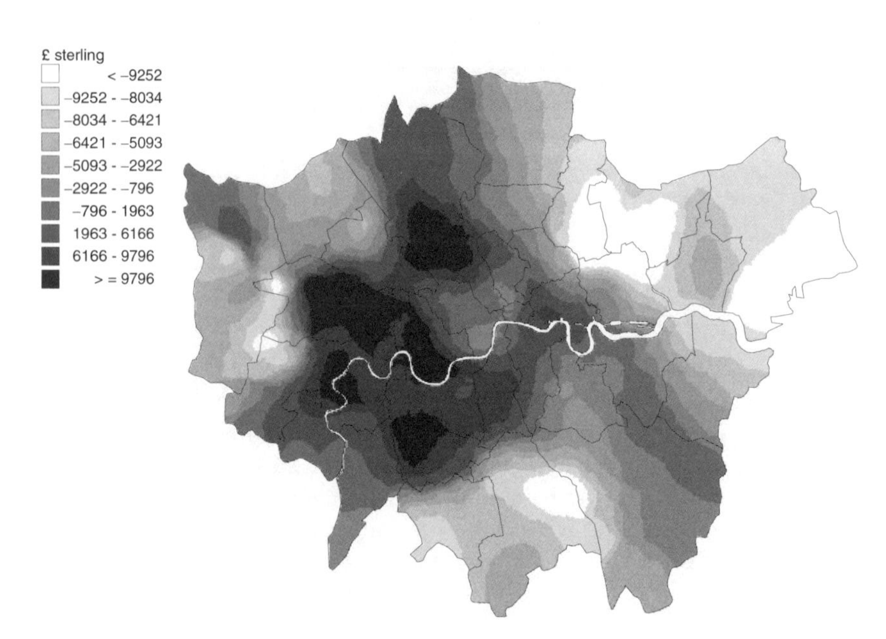

Figure 2.16 *Value of housing built before 1914 compared to housing built 1914–39*

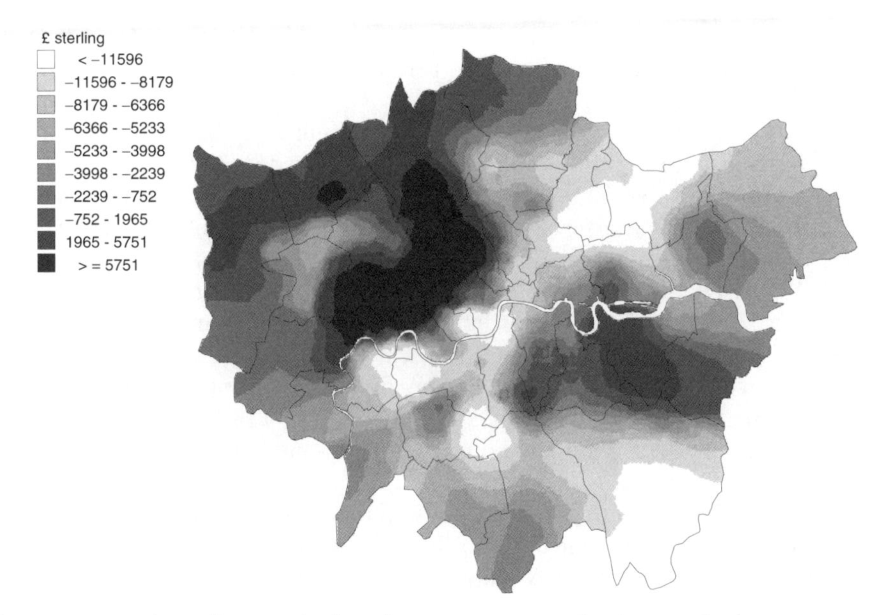

Figure 2.17 *Value of housing built in the 1960s compared to housing built 1914–39*

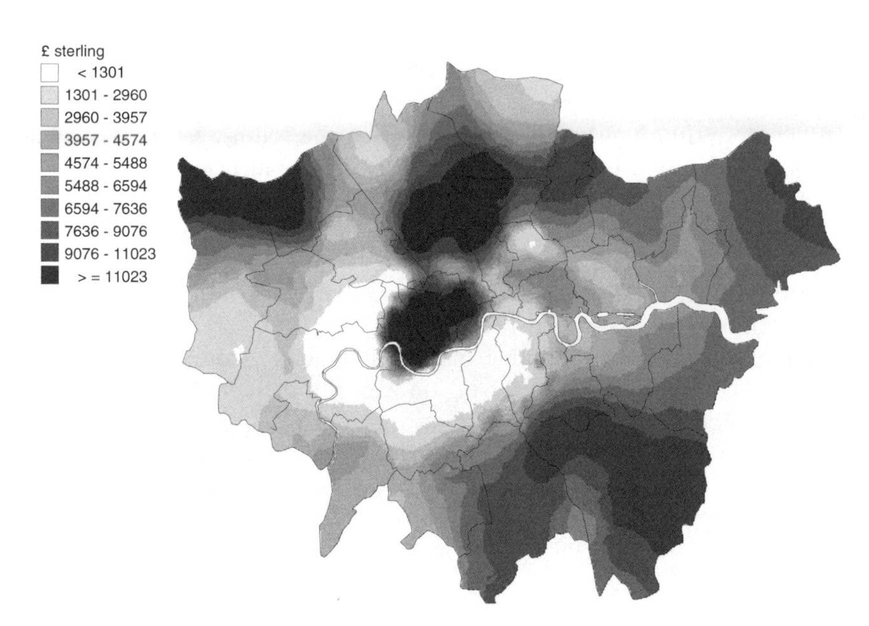

Figure 2.18 *Additional value to property of having garage facilities*

low values in boroughs such as Bexley, Croydon, Brent and Hillingdon where the buildings are of a much more recent period, constructed primarily to house working-class families.

In Figures 2.16 and 2.17, the value of housing built in different time periods is mapped over space. In Figure 2.16 the parameter mapped is the local estimate of α_2 which describes the value of housing built before 1914 compared to housing built in the period 1914–39 (the excluded dummy variable for building period). The global parameter estimate suggests that on average across London a house built before WWI cost £2 340 less than a house with the same attributes built between 1914–39 (see Table 2.4). The local estimates demonstrate how much information is hidden in this global estimate because the local values suggest that equivalent pre-1914 houses would cost over £10 000 *more* in areas such as Barnet, Camden, Ealing, Hammersmith and Merton but would cost over £10 000 *less* in areas such as Havering, Redbridge, Waltham Forest, Croydon and small parts of Hounslow and Hillingdon. There is a general centre–periphery relationship in the spatial pattern exhibited in Figure 2.16: pre-WWI housing towards the centre of the city tends to be valued more highly than housing built between 1914–39, whereas housing built between 1914–39 tends to be valued more highly towards the periphery. As mentioned previously, older housing in the more peripheral parts of London is more likely to have been built to house the working class and has probably aged rather poorly compared to older housing built nearer the centre.

In Figure 2.17, the value of housing built in the 1960s is compared to that of housing built 1914–39. Globally, it was reported that a house built in the 1960s would cost £5 177 less than the equivalent house built between 1914–39 (see Table 2.4). However, the local results demonstrate substantial spatial variation in this relationship. In some parts of London a 1960s' house costs over £12 000 less than its inter-war equivalent whereas in other parts of London, a 1960s' house costs over £6 000 more. In a tight swathe running from the south-west to the north-east across London, housing built during the 1960s has a much lower value than the equivalent housing built between 1914–39. This is also the case for an area of housing centred on the intersection of Merton and Croydon and also for Bromley. In contrast, there are sections of Ealing, Hammersmith, Kensington and Chelsea, Westminster, Camden, Barnet and Greenwich where 1960s' housing is valued more highly than housing built between 1914–39.

The spatial variation in the value of a garage across London is described in Figure 2.18. The global parameter estimate suggests that the average addition to the price of a house resulting from the presence of a garage is £5 956. There are some interesting local variations in this value. The local parameters are always positive (the addition of a garage presumably does not reduce the price of a house as it is not included in the floor space of a property) but the premium ranges from less than £1 000 to over £12 000.[12] The largest premiums are found in the central

[12] There appears to be no end in sight for the astronomical sums people in central London are prepared to pay for garaging space. In a recently published newspaper article (*Sunday Times* 25/03/01) a garage in Kensington and Chelsea was reportedly sold for £200 000.

boroughs of Hammersmith, Kensington and Chelsea and Westminster and also to the north and north-west in Barnet, Haringey, Harrow and north Hillingdon. Interestingly, a large horseshoe-shaped area of very low premiums is found immediately adjacent to the areas of highest premiums in central London. The river appears to act as a sharp divide between areas of high and low premium and the resolution of parameter variation in the GWR surface is particularly noticeable at this point.

These examples serve to demonstrate the ability of GWR to reveal patterns in the data and the processes underlying them. The technique allows detailed spatial variations in relationships to be examined. In doing so, the problem with global parameter estimates is highlighted: global values are nothing more than spatial averages that can hide a great deal of information about the process being studied. One consequence of attempting to model non-stationary processes with a global model is, as is shown in Figure 2.4, that the residual terms exhibit positive spatial autocorrelation (see Chapter 5 for more details). The residuals from the GWR model, on the other hand, exhibit practically no spatial autocorrelation as shown in Figure 2.19 which uses the same key as Figure 2.4 to assist comparison. It can also be seen that the GWR residuals tend to be far less extreme than those of the global model. We now describe in greater detail the methodology of GWR.

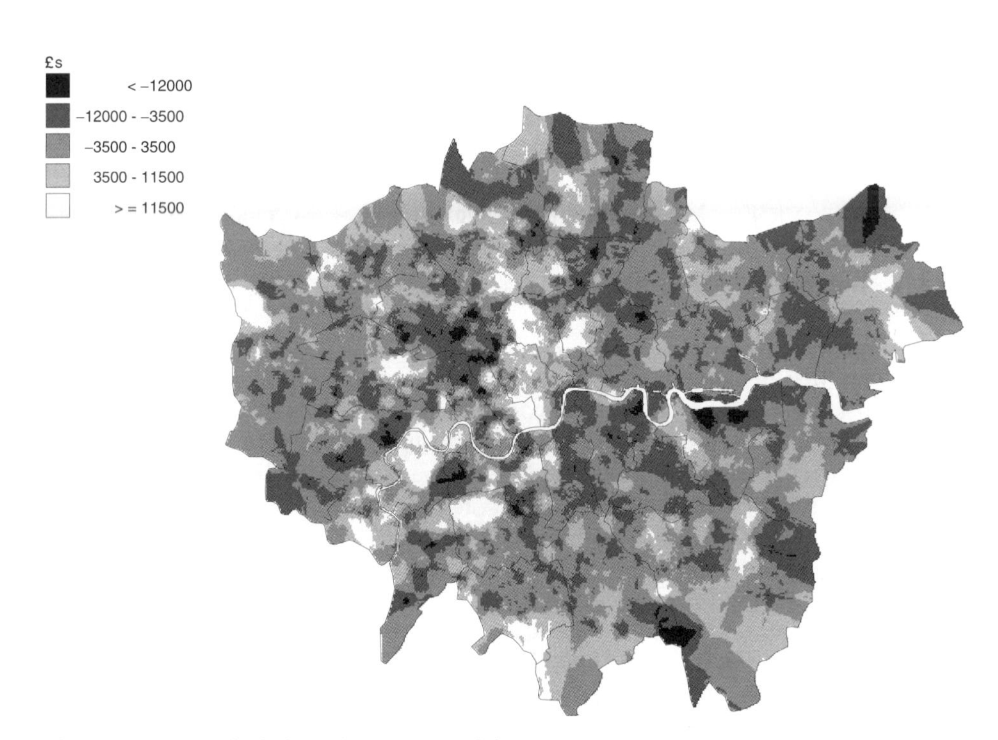

Figure 2.19 *Residuals from the GWR model*

2.7 The Mechanics of GWR in More Detail

2.7.1 The Basic Methodology

Consider a global regression model written as:

$$y_i = \beta_0 + \sum_k \beta_k \, x_{ik} + \varepsilon_i \tag{2.4}$$

GWR extends this traditional regression framework by allowing local rather than global parameters to be estimated so that the model is rewritten as:

$$y_i = \beta_0(u_i,v_i) + \sum_k \beta_k(u_i,v_i)x_{ik} + \varepsilon_i \tag{2.5}$$

where (u_i,v_i) denotes the coordinates of the ith point in space and $\beta_k(u_i,v_i)$ is a realisation of the continuous function $\beta_k(u,v)$ at point i.[13] That is, we allow there to be a continuous surface of parameter values, and measurements of this surface are taken at certain points to denote the spatial variability of the surface. Note that equation (2.4) is a special case of equation (2.5) in which the parameters are assumed to be spatially invariant. Thus the GWR equation in (2.5) recognises that spatial variations in relationships might exist and provides a way in which they can be measured.

As it stands though, there would appear to be problems in calibrating equation (2.5) because there are more unknowns than observed variables. However, models of this kind do occur in the statistical literature and discussions can be found in Rosenberg (1973), Spjotvoll (1977), Hastie and Tibshirani (1990) and Loader (1999). Our approach borrows from the latter two particularly in that we do not assume the coefficients to be random, but rather that they are deterministic functions of some other variables – in our case, location in space. The general approach when handling such models is to note that although an *unbiased* estimate of the local coefficients is not possible, estimates with only a small amount of bias can be provided.[14]

We argue here that the calibration process in GWR can be thought of as a trade-off between bias and standard error. Assuming the parameters exhibit some degree of spatial consistency, then values near to the one being estimated should have relatively similar magnitudes and signs. Thus, when estimating a parameter at a given location i, one can approximate (2.5) in the region of i by (2.4), and perform a regression using a subset of the points in the data set that are close to i. Thus, the $\beta_k(u_i,v_i)$s are estimated for i in the usual way and for the next i, a new subset of 'nearby' points is used, and so on. These estimates will have some degree of

[13] One of the useful properties of GWR is that, as shown subsequently, it allows estimates of the localised parameters to be obtained for any point in space – not just for those points at which data are measured. However, localised residuals are only possible for points in space at which observations of the y variable are available.

[14] Bias here results from inferring the outcome of a non-stationary process at location i from data collected at locations other than i.

bias, since the coefficients of (2.5) will exhibit some drift across the local calibration subset. However, if the local sample is large enough, this will allow a calibration to take place – albeit a biased one. The greater the size of the local calibration subset, the lower the standard errors of the coefficient estimates; but this must be offset against the fact that enlarging this subset increases the chance that the coefficient 'drift' introduces bias. To reduce this effect, one final adjustment to this approach may also be made. Assuming that points in the calibration subset farther from i are more likely to have differing coefficients, a weighted calibration is used, so that more influence in the calibration is attributable to the points closer to i.

As noted above, the calibration of equation (2.5) assumes implicitly that observed data near to location i have more of an influence in the estimation of the $\beta_k(u_i, v_i)$s than do data located farther from i (see Figures 2.11 and 2.13). In essence, the equation measures the relationships inherent in the model *around each location i*. Hence weighted least squares provides a basis for understanding how GWR operates. In GWR an observation is weighted in accordance with its proximity to location i so that the weighting of an observation is no longer constant in the calibration but varies with i. Data from observations close to i are weighted more than data from observations farther away. That is,

$$\hat{\boldsymbol{\beta}}(u_i, v_i) = (\boldsymbol{X}^{\mathrm{T}} \boldsymbol{W}(u_i, v_i) \boldsymbol{X})^{-1} \boldsymbol{X}^{\mathrm{T}} \boldsymbol{W}(u_i, v_i) \boldsymbol{y} \tag{2.6}$$

where the bold type denotes a matrix, $\hat{\boldsymbol{\beta}}$ represents an estimate of $\boldsymbol{\beta}$, and $\boldsymbol{W}(u_i, v_i)$ is an n by n matrix whose off-diagonal elements are zero and whose diagonal elements denote the geographical weighting of each of the n observed data *for regression point i*.

To see this more clearly, consider the classical regression equation in matrix form:

$$\boldsymbol{Y} = \boldsymbol{X\beta} + \boldsymbol{\varepsilon} \tag{2.7}$$

where the vector of parameters to be estimated, $\boldsymbol{\beta}$, is constant over space and is estimated by

$$\hat{\boldsymbol{\beta}} = (\boldsymbol{X}^{\mathrm{T}} \boldsymbol{X})^{-1} \boldsymbol{X}^{\mathrm{T}} \boldsymbol{Y}. \tag{2.8}$$

The GWR equivalent is

$$\boldsymbol{Y} = (\boldsymbol{\beta} \otimes \boldsymbol{X})\mathbf{1} + \boldsymbol{\varepsilon} \tag{2.9}$$

where $\otimes$ is a logical multiplication operator in which each element of $\boldsymbol{\beta}$ is multiplied by the corresponding element of $\boldsymbol{X}$. If there are n data points and k explanatory variables, both $\boldsymbol{\beta}$ and $\boldsymbol{X}$ will have dimensions $n \times (k+1)$ and $\mathbf{1}$ is a $(k+1) \times 1$ vector of 1s. The matrix $\boldsymbol{\beta}$ now consists of n sets of local parameters and has the following structure:

$$\boldsymbol{\beta} = \begin{bmatrix} \beta_0(u_1,v_1) & \beta_1(u_1,v_1) & \dots & \beta_k(u_1,v_1) \\ \beta_0(u_2,v_2) & \beta_1(u_2,v_2) & \dots & \beta_k(u_2,v_2) \\ \dots & \dots & \dots & \dots \\ \beta_0(u_n,v_n) & \beta_1(u_n,v_n) & \dots & \beta_k(u_n,v_n) \end{bmatrix} \tag{2.10}$$

The parameters in each row of the above matrix are estimated by

$$\hat{\boldsymbol{\beta}}(i) = (\boldsymbol{X}^{\mathrm{T}}\boldsymbol{W}(i)\boldsymbol{X})^{-1}\,\boldsymbol{X}^{\mathrm{T}}\boldsymbol{W}(i)\boldsymbol{Y} \tag{2.11}$$

where i represents a row of the matrix in (2.10) and $\boldsymbol{W}(i)$ is an n by n spatial weighting matrix of the form

$$\boldsymbol{W}(i) = \begin{bmatrix} w_{i1} & 0 & \dots\dots\dots\dots & 0 \\ 0 & w_{i2} & \dots\dots\dots\dots & 0 \\ \cdot & \cdot & & \cdot \\ \cdot & \cdot & & \cdot \\ 0 & 0 & \dots\dots\dots\dots & w_{in} \end{bmatrix} \tag{2.12}$$

where w_{in} is the weight given to data point n in the calibration of the model for location i.

The estimator in equation (2.11) is a weighted least squares estimator but rather than having a constant weight matrix, the weights in GWR vary according to the location of point i. Hence the weighting matrix has to be computed for each point i and the weights depict the proximity of each data point to the location of i with points in closer proximity carrying more weight in the estimation of the parameters for location i. Notice, however, that in equations (2.11) and (2.12) there is no reason that i has to be the location of a data point. Local estimates of the parameters can in fact be derived for any point in space, regardless of whether or not that point is one at which data have been observed.

2.7.2 Local Standard Errors

In addition to estimating local parameter estimates, it is also useful to calculate local standard errors in order to account for variations in the data used to compute the estimates. In some cases, for instance, local parameter estimates might be a function of relatively few data points or the data points might have low weights in the local regression because they lie far from the regression point. Local standard errors of GWR parameter estimates are derived in the following manner.

Consider rewriting the estimator of the local parameter estimates given in equation (2.6) as

$$\hat{\boldsymbol{\beta}}(u_i,v_i) = \boldsymbol{C}\boldsymbol{y} \tag{2.13}$$

where,

$$C = (X^{\mathrm{T}} W(u_i, v_i) X)^{-1} X^{\mathrm{T}} W(u_i, v_i) \tag{2.14}$$

The variance of the parameter estimates is given by

$$\mathrm{Var}\,[\hat{\boldsymbol{\beta}}\,(u_i, v_i)] = CC^{\mathrm{T}}\,\sigma^2 \tag{2.15}$$

where σ^2 is the normalised residual sum of squares from the local regression and is defined as

$$\sigma^2 = \sum_i (y_i - \hat{y}_i)/(n - 2v_1 + v_2) \tag{2.16}$$

and where

$$v_1 = \mathrm{tr}(S) \tag{2.17}$$

and

$$v_2 = \mathrm{tr}(S^{\mathrm{T}} S) \tag{2.18}$$

(Loader, 1999). The matrix S is known as the hat matrix (Hoaglin and Welsch, 1978) which maps $\hat{y}$ on to y in the following manner:

$$\hat{y} = Sy \tag{2.19}$$

where each row of S, r_i, is given by:

$$r_i = X_i (X^{\mathrm{T}} W(u_i, v_i) X)^{-1} X^{\mathrm{T}} W(u_i, v_i) \tag{2.20}$$

The term $n - 2v_1 + v_2$ is known as the effective degrees of freedom *of the residual.* The term $2v_1 - v_2$ is equivalent to the number of parameters in a global linear regression model and can be termed the effective number of parameters in the local GWR model. Because the trace of S and the trace of $S^{\mathrm{T}} S$ are generally very similar, the effective number of parameters in the local regression can usually be approximated by v_1 which saves having to compute the trace of $S^{\mathrm{T}} S$.

Once the variance of each parameter estimate is obtained from equation (2.15), the standard errors are obtained from:

$$\mathrm{SE}\,(\hat{\boldsymbol{\beta}}_i) = \mathrm{sqrt}\,[\mathrm{Var}\,(\hat{\boldsymbol{\beta}}_i)] \tag{2.21}$$

where $\boldsymbol{\beta}_i$ is a short-hand notation for $\boldsymbol{\beta}(u_i, v_i)$. An example of the use of local standard errors is provided in Chapter 6 where they are used to compute local t scores. It should be noted that as well as producing local parameter estimates and local standard errors, GWR will also produce local versions of other standard regression diagnostics which can be informative in understanding various aspects of model performance. More details on these functions are provided in Chapter 9 where a user-friendly computer program for GWR is described.

2.7.3 Choice of Spatial Weighting Function

Until this point, it has merely been stated in GWR that $\boldsymbol{W}(u_i, v_i)$ or, in more convenient terms, $\boldsymbol{W}(i)$, is a weighting scheme based on the proximity of the regression point i to the data points around i without an explicit relationship being stated. The choice of such a relationship will be considered here. First, consider the implicit weighting scheme of the OLS framework in equation (2.4). Here

$$w_{ij} = 1 \ \forall \ i,j \tag{2.22}$$

where j represents a specific point in space at which data are observed and i represents any point in space for which parameters are estimated. That is, in the global model each observation has a weight of unity. An initial step towards weighting based on locality might be to exclude from the model calibration observations that are further than some distance d from the regression point. This would be equivalent to setting their weights to zero, giving a weighting function of

$$\begin{aligned} w_{ij} &= 1 \text{ if } d_{ij} < d \\ w_{ij} &= 0 \text{ otherwise} \end{aligned} \tag{2.23}$$

which is the moving window approach described above in Section 2.4. The use of this weighting scheme would simplify the calibration procedure because at every regression point only a subset of the data points would be used to calibrate the model. However, as noted in Section 2.4, this spatial weighting function has the problem of discontinuity. As the regression point changes, the estimated coefficients could change drastically as a data point moves into or out of the window around i. Although sudden changes in the parameters over space might genuinely occur, in this case changes in their estimates would occur as artefacts of the arrangement of data points, rather than necessarily depicting any underlying process in the relationship under investigation. Examples of the use of this type of GWR are given by Fotheringham *et al.* (1996) and by Charlton *et al.* (1997).

One way to combat the problem of discontinuities of weights is to specify w_{ij} as a continuous function of d_{ij}, the distance between i and j. One obvious choice is:

$$w_{ij} = \exp[-{}^{1}/_{2}(d_{ij}/b)^2] \tag{2.24}$$

where b is referred to as the bandwidth. If i and j coincide (that is, i also happens to be a point in space at which data are observed), the weighting of data at that point will be unity and the weighting of other data will decrease according to a Gaussian curve as the distance between i and j increases. In the latter case the inclusion of data in the calibration procedure becomes 'fractional'. For example, in the calibration of a model for point i, if $w_{ij} = 0.5$, then data at point j contribute only half the weight in the calibration procedure as data at point i itself. For data a long way from i the weighting will fall to virtually zero, effectively excluding these observations from the estimation of parameters for location i. The empirical GWR results described above are based on Gaussian or near-Gaussian weighting functions.

An alternative kernel utilises the bi-square function,

$$\begin{aligned} w_{ij} &= [1 - (d_{ij}/b)^2]^2 \text{ if } d_{ij} < b \\ &= 0 \text{ otherwise} \end{aligned} \tag{2.25}$$

which is particularly useful because it provides a continuous, near-Gaussian weighting function up to distance b from the regression point and then zero weights any data point beyond b. Examples of the use of these types of weighting functions are provided by Brunsdon *et al.* (1996; 1997) and by Fotheringham *et al.* (1998).

The above types of spatial kernel are fixed in terms of their shape and magnitude over space and belong to a class of kernels exemplified in Figure 2.11. It could be argued that the kernels should be allowed to vary spatially, with the kernels being smaller in regions where the density of data points is high and larger where the density of data points is low, as shown in Figure 2.13. The rationale for this is twofold: (i) where data points are dense there is more scope for examining changes in relationships over relatively small distances and such changes might be missed with larger kernels; and (ii) in regions where data are scarce, the standard errors of the coefficients estimated in GWR when fixed kernels are used will be high because the number of data points used will be small. In essence the problem of fixed kernels in regions where data are dense is that the kernels are larger than they need be and hence the estimates obtained from them are more likely to suffer from bias. Conversely, the problem with fixed kernels in regions where data are scarce is one of inefficiency: the kernels are smaller than they need be to estimate the parameters reliably. Both problems can be reduced by performing GWR with spatially varying kernels.

At least three methods of producing spatially varying kernels exist. One is to rank the data points in terms of their distance from each point i so that R_{ij} is the rank of the jth point from i in terms of the distance j is from i. The closest data point to i has a weight of 1 and the weights decrease as the rank increases according to some continuous function such as:

$$w_{ij} = \exp(-R_{ij}/b) \tag{2.26}$$

This will automatically reduce the bandwidth of kernels in regions with large amounts of data because the distance to say the 10th nearest data point will be much less than when the regression point is in a region with relatively few data points.

A second, and more complex, method of producing spatially varying kernels is to ensure that the sum of the weights for any point i is a constant, C. In areas where the density of data points is high, the kernel will have to contract to ensure the sum of the weights is equal to C whereas in areas where the density of data points is low, the kernel will have to expand. Formally,

$$\sum_j w_{ij} = \text{C for all } i \tag{2.27}$$

which is a constraint that has to be attached to one of the stationary weighting functions described in equations (2.24) and (2.25). One could simply choose an appropriate value of C and then calibrate the weighting function given this value of C. Alternatively, and perhaps preferably, the following sequence of steps would allow an optimal value of C to be selected:

Step 1: select a value of C.
Step 2: calibrate the weighting function with the selected value of C as a constraint and calculate a goodness-of-fit statistic for the model.
Step 3: select a different value of C.
Step 4: repeat step 2.
Repeat this process and select the value of C that yields the optimal fit.

However, the constraint in equation (2.27) is operationalised, it has the attractive feature of ensuring that the effective sample size is the same for each local regression but the cost is that it can be slightly more complex to calibrate than the other weighting functions.

A third spatially varying weighting method involves a function that is related to the Nth nearest neighbours of point i. That is,

$$\begin{aligned} w_{ij} &= 1 \text{ if } j \text{ is one of the } N\text{th nearest neighbours of } i \\ &= 0 \text{ otherwise} \end{aligned} \tag{2.28}$$

or, given that (2.28) re-introduces discontinuities,

$$\begin{aligned} w_{ij} = {} & [1 - (d_{ij}/b)^2]^2 \text{if } j \text{ is one of the } N\text{th nearest neighbours of } i \text{ and} \\ & b \text{ is the distance to the } N\text{th nearest neighbour.} \\ = {} & 0 \text{ otherwise} \end{aligned} \tag{2.29}$$

If either of the weighting functions in equations (2.28) or (2.29) is used in the GWR, the calibration of the model involves the estimation of N. The value of N is the number of data points to be included within the calibration of the local model and the weighting function determines the weight of each data point up to the Nth one. Weights for all data points beyond the Nth one are set to zero. Hence, the bi-square weighting function, which reaches zero at the Nth data point is a logical one to employ in GWR. The empirical examples of GWR with an adaptive bandwidth described above use nearest neighbour weighting with a bi-square decay function. A comparison of discrete and continuous weighting functions is described by Fotheringham *et al.* (1997).

It should be noted that local kernel bandwidth methods have been applied in the related field of probability density estimation for some time. For example, nearest-neighbour based methods have been used by Loftsgaarden and Quesenberry (1965). Also theoretical work on rules for choosing variable kernel bandwidths has been carried out by Abramson (1982). A useful overview is given by Wand and Jones (1995: 40–2).

Whatever the specific weighting function employed, the essential idea of GWR is that for each regression point i there is a 'bump of influence' around i described by the weighting function such that sampled observations near to i have more influence in the estimation of the parameters than do sampled observations farther away.

2.7.4 Calibrating the Spatial Weighting Function

One aspect of GWR is that the estimated parameters are, in part, dependent on the weighting function or kernel selected. In equation (2.23), for example, as d becomes larger, the closer will be the model solution to that of OLS and when d is equal to the maximum distance between points in the system, the two models will be equal. Equivalently, in equation (2.24) as b tends to infinity, the weights tend to one for all pairs of points so that the estimated parameters become uniform and GWR becomes equivalent to OLS. Conversely, as the bandwidth becomes smaller, the parameter estimates will increasingly depend on observations in close proximity to i and hence will have increased variance. The problem is therefore how to select an appropriate bandwidth or decay function in GWR. There are a number of criteria that can be used for bandwidth selection.

Consider the selection of b in equation (2.24). One possibility is to choose b on a 'least squares' criterion. One way to proceed would be to minimise the quantity

$$z = \sum_{i=1}^{n} [y_i - \hat{y}_i(b)]^2 \tag{2.30}$$

where $\hat{y}_i$ (b) is the fitted value of y_i using a bandwidth of b. In order to find the fitted value of y_i it is necessary to estimate the $\beta_k(u_i, v_i)$s at each of the data points and then combine these with the x-values at these points. However, there is a problem with such a procedure. Suppose b is made very small so that the weighting of all points except for i itself become negligible. Then the fitted values at the sampled points will tend to the *actual* values so that the value of equation (2.30) becomes zero. This suggests that under such an optimising criterion the value of b tends to zero which is not helpful. First, the parameters of such a model will not be defined in this limiting case and, second, the estimates will fluctuate wildly throughout space in order to give locally accurately fitted values at each regression point.

A solution to this problem is a *cross-validation* (CV) approach suggested for local regression by Cleveland (1979) and for kernel density estimation by Bowman (1984). Here, a score of the form

$$\mathrm{CV} = \sum_{i=1}^{n} [y_i - \hat{y}_{\neq i}(b)]^2 \tag{2.31}$$

is used where $\hat{y}_{\neq i}(b)$ is the fitted value of y_i with the observations for point i omitted from the calibration process. This approach has the desirable property of countering the 'wrap-around' effect, since when b becomes very small, the model is calibrated only on samples near to i and not at i itself.

Plotting the CV score against the required parameter of whatever weighting function is selected will therefore provide guidance on selecting an appropriate value of that parameter. If it is desired to automate this process, then the CV score could be maximised using an optimisation technique such as a Golden Section search (Greig, 1980). An example of the use of the cross-validation function to calibrate a GWR model is shown in Figure 2.20 which shows the CV score for the bi-square nearest-neighbour weighting function described in equation (2.29) applied to the London house price data. As can be seen, the plot is a relatively smooth one with a minimum value of the CV score at around 931 nearest neighbours. Thus in the application of the GWR model, only houses up to 930 nearest neighbours of a regression point i have non-zero weights (the weight of the 931st nearest neighbour

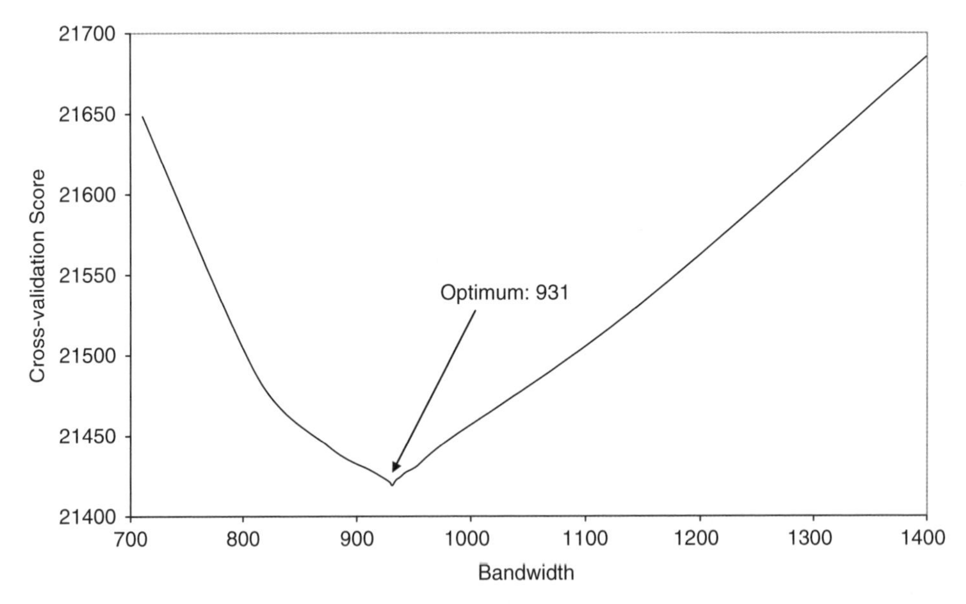

Figure 2.20 Cross-validation scores for the bi-square nearest neighbour weighting function applied to the London housing data

falls to zero using the bi-square function). The weights of these 930 points will decrease the greater is their distance from the regression point according to the bi-square function.

An approximation to the cross-validation statistic that is easier to compute is known as the *generalised cross-validation criterion* (GCV) which is described in Loader (1999) as being first used in the context of smoothing splines by Craven and Wahba (1979). The formula for the GCV score is:

$$\mathrm{GCV} = n \sum_{i=1}^{n} [y_i - \hat{y}_i(b)]^2 / (n - v_1)^2 \tag{2.32}$$

where v_1 is the effective number of parameters in the model as defined by equation (2.17). This term prevents the calibration wrapping itself around the data points because v_1 would then tend to n and the denominator of equation (2.32) would tend to zero.

A similar method of deriving the bandwidth which provides a trade-off between goodness-of-fit and degrees of freedom is to minimise the Akaike Information Criterion (AIC). Following Hurvich *et al.* (1998), we define the AIC for GWR as:

$$\mathrm{AIC}_c = 2n \log_e(\hat{\sigma}) + n \log_e(2\pi) + n\left\{\frac{n + \mathrm{tr}(\boldsymbol{S})}{n - 2 - \mathrm{tr}(\boldsymbol{S})}\right\} \tag{2.33}$$

where n is the sample size, $\hat{\sigma}$ is the estimated standard deviation of the error term, and $\mathrm{tr}(\boldsymbol{S})$ denotes the trace of the hat matrix (see equation 2.19) which is a function of the bandwidth (see Chapters 4 and 9 for further details). The AIC has the advantage of being more general in application than the CV statistics because it can be used in Poisson and logistic GWR (see Chapter 8) as well as in linear models. It can also be used to assess whether GWR provides a better fit than a global model taking into account the different degrees of freedom in the two models (see Chapter 4).

One other bandwidth selection criterion that has been used in the GWR literature by Nakaya (2002) is the Bayesian Information Criterion (BIC), sometimes referred to as the Schwartz Information Criterion (SIC) (Schwartz 1978). This is defined as:

$$\mathrm{BIC} = -2\log_e(L) + k \log_e(n) \tag{2.34}$$

where L is the model likelihood, k is the number of parameters and n is the sample size. This is similar to the AIC although the 'model complexity' penalty differs. Here, the same degree of complexity (that is, the same value of k) carries a higher penalty for larger samples. Consequently, in larger samples the use of the BIC tends to identify models with fewer parameters as optimal.

As its name suggests, the BIC was derived in a Bayesian context. It arises from a Bayesian model in which each of a discrete number of candidate models have equal prior probabilities, but the prior distributions on the model parameters, given the model, are noninformative. Note that unlike the AIC, the BIC is not intended to be an estimator of Kullback-Leibler information distance. Also, it is unclear from Schwartz's derivation of the BIC how the idea may be extended to variable bandwidth non-parametric models with effective degrees of freedom rather than a discrete set of models with whole numbers of parameters. However, given the ease with which the concept of effective degrees of freedom transfers to other model selection criteria, and the success of the criterion in Nakaya's study, the BIC appears to hold some promise in GWR.

2.7.5 Bias–Variance Trade-Off

It is useful at this stage to expand on the ideas discussed briefly above in Section 2.7.1 on the relationship between bias and variance both generally and within the context of GWR. To do this, one needs to discuss some properties of the estimator of $\mathbf{y}$, $\hat{\mathbf{y}}$. At any point in space (u,v), if we are given a set of predictors, $\boldsymbol{X}$, and a set of coefficient estimators $\hat{\boldsymbol{\beta}}$, then $\hat{y} = \boldsymbol{X}^{\mathrm{T}}\hat{\boldsymbol{\beta}}$ is an estimate of $\mathbf{y}$ at that point. However, $\hat{\boldsymbol{\beta}}$ is an estimate of $\boldsymbol{\beta}$ based on a sample of spatially diffuse $\boldsymbol{X}$ and $\boldsymbol{y}$ observations. Due to the randomness of the y terms, $\hat{\boldsymbol{\beta}}$ is random, and therefore so is $\hat{y}$. Two important properties of the distribution of $\hat{y}$ are its standard deviation and its expected value, $\mathrm{SD}(\hat{y})$ and $\mathrm{E}(\hat{y})$, respectively. When, for all $\boldsymbol{X}$, $\mathrm{E}(\hat{y}) = \mathrm{E}(y)$, the estimator is said to be unbiased. In this case, $\mathrm{SD}(\hat{y})$ is a useful measure of the quality of $\hat{y}$ as an estimator of y.

However, zero bias does not in itself guarantee an optimal estimator. Consider Figure 2.21 in which the horizontal line indicates the true value of y and the two box-plots represent the probability distributions of the two estimators of y. Although estimator 1 is a biased estimator (the expected value of y is higher than the true value), its overall variability is far less than that of estimator 2. Thus, the extreme values of potential errors in the prediction of y are less for estimator 1 than for estimator 2 and the only advantage estimator 2 has to offer is that the error

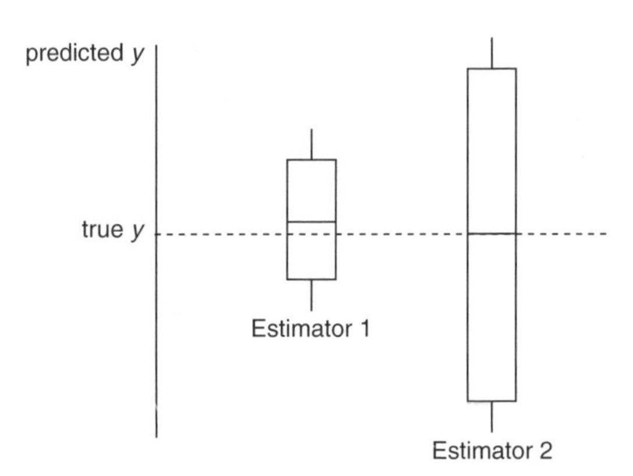

Figure 2.21 Bias and variance in estimators

distribution is centred on zero. If one were to consider the distributions of the prediction squared error (PSE), that for the second estimator would have a much longer tail. Overall, therefore, despite its bias, one might prefer estimator 1 over estimator 2.

This is an example of *bias–variance trade-off* – an issue addressed earlier and one that occurs in many types of statistical modelling – see, for example, work on multi-level modelling (Goldstein, 1987) and ridge regression (Hoerl and Kennard 1970a; 1970b). If regression coefficients vary continuously over space, then using weighted least-squares regression is unlikely to provide a completely unbiased estimate of $\boldsymbol{\beta}(u,v)$ at the given point (u,v) because for each observation there will be a different value of $\boldsymbol{\beta}$ but the regression requires that this value is the same for all observations. The best one can hope is that the values do not vary too much – and this is optimally achieved by only considering observations close to the point (u,v) at which we wish to estimate $\boldsymbol{\beta}(u,v)$. However, since this reduces the effective sample size for the estimate, the standard error of $\hat{\boldsymbol{\beta}}(u,v)$ will increase. Thus, the question arises as to *how close* to (u,v) should points be considered: too close and variance becomes large but the bias is small; too far and the variance is small but the bias is large. At one extreme, if a global model is chosen so that $\boldsymbol{\beta}(u,v)$ is assumed constant for all (u,v), and if there is variability in the true $\boldsymbol{\beta}(u,v)$, then clearly bias will cause problems. At the other extreme, if the local parameter estimates are derived from very small samples of data, they will have very large variances and be increasingly unreliable.

This bias–variance trade-off provides some justification for the use of cross-validation scores as a means of choosing bandwidth. A cross-validation score is essentially the sum of estimated predicted squared errors (PSEs) – the quantity discussed earlier. PSEs can be thought of as a measure of the overall performance of a particular bias/variance combination. We cannot know the exact PSEs (if we did, we would know the true $E(y)$ and β values and would need no statistical prediction), but CV scores provide an estimate which can then be used as a basis for selection.

A final point concerning bandwidth selection is that for a given set of variables – and therefore a given model – optimal bandwidth will change if the sampling strategy is altered. Thus bandwidth choice is *not* a parameter relating to the model itself, but is essentially part of the calibration strategy for a given sample. For example, if further data points were added to the model, although one hopes to achieve better estimates of $\boldsymbol{\beta}(u,v)$, one would expect the optimal bandwidth to decrease. Ultimately, if the sample size is continually increased, $\hat{\boldsymbol{\beta}}(u,v)$ should tend to $\boldsymbol{\beta}(u,v)$ but the bandwidth should tend to zero. However, this does not mean that by altering the bandwidth, and observing the changes in $\hat{\boldsymbol{\beta}}(u,v)$ one cannot gain some insight into the different scales of variation in $\boldsymbol{\beta}(u,v)$.

2.8 Testing for Spatial Non-stationarity

Up to this point the techniques associated with GWR have been predominantly descriptive. However, it is useful to ask the question: 'Does a particular set of local parameter estimates exhibit significant spatial variation?' This question, and others connected with significance testing in GWR, will be examined in greater detail in Chapter 4. Here we simply give the reader a sense of how such a question can be answered.

The variability of the local estimates can be used to examine the plausibility of the stationarity assumption held in traditional regression. For a given relationship k, at a given location i, suppose $\hat{\beta}_k(u_i,v_i)$ is the GWR estimate of $\beta_k(u_i,v_i)$. If we take n values of this parameter estimate (one for each regression point within the region), an estimate of variability in the parameter is given by the standard deviation of the n parameter estimates. This statistic can be defined as s_k.

The next stage is to determine the sampling distribution of s_k under the null hypothesis that the global model holds. Although theoretical properties of this distribution will be discussed in Chapter 7, for the time being a Monte Carlo approach will be adopted. Under the null hypothesis, any permutation of (u_i,v_i) pairs amongst the data points is equally likely. Thus, the observed value of s_k could be compared to the values obtained from randomly rearranging the data in space and repeating the GWR procedure. The comparison between the observed s_k value and those obtained from a large number of randomised distributions can then form the basis of the significance test. Making use of the Monte Carlo approach, it is also the case that selecting a subset of random permutations of (u_i,v_i) pairs amongst the n and computing s_k will also give a significance test when compared with the observed statistics.

To get a good 'feel' for the degree of spatial non-stationarity exhibited by any particular relationship in a model, the set of s_k values can be ranked lowest to highest and the proportion of values exceeding that from the observed data is a measure of the probability of observing such a variation in local parameter estimates from a stationary process. The lower this probability is, the more confidence one has in stating that the process generating the local parameter estimates is non-stationary.

Although the Monte Carlo approach works well in many GWR applications, it has the disadvantage of being computer-intensive, and the time taken to re-run GWR a thousand, or even just a hundred, times when the data set is large can be frustrating even with extremely fast computers. For that reason, theoretical tests have been developed for significance testing in GWR and these are described in Chapter 4. However, it turns out that these tests are also computer-intensive and perhaps the best solution in very large data sets is either to reduce the number of iterations used in the Monte Carlo tests or to use a subsample of the data set to calibrate the GWR model. Both of these options are available in software for GWR developed by the authors and are described in Chapter 9.

2.9 Summary

This chapter has described the basic idea and mechanics of GWR. We now explore some extensions to this basic idea.

3

Extensions to the Basic GWR Model

3.1 Introduction

In the previous chapter, the core idea of GWR models was introduced. However, there are a number of ways in which these models may be extended or enhanced. These will be considered in this chapter. We extend the basic GWR idea initially by specifying that some of the regression coefficients are globally fixed. The second and third extensions are more concerned with the nature of the error term: the second deals with the important problem of outlying observations; the third considers spatial heteroskedasticity, a situation in which the variance of the error term, as well as the regression coefficients, exhibits spatial non-stationarity.

3.2 Mixed GWR Models

In some situations not every regression coefficient in a model varies geographically. In others, the degree of variation for some coefficients might be negligible. It therefore may be helpful to consider mixed GWR models in which some coefficients are global – that is, they do not vary over space. The remaining coefficients are termed local and these are expected to be functions of geographical location, as in the basic GWR model. For example, when predicting house prices it may be hypothesised that the effect of structural attributes such as the number of bedrooms may vary geographically, but socio-economic attributes, such as unemployment level, may have the same effect everywhere. Note that the last statement does not imply that socio-economic variables have no effect, just that they have the same effect across the study region. One line of reasoning for suggesting such a hypothesis is as follows: once buyers have chosen a neighbourhood in which to purchase a house, the value of certain structural attributes will depend to an extent on the neighbourhood housing mix. For example, if few local houses have a garage, the value buyers associate with a garage will be high. Thus, structural coefficients are likely to vary geographically. On the other hand, although socio-economic indicators are neighbourhood characteristics, there is no obvious argument to suggest that their influence on house price will vary with neighbourhoods.

However, it should be noted that the above reasoning is just a hypothesis – although we state that we see no reason for socio-economic variables to have geographically varying effects, this does not imply that there are none. There may be factors that we have overlooked. The aim here is to investigate whether the hypothesis is supported by the data. Here, we use mixed GWR models to achieve this. Thus, we first consider how to specify and calibrate a mixed GWR model, and subsequently how we may compare mixed and basic GWR models and decide which one is the most appropriate. Mixed models may be written in the form

$$y_i = \sum_{j=1,k_a} a_j x_{ij}(a) + \sum_{l=1,k_b} b_l(u_i,v_i)x_{il}(b) + \varepsilon_i \tag{3.1}$$

where for observation i, y_i is the dependent variable, (u_i,v_i) is the geographical location, $\{a_1 \ldots a_{k_a}\}$ are the k_a global coefficients and $\{b_1(u,v) \ldots b_{k_b}(u,v)\}$ are the k_b local coefficient functions. Finally, $\{x_{i1}(a) \ldots x_{ik_a}(a)\}$ are the independent variables associated with global coefficients and $\{x_{i1}(b) \ldots x_{ik_b}(b)\}$ are the independent variables associated with local coefficients. We will refer to these two groups of independent variables as the a-group and the b-group, respectively. Note that one of the variables in either the a-group or the b-group could be a constant giving an intercept term but it is not possible to have intercept terms in *both* groups. If there are no a-group variables, we have a basic GWR model, and if there are no b-group variables, we have a standard linear regression model. Note that if all of $\{a_1 \ldots a_{k_a}\}$ were known, then we could calibrate $\{b_1(u,v) \ldots b_{k_b}(u,v)\}$ using the basic GWR method – by subtracting the first summation term in equation (3.1) from y_i and treating this as the dependent variable. Similarly, we could calibrate $\{a_1 \ldots a_{k_a}\}$ if $\{b_1(u,v) \ldots b_{k_b}(u,v)\}$ were known – this time by subtracting the second summation term from y_i and treating this as the independent variable. In this case we may use OLS to estimate $\{a_1 \ldots a_{k_a}\}$. These observations lead to a method of calibrating the mixed GWR model, adapted from Speckman (1988)[1] also considered in Bowman and Azzalini (1997). First, the mixed GWR model is written in vector-matrix notation as

$$\boldsymbol{y} = \boldsymbol{X_a a} + \boldsymbol{m} + \boldsymbol{\varepsilon} \tag{3.2}$$

where $\boldsymbol{y}$ is the vector of dependent variables, $\boldsymbol{X_a}$ is the matrix of a-group variables, $\boldsymbol{a}$ is the vector of a-group coefficients, $\boldsymbol{\varepsilon}$ is the vector of error terms and the ith

[1] Note, the method we introduce here is a slightly different approach to that adopted in Brunsdon *et al.* (1999) – this newer method is less computationally intensive.

element of vector $\boldsymbol{m}$ is given by $\sum_{l=1,k_b} b_l(u_i,v_i)x_{il}(b)$. Note that $\boldsymbol{m}$ is the geographically weighted term in equation (3.1). Now, recalling the argument above, we subtract the a-group terms from $\boldsymbol{y}$:

$$\boldsymbol{y} - \boldsymbol{X_a a} = \boldsymbol{m} + \boldsymbol{\varepsilon} \tag{3.3}$$

Assuming that $\boldsymbol{a}$ is known, we could then use basic GWR to estimate $\boldsymbol{m}$. Recall from Chapter 2 that the fitted y-values (denoted by $\hat{y}$) for a GWR may be written in the form $\hat{\boldsymbol{y}} = \boldsymbol{Sy}$, where $\boldsymbol{S}$ is termed the hat matrix (Hoaglin and Welsch 1978). Thus, we may estimate $\boldsymbol{m}$ by

$$\hat{\boldsymbol{m}} = \boldsymbol{S}(\boldsymbol{y} - \boldsymbol{X_a a}) \tag{3.4}$$

Substituting the expression for $\hat{\boldsymbol{m}}$ given in (3.4) for $\boldsymbol{m}$ in equation (3.2) and rearranging slightly gives the expression

$$(\boldsymbol{I} - \boldsymbol{S})\boldsymbol{y} = (\boldsymbol{I} - \boldsymbol{S})\boldsymbol{X_a a} + \boldsymbol{\varepsilon} \tag{3.5}$$

If we write $\boldsymbol{z} = (\boldsymbol{I} - \boldsymbol{S})\boldsymbol{y}$ and $\boldsymbol{Q} = (\boldsymbol{I} - \boldsymbol{S})\boldsymbol{X_a}$ then we have:

$$\boldsymbol{z} = \boldsymbol{Qa} + \boldsymbol{\varepsilon} \tag{3.6}$$

This is an ordinary (OLS) regression problem and $\boldsymbol{a}$ may be estimated using the standard approach:

$$\hat{\boldsymbol{a}} = (\boldsymbol{Q}^{\mathrm{T}}\boldsymbol{Q})^{-1}\boldsymbol{Q}^{\mathrm{T}}\boldsymbol{z} = (\boldsymbol{X_a}^{\mathrm{T}}(\boldsymbol{I} - \boldsymbol{S})^{\mathrm{T}}(\boldsymbol{I} - \boldsymbol{S})\boldsymbol{X_a})^{-1}\boldsymbol{X_a}^{\mathrm{T}}(\boldsymbol{I} - \boldsymbol{S})^{\mathrm{T}}(\boldsymbol{I} - \boldsymbol{S})\boldsymbol{y} \tag{3.7}$$

Thus, we may obtain an estimate of $\boldsymbol{a}$, say $\hat{\boldsymbol{a}}$. Returning to the original representation of the model in equation (3.1), we have estimates for $\{a_1 \ldots a_{k_a}\}$. As suggested earlier, subtracting this term (which we now treat as known) from the left-hand side of the equation allows the b-group coefficient functions to be estimated using the basic GWR procedure. In this way, the mixed GWR model may be calibrated.

It is worth observing that pre-multiplying by the matrix $\boldsymbol{I} - \boldsymbol{S}$, which occurs in the definitions of both $\boldsymbol{z}$ and $\boldsymbol{Q}$, gives the residuals from a GWR model. To see this, note that $(\boldsymbol{I} - \boldsymbol{S})\boldsymbol{y} = \boldsymbol{y} - \boldsymbol{Sy} = \boldsymbol{y} - \hat{\boldsymbol{y}}$. Thus, $\boldsymbol{z}$ is the vector of residuals obtained when regressing the b-group variables against $\boldsymbol{y}$ using basic GWR and the columns of $\boldsymbol{Q}$ are the vectors of residuals obtained when regressing the b-group variables against the a-group variables. Denoting the matrix of b-group variables as $\boldsymbol{X_b}$, the algorithm for mixed GWR may be set out as below:

1. For each column of $\boldsymbol{X_a}$:
   (i) Regress the column against $\boldsymbol{X_b}$ using basic GWR.
   (ii) Compute the residuals from the above regression.
2. Regress $\boldsymbol{y}$ against $\boldsymbol{X_b}$ using basic GWR.
3. Compute the residuals from the above regression.

4. Regress the $\boldsymbol{y}$-residuals against the $\boldsymbol{X_a}$-residuals using OLS. This gives the estimate $\hat{\boldsymbol{a}}$.
5. Subtract $\boldsymbol{X_a}\hat{\boldsymbol{a}}$ from $\boldsymbol{y}$. Regress this against $\boldsymbol{X_b}$ using basic GWR to obtain the geographically varying (b-group) coefficient estimates.

Before considering a practical example, it is worth making a few notes about this process. First, writing $\boldsymbol{W} = (\boldsymbol{I} - \boldsymbol{S})^{\mathrm{T}}(\boldsymbol{I} - \boldsymbol{S})$, equation (3.7) becomes

$$\hat{\boldsymbol{a}} = (\boldsymbol{X_a}\boldsymbol{W}\boldsymbol{X_a})^{-1}\boldsymbol{X_a}^{\mathrm{T}}\boldsymbol{W}\boldsymbol{y} \tag{3.8}$$

Further noting that $\boldsymbol{W}$ is a positive semi-definite symmetric matrix (as it is a product of a matrix and its transpose), equation (3.8) is identical to the estimation equation of $\boldsymbol{a}$ when the error terms are correlated. However, unlike existing models for regression in which the error terms are correlated, such as Kriging, the model for correlation will depend on $\boldsymbol{X_b}$. Second, it is possible to identify a hat matrix for mixed GWR. Substituting the expressions for $\hat{\boldsymbol{m}}$ and $\hat{\boldsymbol{a}}$ into equation (3.2), we note that

$$\hat{\boldsymbol{y}} = \boldsymbol{X_a}(\boldsymbol{X_a}\boldsymbol{W}\boldsymbol{X_a})^{-1}\boldsymbol{X}^{\mathrm{T}}\boldsymbol{W}\boldsymbol{y} + \boldsymbol{S}\boldsymbol{y} = \boldsymbol{S}^{*}\boldsymbol{y} \tag{3.9}$$

where $\boldsymbol{S}^{*} = \boldsymbol{X_a}(\boldsymbol{X_a}\boldsymbol{W}\boldsymbol{X_a})^{-1}\boldsymbol{X_a}^{\mathrm{T}}\boldsymbol{W} + \boldsymbol{S}$. This will prove useful for developing inferential approaches, such as those in Chapter 4. Finally, we note that if there are k_a a-group variables, then using the algorithm set out above it will be necessary to run $k_a + 2$ basic GWR calibrations each with k_b variables. This information is helpful in estimating run times for large mixed GWR model calibrations.

3.3 An Example

To illustrate the method described above, we consider a subset of the Nationwide Building Society house price data set for London described in Chapter 2. Here, we randomly select 500 houses from this set, mainly to reduce the computational effort required to calibrate the mixed GWR. Recall that mixed GWR requires $k_b + 2$ basic GWR models to be calibrated and in addition to this, computation of the hat matrix will require manipulation of several n by n matrices. The locations of the 500 sampled houses are shown in Figure 3.1.

As suggested in the previous section, two competing models for house price are:

1. Model A: a basic GWR in which all coefficients vary geographically.
2. Model B: a mixed GWR in which structural coefficients vary locally but socio-economic variables are the same everywhere.

A full listing of the variables in the two models is given in Table 3.1. Two other models (C and D) also used in the analysis are listed; these latter two models have no geographically varying terms. Note that as well as the two socio-economic

Figure 3.1 *Locations of houses in sample*

variable coefficients, the coefficient for the log distance from the centre of London is also treated as fixed in model B. The intention here is to preserve the isotropic nature of the term in the model. Effectively, in the mixed model the distance variable is forced to take the functional form $\frac{1}{2}\, a \log_e (u_l^2 + v_l^2)$ where (u_l, v_l) are locational coordinates centred on Nelson's column and a is the corresponding coefficient in the regression model. Note that this function has directional symmetry. However, in the basic model the term takes the more general form $\frac{1}{2}\, b^*(u_l, v_l) \log_e (u_l^2 + v_l^2)$ where b^* is the geographically varying coefficient function for the log distance term, with coordinates re-centred on Nelson's column. In this case, as b^* is an arbitrary two-dimensional function, we have no guarantee that the term exhibits directional symmetry.

It is important to note that here we are interested in deciding which model is the more appropriate – we do not claim that either is faultless, and do not expect that the data are generated by *exactly* the stochastic process described by either model A or model B. However, we expect that these models can potentially perform well and approximate reality reasonably accurately. The issue here is which of the two models approximates reality more closely. For this reason, we adopt the Akaike Information Criterion (AIC) approach to statistical inference (Akaike 1973). This is described in Chapters 2 and 4 so here we simply state that in general an AIC is computed for each of a number of competing models fitted to a given data set, and the model with the smallest AIC is deemed to be the best fit to the data. As a rule of thumb, in cases where the difference between AICs is less than around 3, the competition between models is regarded as 'too close to call', i.e. there is no clear evidence as to which of the two models is better.

Here, four competing models are considered – the full GWR model (A), the mixed GWR model (B), and the stationary models (C) and (D) as described in Table 3.1. As we move from model A to model D we reduce model generality. In model A, all coefficients are assumed to vary geographically. In model B, the socio-economic and distance coefficients are assumed fixed; in model C all coefficients are assumed fixed; and in model D the socio-economic and distance-based variables are dropped entirely.

The GWR models were fitted using an adaptive method with a bi-square kernel with a span of 30% of the data set. The AICs for the four models are listed in

Table 3.1 *Variables included in models A, B, C and D*

Variable	Model A	Model B	Model C	Model D
INTERCEPT	Non-stationary	Non-stationary	Stationary	Stationary
FLRAREA	Non-stationary	Non-stationary	Stationary	Stationary
BLDPWW1	Non-stationary	Non-stationary	Stationary	Stationary
BLDPOSTW	Non-stationary	Non-stationary	Stationary	Stationary
BLDP60S	Non-stationary	Non-stationary	Stationary	Stationary
BLDP70S	Non-stationary	Non-stationary	Stationary	Stationary
BLDP80S	Non-stationary	Non-stationary	Stationary	Stationary
TYPDETCH	Non-stationary	Non-stationary	Stationary	Stationary
TYPTRRD	Non-stationary	Non-stationary	Stationary	Stationary
TYPBNGLW	Non-stationary	Non-stationary	Stationary	Stationary
TYPFLAT	Non-stationary	Non-stationary	Stationary	Stationary
GARAGE	Non-stationary	Non-stationary	Stationary	Stationary
CENTHEAT	Non-stationary	Non-stationary	Stationary	Stationary
BATH2	Non-stationary	Non-stationary	Stationary	Stationary
FLRDETCH	Non-stationary	Non-stationary	Stationary	Stationary
FLRFLAT	Non-stationary	Non-stationary	Stationary	Stationary
FLRBNGLW	Non-stationary	Non-stationary	Stationary	Stationary
FLRTRRD	Non-stationary	Non-stationary	Stationary	Stationary
PROF	Non-stationary	Stationary	Stationary	Excluded
UNEMPLOY	Non-stationary	Stationary	Stationary	Excluded
ln(DISTCL)	Excluded	Stationary	Stationary	Excluded

Note: Variable names are those defined in Chapter 2. Here stationary variables constitute the *a*-group and non-stationary variables constitute the *b*-group in mixed GWR.

Table 3.2 *Akaike information criteria for models A,B,C and D*

Model	A	B	C	D
AIC	11 349	11 342	11 389	11 430
Effective number of parameters	110.4	99.23	21	18

Table 3.2. These suggest that of the four models considered, the best is model B. This is the model with fixed socio-economic coefficients but geographically varying structural coefficients. The second row lists the effective numbers of parameters – another concept discussed in Chapters 2 and 4. For OLS regressions, this is simply the number of linear coefficients in the model. For the GWR-based models, it is a more general measure of model complexity. The values listed in Table 3.2 reflect the relative reduction in complexity from models A to D.

Thus we conclude from the above analysis that there is some support for the hypothesis that the effect of socio-economic variables on house price is constant over London, while the effect of structural variables is non-stationary – at least for our analysis of the 500 item subset of the data. Comparing the maps of coefficients for models A and B, for example the INTERCEPT term (see Figures 3.2 and 3.3), the general patterns are similar although there are some notable differences. First, a glance at the shading of the two maps suggests that the coefficient has a greater range in model A than in model B. Although the general trends in the two maps are similar, there are marked peaks to the east and west of model A.

Likewise, looking at the coefficient for BLD70S (the value of a 1970s-built house relative to an inter-war period house) in Figures 3.4 and 3.5, the two distributions

Figure 3.2 *Map of the INTERCEPT term (model A)*

Figure 3.3 *Map of the INTERCEPT term (model B)*

Figure 3.4 *Map of the BLD7OS term (model A)*

exhibit the same general trends but there are some interesting differences in the detail. There is also the same range inflation of the coefficient for model A.

In general, we conclude that although a slightly better model is achieved using mixed GWR, basic GWR still gives a fair approximation of the key trends. However, working with basic GWR may lead to some misinterpretation of local details.

Figure 3.5 Map of the BLD7OS term (model B)

One possible explanation for this is that including geographically varying terms in the model when in reality the term is fixed will result in a degree of random local fluctuation in the socio-economic coefficients. Since the structural characteristics of the houses will most likely be correlated to the socio-economic variables (for example, houses with two bathrooms tend to be found in affluent areas), this in turn may cause some extra fluctuation in the GWR estimates of the structural variables. This may explain the spurious features and greater range of values of the GWR coefficients in model A.

In summary, although mixed GWR requires a greater computational effort than basic GWR, the gain is that more stable estimates of the varying coefficients are achieved if some of the coefficients are stable over space.

3.4 Outliers and Robust GWR

The influence of outliers is a key issue in ordinary regression and is no less of an issue for GWR. In the bivariate case, the problem is illustrated in Figure 3.6. Here the point labelled 'outlier' stands clear of the other (x,y) pairs in the scatterplot. The effect of the outlier on calibrating (a,b) in the regression model $y_i = a + bx_i + \varepsilon_i$ with $\varepsilon_i \sim N(0,\sigma^2)$ can also be seen in Figure 3.6. The solid line shows the fitted regression when the outlier is included in the analysis; the dotted line shows the fitted regression when it is excluded. There is a clear difference between the two lines. In fact, the points in Figure 3.6 were simulated from a model in which $(a,b) = (0,1)$. The dotted line corresponds to an estimate of $(a,b) = (0.01, 0.95)$ whereas the solid line corresponds to $(a,b) = (0.12, 0.77)$.

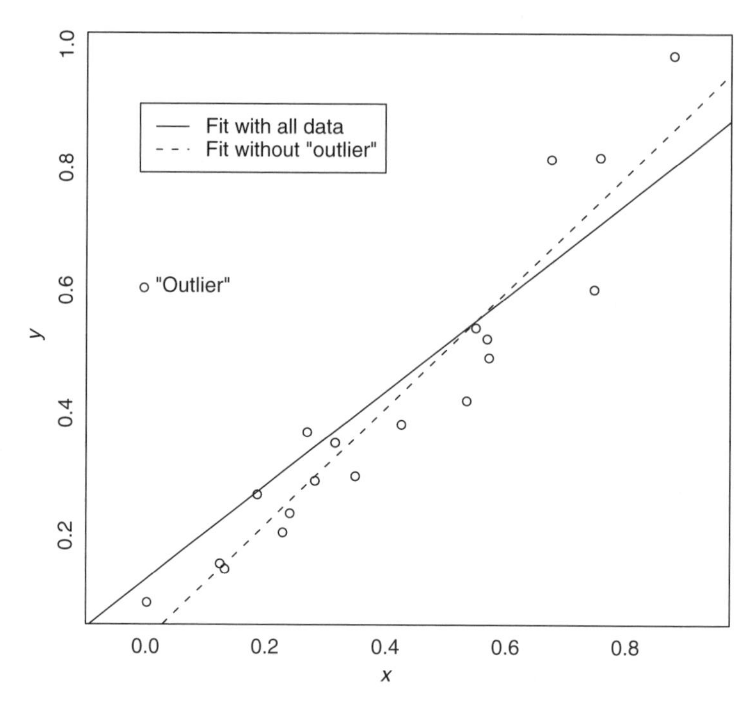

Figure 3.6 *The effect of an outlier on regression*

Thus, the inclusion of the outlying unusual observation in the model calibration gave considerably less accurate parameter estimates.

The above situation illustrates the importance of allowing for outliers in regression models. There are situations in which 'weeding out' problematic observations can improve model calibration; this is true of GWR. The effect of an outlier can be to distort local parameter estimates, and consequently to distort local parameter surface estimates. Unfortunately, outliers are often harder to detect in the GWR situation. This is mainly because an observation has only to be unusual *in its locality* for the distorting effect to occur. Since, computationally at least, calibrating GWR at a regression point is equivalent to fitting a regression model in a window around that point, one only needs an observation to be unusual relative to other observations inside the window for problems to occur. Even in the bivariate case, scatterplots such as Figure 3.6 may be unhelpful: although they identify global outliers, a local outlier may be impossible to spot since it may not be an unreasonable observation in a global context. Furthermore, scatterplots run into problems even in standard regressions when there is more than one independent variable since they are essentially two-dimensional graphical tools.

All of this implies that some means of identifying and handling outliers would be a useful tool for GWR analysis but that simple inspection of scatterplots is unlikely to suffice. Here we consider two approaches, both based on existing approaches for ordinary linear regression. In both cases, the residual $y_i - \hat{y}_i = e_i$ plays a central role. In cases where an outlier is present, this observation will lie a long way from the general trend (even in the GWR case where such a trend varies locally), and thus one expects the residual at observation i to be large.[2] Thus, examining residuals appears a useful alternative to scatterplots as an outlier identification tool. However, 'raw' residuals need to be treated with some care, since they do not all have the same variance. To see this, consider in vector notation

$$\hat{y} = \boldsymbol{S}\boldsymbol{y} \tag{3.10}$$

(with the variables defined as above) so that

$$\boldsymbol{e} = \boldsymbol{y} - \boldsymbol{S}\boldsymbol{y} = (\boldsymbol{I} - \boldsymbol{S})\boldsymbol{y} \tag{3.11}$$

where $\boldsymbol{e}$ is the vector of e_is. Now we can see that

$$\mathrm{Var}(\boldsymbol{e}) = (\boldsymbol{I} - \boldsymbol{S})(\boldsymbol{I} - \boldsymbol{S})^{\mathrm{T}}\mathrm{var}(\boldsymbol{y}) = (\boldsymbol{I} - \boldsymbol{S})(\boldsymbol{I} - \boldsymbol{S})^{\mathrm{T}}\sigma^2 \tag{3.12}$$

Thus, the variances of the e_is are the leading diagonal elements of $(\boldsymbol{I} - \boldsymbol{S})(\boldsymbol{I} - \boldsymbol{S})^{\mathrm{T}}\sigma^2$, which in general are not all equal. The residuals may be standardised (*Studentised*) by dividing by the square root of this quantity:

$$r_i = \frac{e_i}{\hat{\sigma}\sqrt{q_{ii}}} \tag{3.13}$$

where q_{ii} is the ith element of the leading diagonal elements of $\boldsymbol{Q} = (\boldsymbol{I} - \boldsymbol{S})(\boldsymbol{I} - \boldsymbol{S})^{\mathrm{T}}$. For ordinary regression, $q_{ii} = 1 - s_{ii}$. A further refinement is possible by replacing the term $\hat{\sigma}^2$ by $\hat{\sigma}_{-i}{}^2$, the estimate of σ^2 obtained when the ith observation is omitted from the calibration data set:

$$r^*_i = \frac{e_i}{\hat{\sigma}_{-i}\sqrt{q_{ii}}} \tag{3.14}$$

The idea here is that if point i is an outlier, including it in the estimate of σ^2 is likely to give an upward bias. The term r_i is called the *internally Studentised residual* and r_i* is called the *externally Studentised residual*. Chatfield (1995) suggests tagging observations for which $|r_i^*|$ exceeds 3 as potential outliers. For the housing data subsample used earlier in this chapter, the externally Studentised residuals are shown in Figure 3.7.

From Figure 3.7, a number of potential outliers are visible. Their locations are shown in Figure 3.8. There are more dwellings that are unusually expensive rather than unusually cheap and a number of these appear to be close to the River Thames. It is possible that these houses have a good view across the river – a factor likely to inflate price but one not recorded in our data set.

[2] One problem with this method occurs when the outlier has such a high leverage that it pulls the regression line very close to it. However, we anticipate such occurrences to be rare.

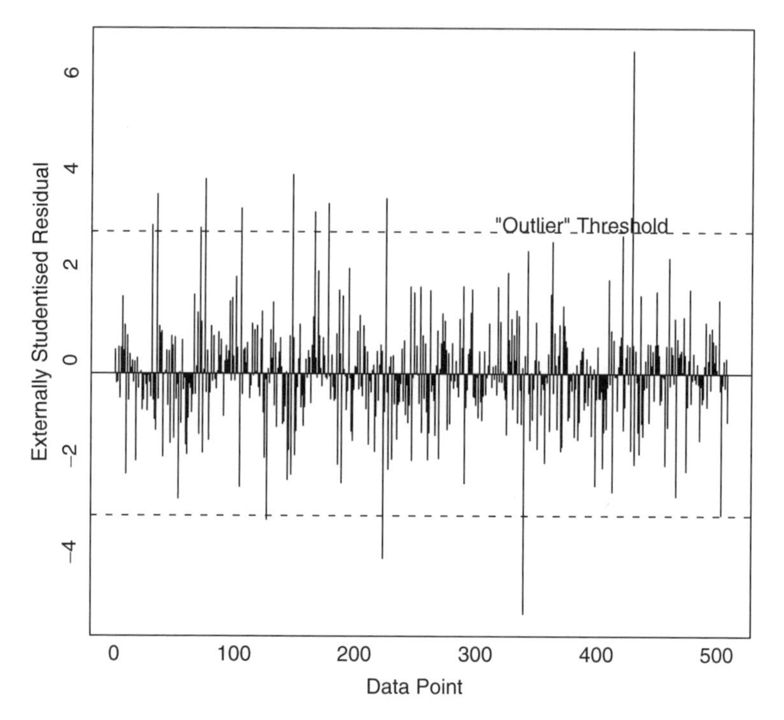

Figure 3.7 *Externally Studentised residuals for housing subsample data*

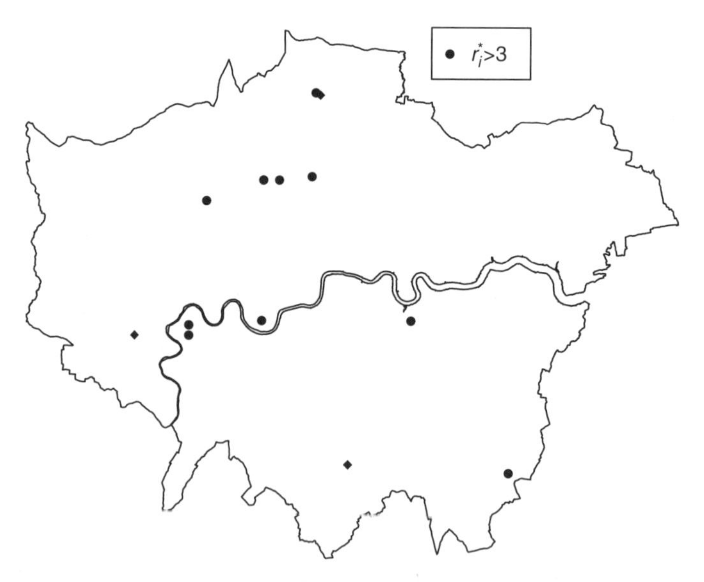

Figure 3.8 *Locations of observations with Studentised residuals greater than 3*

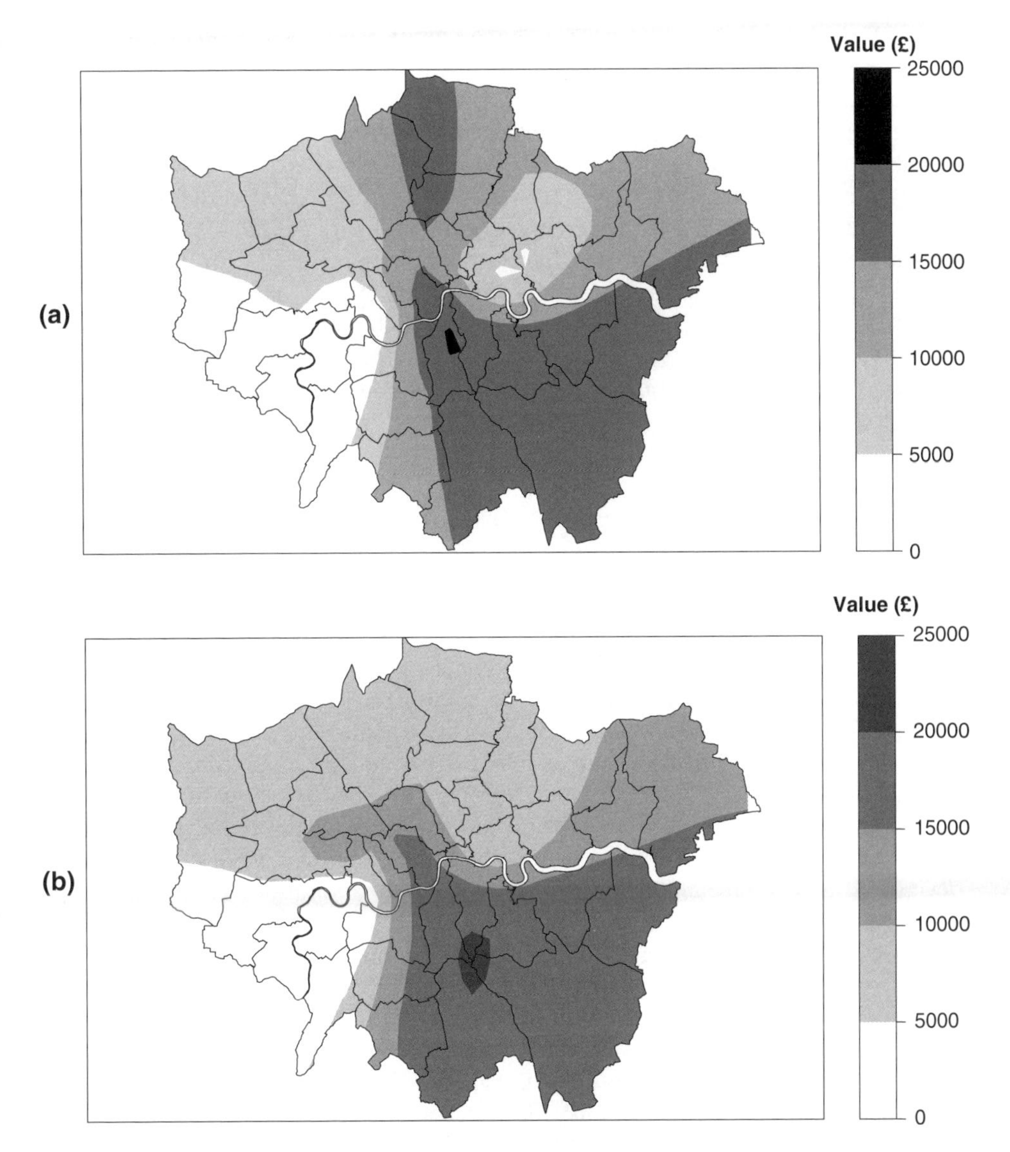

Figure 3.9 *Fitted surfaces for GARAGE: (a) all data; (b) data with outliers removed*

The usual approach at this point is to remove the outliers from the data set and then re-fit the model. To appreciate the extent to which the outliers have distorted the model fitting process, one should compare results before and after the outliers have been filtered. In practice, for the GWR case, one would compare all coefficient surfaces of interest for both fits. Here, we will just consider the surface associated with the variable GARAGE. It is interesting to note that for all of the outlying data items near to the river, the properties concerned do not have garages, despite their unusually high selling prices. The two surfaces are shown in Figure 3.9. Here, we see that although the surfaces are similar, the data with the outliers contains an area of apparently high GARAGE coefficients in north-central London which disappears when the model is recalibrated without the outliers.

The 'weeding out' approach using externally Studentised residuals seems to be a useful way of identifying outliers and assessing their effect. However, there are computational problems for large data sets. The term r_i* relies on the computation of q_{ii} which in turn needs the computation of $\boldsymbol{Q} = (\boldsymbol{I} - \boldsymbol{S})(\boldsymbol{I} - \boldsymbol{S})^{\mathrm{T}}$. The latter expression is an n by n matrix which, although manageable for the 500-point dataset used in this example, would be unwieldy for large data sets (say, $n = 10\,000$). In this case an alternative approach based on *Robust Regression* is suggested – see, for example, Huber (1981) or Hampel *et al.* (1986). Robust regression works by downweighting observations having large residuals when an ordinary regression model is fitted. Typically a weighting function might be

$$w_r(e_i) = \begin{cases} 1 & \text{if } |e_i| \leq 2\hat{\sigma} \\ [1 - (|e_i| - 2)^2]^2 & \text{if } 2\hat{\sigma} < |e_i| < 3\hat{\sigma} \\ 0 & \text{otherwise} \end{cases} \tag{3.15}$$

where w_r is a weighting function depending on the residual e_i. This weighting function provides weights which are used in a re-fitting of the model. In simple terms, the re-fitted model downplays the effect of observations that did not agree well with the original model. This function is shown in Figure 3.10. When the fitted residual is between $-2\hat{\sigma}$ and $2\hat{\sigma}$, the data for that point have unit weighting. Thus, if the original model has no residuals of greater than two standard deviations, the re-fitted model will be no different. However, between two and three standard deviations the weighting tapers off. Thus, in a re-fitting, observations with residuals in this band will have less influence. Finally, if a residual is outside three standard deviations, it is completely dropped from the second calibration. Thus, the method works by applying a 'trial fit' and identifying high residuals, and then re-calibrating but this time downweighting the effect of these residuals. Note that in the GWR situation each local regression fit is already weighted (by the kernel function) and so the existing weight has to be multiplied by $w_r(e_i)$ in the re-fit. This process may be cycled through a number of times: after each weighted model fit, the latest set of residuals are computed, the weights re-calculated and the model updated once again using the new weights. Thus, the process provides an automated means of allowing for outliers when calibrating a GWR model. Computationally, this has much less overhead than the Studentised residual approach, since each cycle only requires a set of n residuals to be computed – one does not have to deal with n by n hat matrices. In many cases, a single cycle is sufficient to remove the worst distortions of any outliers.

However, the increased computational efficiency of the weighted residual approach comes at a cost. The method relies on raw residuals and so overlooks the point made earlier in this section that all residuals do not have the same variance – indeed it is compensating for the differences in variance that requires much of the computation when using Studentised residuals. Also, while the Studentised residual approach lends itself to identification and human examination of unusual cases, the

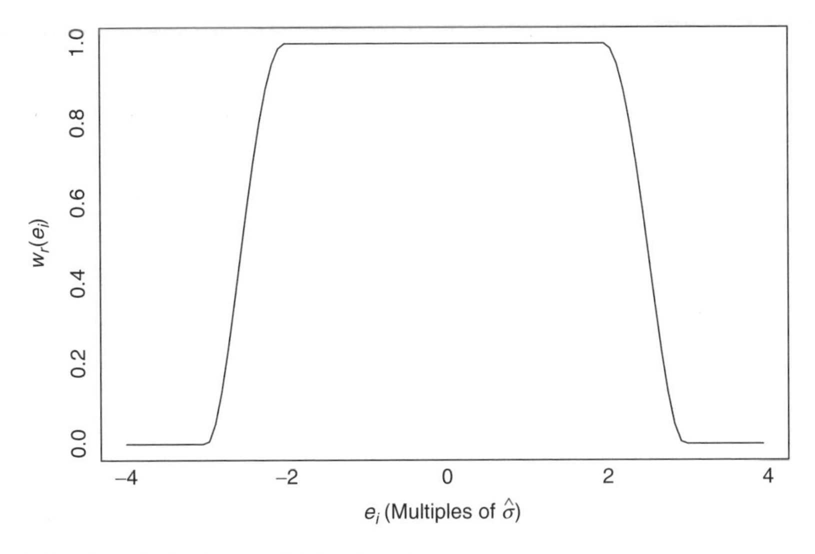

Figure 3.10 *A typical robust weighting function*

Figure 3.11 *A robust fitting of the GARAGE parameter surface*

robust regression relies much more on a 'black box' approach. This could be considered a disadvantage – human examination of exceptional cases may give rise to a greater understanding of the data under study and of the underlying process giving rise to those data. Black box approaches tend to sweep such exceptional cases under the carpet and therefore possibly prevent some interesting discoveries.

However, in some situations when the data set is very large, practicality dictates that the robust method is the only real option.

The robust method for the housing data is shown in Figure 3.11, again for the surface of the GARAGE coefficient. Here we see a similar pattern to that in Figure 3.9 – the surface is closer to the surface generated with the outliers removed than the full data surface, although in general the values of the surface are slightly higher. This is due to the fact that the attenuation of outliers is more extreme than in the Studentised-residuals based method; as well as removing some observations completely, the robust method also downweights some less extreme 'near-outliers', due to the shape of the w_r function.

3.5 Spatially Heteroskedastic Models

In the basic GWR model, although the regression coefficients vary geographically, the variance of the error term is assumed fixed. In this section we introduce a generalisation of this model in which this variance may also vary geographically. Thus, the model still takes the form

$$y_i = \sum_{j=1}^{m} \beta(u_i, v_i) x_{ij} + \varepsilon_i \tag{3.16}$$

but now the distribution of ε_i is $N(0,\sigma^2(u_i,v_i))$ rather than simply $N(0,\sigma^2)$. Thus σ^2 is now a function of (u_i,v_i) instead of a constant. This is a form of heteroskedasticity, as the variance of the error term varies from place to place. When the heteroskedasticity is of a known form, and the variance depends on the expected value of the dependent variable, the situation is often resolved by transforming the y-variable. The square root transform for Poisson variables is an example of this (Bartlett 1936). However, in equation (3.16) it is possible for the expected value of y to be the same in two places, while the variances differ in those places. Thus an alternative approach is needed here.

One superficially appealing solution may be to estimate the regression σ^2 at each local regression surface and to draw the surface of these. However, strictly this is not the correct approach since the regression coefficients would be calibrated on the assumption that σ^2 is constant. The key characteristic of equation (3.15) is that this assumption is *not* true. An alternative approach is to weight the observations to allow for the different levels of variability. Since the error term is normally distributed, we can allow for heteroskedasticity by weighting according to the inverse variance of each observation. For the GWR situation, weighting is applied in the same way as for robust regression – that is, the spatial kernel weight is multiplied by the inverse variance weight. One obvious problem with the above method is that it assumes that the functional form of the variance of the error term $\sigma^2(u_i,v_i)$ is already known – which is unlikely to be true in practice. A tentative method of overcoming this problem is proposed here. First, we note that

$$E(\varepsilon_i^2) = \sigma^2(u_i, v_i) \tag{3.17}$$

Recall that ε_i is the error term at point i. Although we do not know ε_i we may estimate it using e_i, the residual at point i, then an ordinary GWR model is fitted. This assumes that the basic GWR fit has a negligible bias – the method should not be used if there is some suspicion that the basic model is highly biased. If we assume that $\sigma^2(u_i, v_i)$ is a continuous function over space, we may reasonably estimate it by applying a mean smoother over the e_i^2s. Call this estimate $\hat{\sigma}^2(u_i, v_i)$. Using this, we then re-estimate the GWR model applying the weight $1/\hat{\sigma}^2(u_i, v_i)$ to observation i in addition to the usual geographical kernel weights. This will give us an improved set of coefficient surface estimates, which in turn leads to an updated set of residuals, and finally an updated estimate $\sigma^2(u_i, v_i)$. There are parallels between this method and the robust estimation algorithm introduced in the previous section. In both cases, the starting point is a trial estimation based on standard GWR, followed by a re-weighted model fit where the weights are determined by the residuals. The main difference is that here the re-weighting is based on a smoothing of the residuals, whereas the robust technique is based on re-weighting isolated points. This reflects the nature of the underlying models in each case. In the robust model, we assume all observations follow the standard GWR model (where error variance is stationary) with the exception of a handful of isolated 'rogue' points. On the other hand, in the heteroskedastic model we assume that there is a continuously varying geographical trend in the error term.

Another characteristic of the above procedure that is shared with the robust algorithm is that it may be regarded as cyclic. In each cycle, a new set of residuals are produced, giving rise to a new set of estimates of $\hat{\sigma}^2(u_i, v_i)$s. These enable a new weighted GWR to be calibrated and so the cycle repeats. Experimentally, it has been found that this iterative approach generally converges. Thus, the iterative method may be used to calibrate the model in equation (3.16). However, it should be noted that little is known about the distribution of the estimates, making it difficult to calculate their standard errors. This is mainly because the surface standard errors are dependent on $\hat{\sigma}^2(u_i, v_i)$ which itself is subject to random variation. For the time being then, this approach should perhaps be regarded as an exploratory method rather than as an inferential tool. It could also be used as a tool to investigate the stability of GWR against heteroskedasticity. If the above algorithm gives a very different surface than that from ordinary GWR for some variable, this suggests that further investigation is needed into the model of the error term and that the results from ordinary GWR are suspect.

The algorithm was applied to the housing subset data, and the results are shown in Figure 3.12. Here, the estimate of local error variance was carried out using an adaptive bi-square kernel moving window average, where in each case the bandwidth was selected to cover 30% of the data. The estimate of GARAGE is similar to the earlier estimates in Figures 3.9 and 3.11 (perhaps having more in common with the robust estimates), although the peak in the coefficient in the south-east appears less marked. In terms of error variance there appears to be a band of higher values running down the centre of the study area from the north to the south-east.

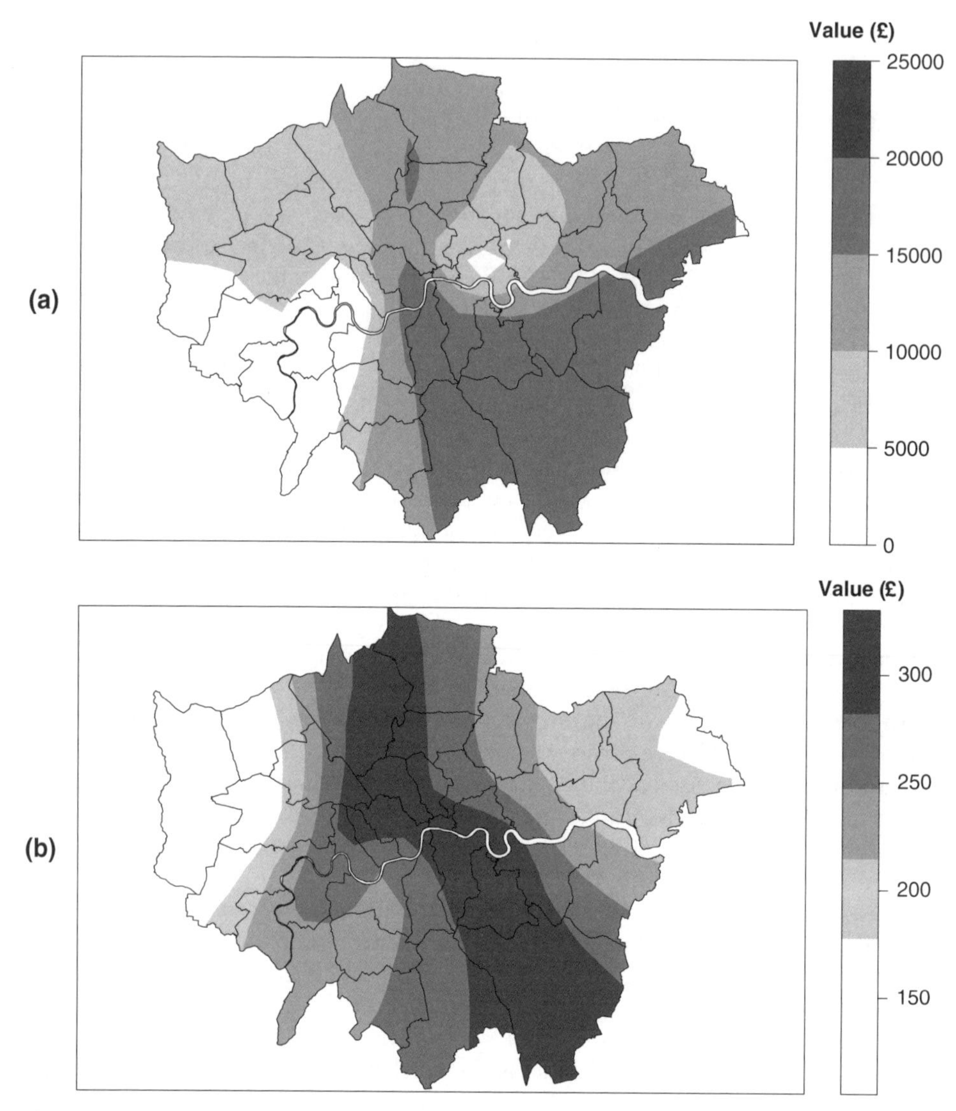

Figure 3.12 Heteroskedastic model: (a) surface for GARAGE; (b) $\hat{\sigma}^2(u_i,v_i)$

3.6 Summary

In this chapter, a number of extensions of the basic GWR method have been introduced. All of these may be regarded as generalisations of the basic method where the core notion of a spatially non-stationary OLS regression model is enhanced, by specifying that some coefficients are fixed, or that the error variance is not fixed, or by fine-tuning the calibration method to be more robust to outliers. This may be contrasted with Chapter 7 where geographically weighted extensions of entirely different models and methods are introduced.

4

Statistical Inference and Geographically Weighted Regression

4.1 Introduction

The use of GWR as a visual technique can reveal some interesting patterns in geographical data. However, a quite reasonable criticism is that one cannot be sure that the patterns viewed could have occurred by chance. That is, even with a completely random data set, it is possible that some degree of spatial pattern in the estimated parameters may be observed when applying a GWR algorithm. For example, geographically weighted regression was carried out on a 21 × 21 rectangular grid of points covering the region $[-10,10] \times [-10,10]$. The independent variable u was set equal to the x-coordinate of the grid point, and the dependent variable v was simulated using the global relationship $v = 10u + \varepsilon$ where ε is a standard normal variate. Although there was no local variation in the relationship between u and v, carrying out GWR on this data set with a bandwidth of 6 and a bi-square kernel gives the result seen in Figure 4.1. Clearly, this does show some form of spatial pattern despite the global nature of the model.

This is perhaps not particularly worrying if one looks at the scale of the graph – the variation in the estimate of the slope term lies within the range [9.85,10.20] which is reasonable given the true global value of 10. Also, the fact that a pattern occurs is not very surprising; given there is some random variation in the data, it is highly unlikely that GWR would produce a perfectly flat surface of parameter estimates. However, although these comments shed some light on the interpretation of Figure 4.1, a number of questions remain unanswered. In particular, how can one decide whether an observed pattern could have arisen due to random variation in a global model as in Figure 4.1, or whether it reflects a true geographical trend

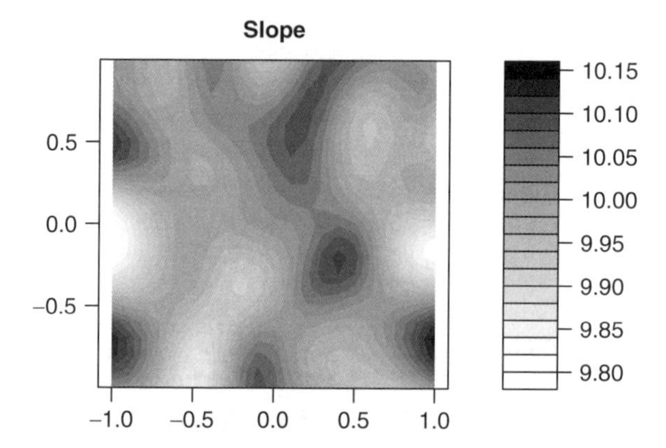

Figure 4.1 GWR estimates of local slope parameters

in the model? This is a question of statistical inference and in this chapter we set out to choose a set of inferential approaches for GWR, and to provide a justification for our choice. In addition, we also provide some practical examples to show how the methods advocated here may be used in practice. In order to do this, we first consider the notion of inference in greater depth (Section 4.2). Following from this discussion we suggest two possible inferential approaches, a classical, confidence interval-based approach and an approach based on the Akaike Information Criterion (AIC) for model selection (Akaike 1973). It is expected that readers will be more familiar with the former approach, and so some effort has been made to introduce AIC methods both here and in Chapter 2.

4.2 What is Meant by 'Inference' and How Does it Relate to GWR?

Questions of inference are fundamental to statistical analysis. However, before considering how inferential methods may be applied to GWR, it is important to discuss the meaning of the term 'inference' in a statistical context. As its name suggests, statistical inference is concerned with the process of inferring information from the analysis of statistical data sets. Typically, the process sets out to answer one of three kinds of questions, as set out in Table 4.1.

Although data exploration and visualisation are important tools for uncovering the structure of data and identifying potential spatial patterns, there are many pitfalls in attempting to answer the above questions using only these kinds of approach. In many instances, difficulties arise in much the same way as in the example seen in Section 4.1; there is always some degree of random variation in data collection, and observed features in some representation or transformation of a data set may simply be a consequence of this. To decide whether this is the case, or if the observed pattern is attributable to some geographical trend, we must attempt to answer at least one of the questions listed in Table 4.1.

Table 4.1 General questions underlying statistical hypotheses

1	Is some fact true on the basis of the data?
2	Within what interval does some model coefficient lie?
3	Which one of a series of potential mathematical models is 'best'?

In fact, in the GWR case, we can use any one of the three questions in Table 4.1 to investigate aspects of non-stationarity. In each case, the problem would be approached in a different way, but all would shed some light on whether the pattern observed in Figure 4.1 is due to a geographical trend. The following three subsections will introduce each approach and consider how each could be used in a GWR context.

4.2.1 How Likely is it that Some Fact is True on the Basis of the Data?

This form of inference is commonly known as the hypothesis test or significance test. The method here is to ask how likely the observed results are, given some hypothesis. Typically, this is denoted as H_0, the null hypothesis. In the GWR context, we could ask how likely is an observed pattern (such as the one in Figure 4.1) if H_0 is true, given that H_0 is that the data are generated by a global model. If the observed pattern is extremely unlikely under this hypothesis, we have evidence that H_0 is unlikely to be true. The so-called p-value of a test is the probability of obtaining an observed pattern (or one that is more extreme – see Section 4.3) given that H_0 is true. A significance test is carried out if one checks whether the p-value falls below some threshold – typically 0.01 or 0.05. Note that these are not probabilities of whether H_0 is true, they are probabilities of incorrectly rejecting the hypothesis given that it is actually true. Strictly speaking, the approach does not address the question at the head of this section. This, and some other arguments (Chatfield 1995; Nester 1996) suggest that there are problems with this approach under some circumstances. Although p-values and significance tests are of some use, their utility is not as great as some practitioners appear to believe. In Section 4.2.3 we will discuss this in more detail.

4.2.2 Within What Interval Does Some Model Coefficient Lie?

In a standard global regression, it is possible to estimate standard errors for the regression coefficients. These are the sampling standard deviations of the coefficient estimates. These can be used to give confidence intervals for the coefficients: for example, when the sample size n is sufficiently large, then the interval defined by the coefficient estimate plus or minus approximately 1.96 times the standard error will contain the true coefficient value 95% of the time. In GWR we are estimating coefficient *surfaces*, or regression coefficient values for a set of geographical locations. It is possible to compute standard errors for these coefficient values at any of the geographical locations.

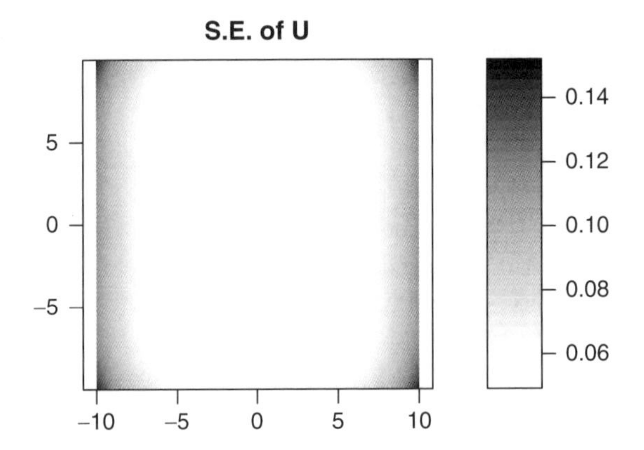

Figure 4.2 *GWR estimate of standard error of local slope parameters*

Having computed standard errors for all points on a GWR coefficient surface, we may then consider the degree to which the surface varies within the study area, and compare this to the point-based standard errors. For example, the standard error for the slope coefficient from the example in Section 4.1 is shown in Figure 4.2. Here, it can be seen that the standard error ranges from around 0.05 to 0.15. Thus, the minimum 95% confidence interval width is around four times 0.05 (around 0.2), and at places is around 0.6. Given the limited range over which the slope estimate itself varies (see Figure 4.1), from an informal basis it would seem that the degree of variation displayed in Figure 4.1 is unlikely to be due to a genuine geographical pattern.

Using a more formal approach, we could consider the difference between the coefficient estimate at two distinct locations. Again, we could work out the standard error (and subsequently a confidence interval) for this difference, and decide whether the estimated value of this difference was sufficiently 'important' to justify using a geographically non-stationary model. This latter approach differs subtly from the first; we are not asking whether the difference between coefficients is zero (as we would with H_0 as used in Section 4.2.1) but are attempting to *measure* the difference and provide some kind of upper and lower bounds. On the basis of this, we must then decide whether the difference is important *in the context of the situation to which GWR is applied.*

4.2.3 Which One of a Series of Potential Mathematical Models is 'Best'?

The previous two subsections have made use of classical statistical inference techniques. However, the approach considered in this section is quite different in character. In these subsections we have, at some stage, proposed the notion of a 'true' model for the data. In Section 4.2.1 we put a hypothesised true model, H_0, to the test. In Section 4.2.2 confidence interval calculations were based on the assumption that the fitted model was correct. However, this is at variance with some philosophies of modelling. In the words of G.E.P. Box, cited in Chatfield (1995: 26): 'All models are wrong but some are useful.'

The approach set out in this subsection attempts to find out which of a collection of 'wrong' models is most useful. In a GWR context, we can ask whether a global regression model or GWR is most useful; as well as some more sophisticated questions. Here, the measurement of utility is the Akaike Information Criterion (AIC) or the corrected Akaike Information Criterion (AIC_c) (Akaike 1973). The AIC makes use of the notion that a 'true' model may exist, but is not directly verifiable. However, it is possible to estimate how close a proposed model is to the true model. The AIC is the measure of this closeness, and the closest model is nominated as 'best'. Note that the AIC is not simply a measure of 'goodness of fit' such as a sum of squared errors, but also takes model complexity into account. A more detailed discussion will be given in Section 4.6.

4.3 GWR as a Statistical Model

In the previous subsections, brief sketches of some inferential models for GWR were given. In the sections following this one, these ideas will be considered in greater detail. However, in order to do this, we must first consider GWR as a statistical model, rather than as an *ad hoc* data exploration tool. Such a model will be set out in this section, and used throughout the remainder of the chapter.

Suppose we have a set of observations $\{x_{ij}\}$ for $i = 1, \dots, n$ cases and $j = 1, \dots, k$ explanatory variables, and a set of dependent variables $\{y_i\}$ for each case. This is a standard data set for a global regression model. Now suppose that in addition to this we have a set of location coordinates $\{(u_i, v_i)\}$ for each case. The underlying model for GWR is

$$y_i = \beta_0(u_i, v_i) + \sum_{j=1}^{k} x_{ij}\beta_j(u_i, v_i) + \varepsilon_i \tag{4.1}$$

where $\{\beta_0(u,v), \dots \beta_k(u,v)\}$ are $k + 1$ continuous functions of the location (u,v) in the geographical study area. The ε_is are random error terms. In the basic GWR model we assume that these are independently normally distributed with mean zero and common variance σ^2. The aim of GWR is to obtain non-parametric estimates of these functions. A related technique, the expansion method (Casetti 1972), attempts to obtain parametric estimates. Both are special cases of the very general varying coefficient model of Hastie and Tibshirani (1993). More specifically, GWR attempts to obtain estimates of the functions using kernel-based methods.

The log-likelihood for any particular set of estimates of the functions may be written as

$$L(\beta_0(u,v) \dots \beta_k(u,v)|D) = -\frac{\sigma^{-2}}{2}\sum_{i=1}^{n}\left(y_i - \beta_0(u_i, v_i) - \sum_{j=1}^{k} x_{ij}\beta_j(u_i, v_i)\right)^2 \tag{4.2}$$

where D is the union of the sets $\{x_{ij}\},\{y_i\}$ and $\{(u_i,v_i)\}$. As with many situations involving non-parametric regression, choosing function estimates to maximise this expression is not very helpful. With the distribution assumptions for the error terms above, this maximum likelihood approach is equivalent to choosing the functions $\beta_k(u_i,v_i)$ using least squares. However, since the functions are arbitrary, and therefore they may take any value at (u_i,v_i), we can simply choose them to obtain a residual sum of squares of zero, with an associated and rather unconvincing estimate of σ^2 of zero. For example, consider the rather simplistic data set in Table 4.2.

Suppose we wish to fit the model $y_i = \beta_0(u_i,v_i) + \beta_1(u_i,v_i)x_i + \varepsilon_i$. One possibility giving a 'perfect' fit at the four observation points is set out in Table 4.3. Concentrating on $\beta_1(u_i,v_i)$, Table 4.3 tells us that $\beta_1(1,0) = 0$, $\beta_1(0,1) = 2$, $\beta_1(1,1) = 2$ and $\beta_1(0,0) = 1$. Clearly there are a number of β_1 functions that satisfy this. Two are shown in Figures 4.3a and b. Although both functions agree at the four corners of the surfaces shown – which correspond to the four data points – they take different values at other points. Since these two functions provide a perfect fit to the data in Table 4.2, they are both maximum likelihood (ML) estimates. Thus, ML does not provide us with a unique solution. In fact there is a further level of ambiguity, since Table 4.3 does not provide a unique set of pointwise β_0 and β_1 functions that give a perfect fit. The reader may wish to verify that Table 4.4 provides an alternative perfect fit.

Table 4.2 *Example data set*

i	u_i	v_i	x_i	y_i
1	1	0	1	0
2	0	1	2	4
3	1	1	1	3
4	0	0	3	2

Table 4.3 *Pointwise 'perfect' GWR fit*

i	1	2	3	4
$\beta_0(u_i,v_i)$	1	0	1	−1
$\beta_1(u_i,v_i)$	0	2	2	1

Table 4.4 *Alternative pointwise 'perfect' GWR fit*

i	1	2	3	4
$\beta_0(u_i,v_i)$	0	4	3	2
$\beta_1(u_i,v_i)$	0	0	0	0

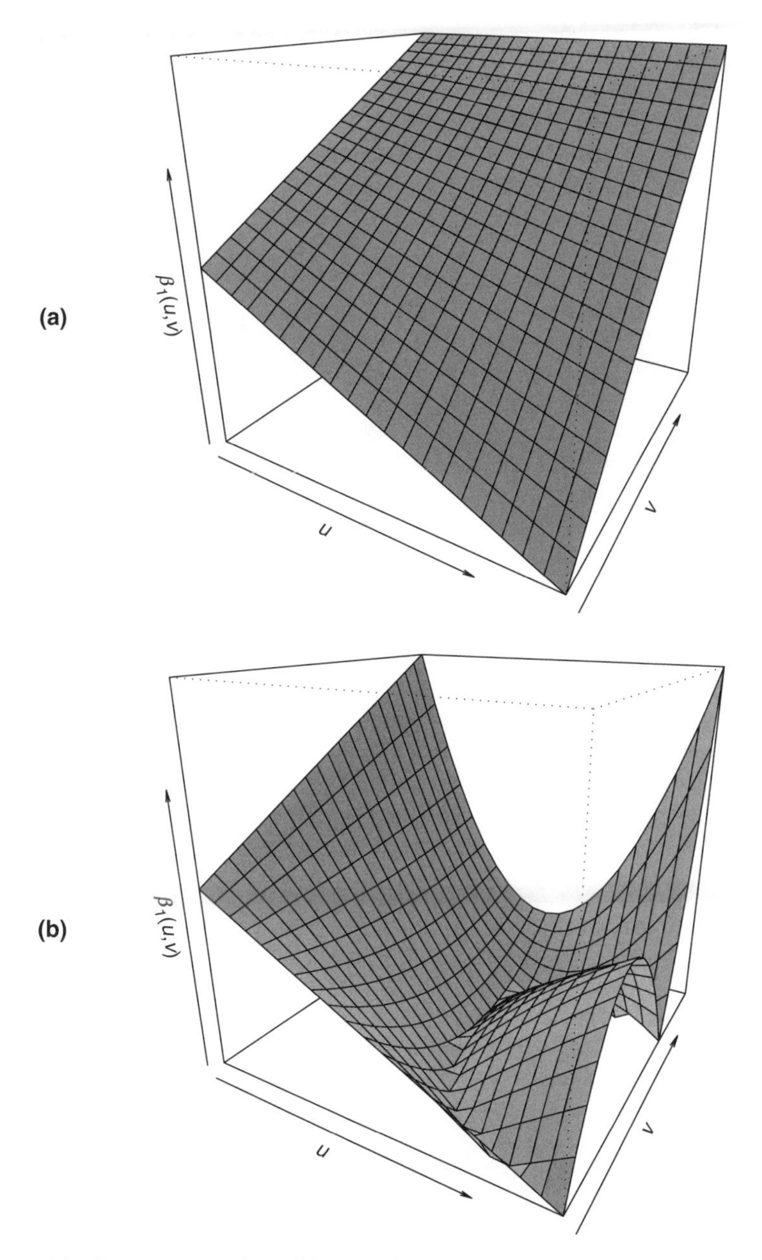

Figure 4.3 *(a) a linear Beta surface; (b) a non-linear Beta surface*

There are two alternative strategies to avoid this ambiguity – first, that we restrict the functions $\beta_0(u_i,v_i)$ and $\beta_1(u_i,v_i)$ to some prescribed parametric form (say, a polynomial in u and v) and then estimate the parameters using ML techniques,[1] or, second, that some method other than basic ML is used to calibrate the model in equation (4.1). The first approach is reasonable if we have some prior reason to believe that a particular functional form is appropriate, but in many geographical examples this is not the case. The following section therefore considers the second alternative.

4.3.1 Local Likelihood

A more useful way forward is to consider local likelihood. Rather than attempting to minimise equation (4.2) globally, we consider the problem of estimating $\{\beta_0(u,v), \ldots \beta_k(u,v)\}$ on a pointwise basis. That is, given a specific point in geographical space (u_0, v_0) (which may or may not correspond to one of the observed $\{(u_i, v_i)\}$s) we attempt to estimate $\{\beta_0(u,v), \ldots \beta_k(u,v)\}$. If these functions are reasonably smooth, we can assume that a simple regression model

$$y_i = \gamma_0 + \sum_{j=1}^{k} x_{ij}\gamma_j + \varepsilon_i \tag{4.3}$$

holds close to the point (u_0, v_0), where each γ_j is a constant valued approximation of the corresponding $\beta_j(u,v)$ in model 4.1. We can calibrate a model of this sort by considering observations close to (u_0, v_0). An obvious way to do this is to use weighted least squares, that is to choose $\{\gamma_0 \ldots \gamma_k\}$ to minimise

$$\sum_{i=1}^{n} w(d_{0i})\left(y_i - \gamma_0 - \sum_{j=1}^{k} x_{ij}\gamma_j\right)^2 \tag{4.4}$$

where d_{0i} is the distance between the points (u_0, v_0) and (u_i, v_i). This gives us the standard GWR approach. We simply set $\hat{\beta}_j(u_0, v_0)$ as $\hat{\gamma}_j$ to obtain the familiar GWR estimates. At this stage it is worth noting that equation (4.4) may be multiplied by $-\sigma^{-2}$ and be considered as a local log-likelihood expression:

$$WL(\gamma_0 \ldots \gamma_k | D) = \sum_{i=1}^{n} w(d_{0i}) L(\gamma_0 \ldots \gamma_k | D) \tag{4.5}$$

The properties of such estimators have been studied fairly comprehensively over the past decade or so. Typically this is in the context where the weighting function is applied to the $\{x_{ij}\}$s, but this does not have to be the case (see, for example, Hastie and Tibshirani (1993)). In particular, Staniswalis (1987a) notes that if $w(\ )$ is scaled to sum to unity (which it may be without loss of generality), then

[1] Essentially, this is the expansion method.

$WL(\gamma_0 \ldots \gamma_k|D)$ is an empirical estimate of the expected log-likelihood (not the local log-likelihood) at the point of estimation. Further work by Staniswalis (1987b) shows that under certain conditions – which will apply for any bounded $\beta_j(u,v)$ functions with bounded first, second and third derivatives – the γ_js do provide pointwise consistent estimators for the $\beta_j(u_0,v_0)$s. Furthermore, the distribution of the estimates for the γ_js is asymptotically normal and asymptotically unbiased. For a more in-depth view of this body of work, see, for example, Bowman and Azzalini (1997).

Thus, GWR does provide a reasonable calibration technique for model 4.1. On a historical note, it must be admitted that GWR was first devised as an exploratory technique, and not as an explicit attempt to fit model 4.1, but with hindsight it now seems that the approach does have a more formal interpretation. What is interesting to note, however, is that although the model in equation (4.1) has non-stationary regression coefficients, the variance of the error term, σ^2, is a global constant.

4.3.2 Using Classical Inference – Working with *p*-values

In a global regression model, 'goodness of fit' is measured in terms of the sum of squared residuals usually referred to as the residual sum of squares (RSS). Under the usual global model assumptions, for a global model with k linear parameters, the expected residual sum of squares, E(RSS), is given by

$$\mathrm{E(RSS)} = (n-k)\sigma^2 \tag{4.6}$$

where k denotes the number of parameters estimated in the model, n is the number of observations and $n-k$ is referred to as the degrees of freedom of the residual. This gives the usual estimate for σ^2

$$\hat{\sigma}^2 = \frac{\mathrm{RSS}}{n-k} \tag{4.7}$$

In the non-parametric framework of GWR, the concepts of 'number of parameters' and 'degrees of freedom' are fairly meaningless. However, as is the case with many non-parametric regression problems, the related ideas of 'effective number of parameters' and 'effective degrees of freedom' can be considered. Note that from equation (4.6), the number of degrees of freedom is linked to the expected value of the residual sum of squares (RSS) of the model. Now consider the distribution for the RSS in the GWR situation. We have noted earlier that the fitted values for the y_is, denoted by $\{\hat{y}_i\}$ can be expressed as a matrix transform of the raw y_is. In matrix form we write this as

$$\hat{\boldsymbol{y}} = \boldsymbol{S}\boldsymbol{y} \tag{4.8}$$

for some n by n matrix $\boldsymbol{S}$. Thus, fitted residuals are just $(\boldsymbol{I}-\boldsymbol{S})\boldsymbol{y}$, where $\boldsymbol{I}$ is the identity matrix, and

$$\mathrm{RSS} = \boldsymbol{y}^{\mathrm{T}}(\boldsymbol{I} - \boldsymbol{S})^{\mathrm{T}}(\boldsymbol{I} - \boldsymbol{S})\boldsymbol{y} \tag{4.9}$$

Following Cleveland (1979) and Tibshirani and Hastie (1987), we then note that

$$\mathrm{E(RSS)} = (n - [2\mathrm{tr}(\boldsymbol{S}) - \mathrm{tr}(\boldsymbol{S}^{\mathrm{T}}\boldsymbol{S})])\sigma^2 + \mathrm{E}(\boldsymbol{y})^{\mathrm{T}}(\boldsymbol{I} - \boldsymbol{S})^{\mathrm{T}}(\boldsymbol{I} - \boldsymbol{S})\mathrm{E}(\boldsymbol{y}) \tag{4.10}$$

where $\mathrm{tr}(\boldsymbol{S})$ is the trace of matrix $\boldsymbol{S}$. The first term of this expression relates to the variance of the fitted values, and the second to the bias. However, if we assume that the bandwidth for the GWR is chosen so that bias is negligible, which is reasonable for a large sample as the asymptotic results suggest, then we have the approximation

$$\mathrm{E(RSS)} = (n - [2\mathrm{tr}(\boldsymbol{S}) - \mathrm{tr}(\boldsymbol{S}^{\mathrm{T}}\boldsymbol{S})])\sigma^2 \tag{4.11}$$

which is analogous to equation (4.6). As noted in Chapter 2, the effective number of parameters in GWR is given by $2\mathrm{tr}(\boldsymbol{S}) - \mathrm{tr}(\boldsymbol{S}^{\mathrm{T}}\boldsymbol{S})$ and the effective degrees of freedom are given by $n - 2\mathrm{tr}(\boldsymbol{S}) - \mathrm{tr}(\boldsymbol{S}^{\mathrm{T}}\boldsymbol{S})$. The effective number of parameters in a GWR is often not an integer but varies between k (when the bandwidth tends to infinity) and n (when the bandwidth tends to zero). In many cases, $\mathrm{tr}(\boldsymbol{S})$ is very close to $\mathrm{tr}(\boldsymbol{S}^{\mathrm{T}}\boldsymbol{S})$ so an approximate value for the effective number of parameters is $\mathrm{tr}(\boldsymbol{S})$.

Cleveland (1979) and Hastie and Tibshirani (1990) suggest that the distribution of the RSS divided by the effective number of parameters may be reasonably approximated by a χ^2 distribution with effective degrees of freedom equal to the effective number of parameters.[2] Given this, we have the basis for an approximate likelihood ratio test, based on the F-test, which can be used to compare the abilities of the GWR and global models to replicate the observed data set. Simply divide the residual sum of squares for the standard OLS model by that for the GWR model and then carry out an F-test on this ratio with (d_1, d_2) degrees of freedom (DF) where d_1 is the DF for the OLS model and d_2 is the DF for the GWR model.

4.3.3 Testing Individual Parameter Stationarity

An earlier, perhaps more pragmatic, approach to inference about GWR models was outlined in Brunsdon *et al.*, 1996 and is described in Chapter 2. This allows testing the stationarity of individual parameters based on measuring their variability over space when estimated using GWR. The method is carried out as follows: a GWR estimate of the coefficient of interest is taken at each of the n data points and the variance (or standard deviation) of these estimates is computed. If the variance for parameter k is termed V_k then

[2] Note these degrees of freedom may not be an integer but the χ^2 distribution is well defined for *any* positive DF.

$$V_k = \frac{1}{n}\sum_{i=1}^{n}\left(\hat{\beta}_{ik} - \frac{1}{n}\sum_{i=1}^{n}\hat{\beta}_{ik}\right)^2 \tag{4.12}$$

Of course, even if the parameter of interest did not vary geographically, one would expect to see some variation in the estimated local values of the parameter. The question here is whether the observed variation is sufficient to reject the hypothesis that the parameter is globally fixed. To do this, consider the null distribution of the variance under this hypothesis. If there is no spatial pattern in the parameter, then any permutation of the regression variables against their locations is equally likely and on this basis we can model the null distribution of the variance.

However, it would be difficult, if not impossible, to compute this distribution analytically and therefore a Monte Carlo approach is adopted. For a given number of times, say, n, the geographical coordinates of the observations are randomly permuted against the variables. Note that since we are only questioning the geographical variability of the observations we do not permute the independent variables against the dependent variable. Thus, we have n values of the variance of the coefficient of interest which we use as an experimental distribution. We compare the actual value of the variance against this list to obtain an experimental significance level.

This approach allows significance testing for the variability of individual coefficients. However, the method is computationally intensive. Subsequently, Leung *et al.* (2000a, 2000b) have suggested an analytical approach to testing the variability of the variance under a null hypothesis of a stationary coefficient. To do this they first note that defining $\hat{\boldsymbol{\beta}}_k$ as the column vector containing elements $[\hat{\beta}_{1k} \ldots \hat{\beta}_{nk}]$ it is possible to write

$$V_k = 1/n\hat{\boldsymbol{\beta}}_k{}^{\mathrm{T}}[\boldsymbol{I} - (1/n)\boldsymbol{J}]^{\mathrm{T}}[\boldsymbol{I} - (1/n)\boldsymbol{J}]\hat{\boldsymbol{\beta}}_k \tag{4.13}$$

where $\boldsymbol{I}$ is the $n \times n$ identity matrix and $\boldsymbol{J}$ is an $n \times n$ matrix whose elements are all equal to 1. To see more clearly how equation (4.13) is derived, note that $(1/n)\boldsymbol{J}\hat{\boldsymbol{\beta}}_k$ is the mean value of $\hat{\boldsymbol{\beta}}_k$ repeated n times in a column vector. Note also that $\left(\boldsymbol{I} - \frac{1}{n}\boldsymbol{J}\right)^{\mathrm{T}}\left(\boldsymbol{I} - \frac{1}{n}\boldsymbol{J}\right) = \left(\boldsymbol{I} - \frac{1}{n}\boldsymbol{J}\right)$ so equation (4.13) can be simplified to

$$V_k = 1/n\hat{\boldsymbol{\beta}}_k{}^{\mathrm{T}}[\boldsymbol{I} - (1/n)\boldsymbol{J}]^{\mathrm{T}}\hat{\boldsymbol{\beta}}_k \tag{4.14}$$

Also, Leung *et al.* (2000a) note that $\hat{\boldsymbol{\beta}}_k$ can be written as a linear transform of the dependent variable $\boldsymbol{y}$, say, $\hat{\boldsymbol{\beta}}_k = \boldsymbol{B}\boldsymbol{y}$, where

$$\boldsymbol{B} = \begin{bmatrix} \boldsymbol{e}_k^{\mathrm{T}}[\boldsymbol{X}^{\mathrm{T}}\boldsymbol{W}_1\boldsymbol{X}]^{-1}\boldsymbol{X}^{\mathrm{T}}\boldsymbol{W}_1 \\ \vdots \\ \boldsymbol{e}_k^{\mathrm{T}}[\boldsymbol{X}^{\mathrm{T}}\boldsymbol{W}_n\boldsymbol{X}]^{-1}\boldsymbol{X}^{\mathrm{T}}\boldsymbol{W}_n \end{bmatrix} \tag{4.15}$$

and $\boldsymbol{W}_1 \ldots \boldsymbol{W}_n$ are the diagonal weighting matrices associated with kernels centred around each of the case locations, and $\boldsymbol{e}_k$ is an m-element column vector with 1 for the kth element and zero for the others. Then, it is possible to write

$$V_k = \mathbf{y}^{\mathrm{T}} \frac{1}{n} \mathbf{B}^{\mathrm{T}} \left(\mathbf{I} - \frac{1}{n} \mathbf{J} \right) \mathbf{B} \mathbf{y} \tag{4.16}$$

However, this is simply a quadratic form in the random variable $\mathbf{y}$. Applying similar logic to the last section, we note that, approximately,

$$\mathrm{E}(V_k) = \mathrm{tr}\left(\frac{1}{n} \mathbf{B}^{\mathrm{T}} (\mathbf{I} - \frac{1}{n} \mathbf{J}) \mathbf{B} \right) \sigma^2 \tag{4.17}$$

In the original paper the authors use a more accurate approximation although we suggest this one to reduce computational complexity. We also note that $\mathrm{E}(V_k)/\sigma^2$ is approximately χ^2 with DF equal to $\mathrm{tr}\left(\frac{1}{n}\mathbf{B}^{\mathrm{T}}(\mathbf{I} - \frac{1}{n}\mathbf{J})\mathbf{B}\right)$. This approximately chi-squared quantity may be divided by the estimate of σ^2 given in the last section, which is also approximately chi-squared with $n - 2\mathrm{tr}(\mathbf{S}) - \mathrm{tr}(\mathbf{S}^{\mathrm{T}}\mathbf{S})$ degrees of freedom. Hence, the ratio of these two chi-squared distributed quantities yields a statistic with an F distribution with degrees of freedom equal to $\mathrm{tr}\left(\frac{1}{n}\mathbf{B}^{\mathrm{T}}(\mathbf{I} - \frac{1}{n}\mathbf{J})\mathbf{B}\right)$ and $n - 2\mathrm{tr}(\mathbf{S}) - \mathrm{tr}(\mathbf{S}^{\mathrm{T}}\mathbf{S})$. This F-test examines whether the kth parameter is stationary.

We propose one further adjustment to this statistic. The derivation applied above does not require the points at which the kth coefficient is estimated (the regression points) to be the same as the points at which data are observed. There is no reason why estimates could not be made at a subset of the locations of cases, or indeed of an entirely different set of points – for example, a regular grid covering the study area. The $\mathbf{B}$ matrix would need to be adjusted, by changing the number of rows to the number of points at which coefficients are estimated, and by choosing $\mathbf{W}$-matrices centred on the new regression points. Once this is done, the above results would still apply and the F-test could still be carried out. Note, however, that as yet no work has been done to investigate how a choice of regression points affects the power of the statistical test.

We finally note that in practice the use of the Leung test has proven as computationally intensive as the Monte Carlo test. Both tests are available in the GWR software described in Chapter 9.

4.4 Confidence Intervals

Here inference will be regarded more in terms of confidence intervals for estimated values than in terms of significance tests. This simply reflects trends in the statistical community over the past few years (Nester 1996). To establish pointwise confidence intervals for the regression coefficients, we need to know the form for the asymptotic variance-covariance matrix. In a GWR context, this is given by inverting the local information matrix. The expression may be found by re-casting a result from Staniswalis (1987b):

$$\mathbf{I}(\gamma_0 \ldots \gamma_k) = \mathrm{outer}\left(E\left(\left\{ \frac{\partial L(\gamma_0 \ldots \gamma_k)}{\partial \gamma_i} | u_0, v_0 \right\} \right) \right) \tag{4.18}$$

where outer( ) denotes a multiplicative outer product, $L(\gamma_0, \ldots, \gamma_k)$ is the global likelihood of $\gamma_0, \ldots, \gamma_k$ at the point (u_0, v_0) and $\boldsymbol{I}$ is the information matrix associated with the estimates of $\gamma_0, \ldots, \gamma_k$ at the point (u_0, v_0). Note that although the estimates of the $\beta_j(u_0, v_0)$ are the local ones for (u_0, v_0), the likelihood function is the global one. Since we do not know the true values of these partial derivatives, we could use the fact that the local likelihood is an estimator of the expected global likelihood and 'plug in' the local likelihood estimates of the functions to the likelihood expression. In fact, although this is a general result which could be applied to a variety of models, there is a more direct approach for the GWR model in equation (4.1). To see this, we note that for any pointwise model calibration, we may write in matrix form

$$\hat{\boldsymbol{\gamma}} = (\boldsymbol{X}^{\mathrm{T}}\boldsymbol{W}\boldsymbol{X})^{-1}\boldsymbol{X}^{\mathrm{T}}\boldsymbol{W}\boldsymbol{y} \quad (4.19)$$

where the matrix $\boldsymbol{X}$ and the vectors $\boldsymbol{y}$ and $\boldsymbol{\gamma}$ correspond to the xs, ys and γs used previously, and $\boldsymbol{W}$ is the diagonal matrix of the local weights around (u_0, v_0). Thus, the vector of local coefficient estimates $\hat{\boldsymbol{\gamma}}$ is a linear function of the observed dependent variable vector, $\boldsymbol{y}$. Thus, equation (4.19) could be simplified to $\hat{\boldsymbol{\gamma}} = \boldsymbol{C}\boldsymbol{y}$, where $\boldsymbol{C} = (\boldsymbol{X}^{\mathrm{T}}\boldsymbol{W}\boldsymbol{X})^{-1}\boldsymbol{X}^{\mathrm{T}}\boldsymbol{W}$. But, in model (4.1) we have assumed that the y_is are independently distributed with the same variance σ^2. Thus, $\mathrm{var}(\boldsymbol{y}) = \sigma^2\boldsymbol{I}$ and the pointwise variance of the vector $\hat{\boldsymbol{\gamma}}$ is just $\boldsymbol{C}\boldsymbol{C}^{\mathrm{T}}\sigma^2$. Consequently, we can obtain pointwise confidence intervals for the surface estimates if we use the way of estimating σ^2 outlined in equation (4.7) in section 4.4.[3]

It is important to note here that although the pointwise confidence intervals are local, the value of σ is global, in keeping with equation (4.2). It follows that these confidence intervals are not those obtained from the output of any of the local weighted regression models. This is an important practical point – although local parameter estimates are obtained by moving a kernel through the study area and carrying out weighted regressions, local parameter standard errors are not simply the standard errors obtained from these weighted regressions. These latter figures are based on *localised* estimates of σ, rather than the global value implied above.

4.5 An Alternative Approach Using the AIC

One useful approach to model selection theory is the use of the Akaike Information Criterion (AIC) (Akaike 1973) introduced in Chapter 2. The underlying idea is to estimate the quantity

$$\int f(y)\log_e(f(y)/g(y))\mathrm{d}y \quad (4.20)$$

[3] Because a small amount of bias is inevitable in the calibration of the local model, Fox (2000a) prefers the term 'variability band or interval' to that of 'confidence interval'. However, we use the latter to avoid confusion in the discussion – we assume the term 'confidence interval' makes it immediately clear to the reader what is being measured.

which measures the information distance between the model distribution g and the true distribution f (Kullback and Leibler 1951). By comparing this quantity for a number of competing models $g_1 \ldots g_l$, we can decide which is closest to reality (Burnham and Anderson 1998). Unlike classical statistical inference, this does not involve a decision as to whether a hypothesis is 'true'. Such an assertion of absolute truth is certainly dubious in many social science situations, and arguably so in many other areas of study. Instead, this approach assumes that all models are 'wrong' in a strict sense, but that some are closer approximations to reality than others.

In general hat matrix situations such as equation (4.8) where $\hat{\boldsymbol{y}}$ (often called '$\boldsymbol{y}$-hat') is obtained by pre-multiplying $\boldsymbol{y}$ by a matrix $\boldsymbol{S}$, some lengthy workings show that the AIC can be reasonably estimated by the expression

$$\mathrm{AIC}_c = 2n\log_e(\hat{\sigma}) + n\log_e(2\pi) + n\left\{\frac{n + \mathrm{tr}(\boldsymbol{S})}{n - 2 - \mathrm{tr}(\boldsymbol{S})}\right\} \tag{4.21}$$

where $\hat{\sigma}$ is the estimated standard error of the error term (Hurvich *et al.*, 1998). The c subscript is used to denote that this is the 'corrected' AIC estimate. The more simplistic form

$$\mathrm{AIC} = 2n\log_e(\hat{\sigma}) + n\log_e(2\pi) + n + \mathrm{tr}(\boldsymbol{S}) \tag{4.22}$$

is used in some other situations. Note that direct comparisons should not be made between AIC and AIC_c. Note also that in both the AIC and the AIC_c the estimate of sigma used is not that given in equation (4.7) but one based on the maximum likelihood estimate:

$$\hat{\sigma}^2 = \frac{\mathrm{RSS}}{n} \tag{4.23}$$

To choose between a number of competing models, we compute the residual sum of squares and the hat matrix for each model, and then compute AIC_c. The best model is then the one with the smallest AIC_c value. Note that, as a rule of thumb, a 'serious' difference between two models is generally regarded as one in which the difference in AIC_c values between the models is at least 3. The AIC approach thus provides a useful tool for choosing models in the GWR context. First, one could compare a pair of GWR models with different explanatory variables. Second, one could compare a GWR model with an ordinary least squares regression – providing another way of testing whether the observed patterns in GWR coefficient surfaces are attributable to chance. If the OLS model has a lower AIC than the GWR model, this suggests that the extra detail of the GWR model is unjustified. Finally, one may compare GWR models having different bandwidths. Two different bandwidths will give two different sets of GWR surfaces and the task of the AICs here is to determine which set of surfaces gives the model that is closest to reality.

Indeed, it is possible to use the AIC as an alternative criterion for bandwidth selection as discussed in Chapter 2.

4.6 Two Examples

In this section, two practical examples of inference in GWR will be given using data on educational attainment from the 1990 US Census for counties in the southern United States. The first example demonstrates pointwise standard error estimation and the second the AIC approach. A total of six variables were extracted for each of the counties, together with coordinates for the county centroids. These variables are listed in Table 4.5. The variable PCTBACH (percentage of the employable population with a Bachelor's degree) is used as a measure of educational attainment dependent variable and a GWR model is fitted using the other five variables as predictors. It is felt that educational attainment might be inversely related to the percentage of people classed as rural, the percentage elderly, and the percentage below the poverty line. It is not clear what, if any, relationship there might be between educational attainment and percentage black and percentage foreign born. The latter variable consists of a potentially heterogeneous group and might provide a good example of the utility of GWR: if immigrants from the same country tend to be located in spatial clusters and if there is a difference in educational aspirations amongst immigrants from different countries, this would manifest itself in spatially varying parameter estimates for this variable.

4.6.1 Basic Estimates

First, a basic GWR analysis was run on the southern US data. The results are shown in Figures 4.4a–f. These local parameter estimates were obtained using an adaptive bandwidth method where at each regression point 5.5% of the entire data set was used in the locally weighted regression. The regression points were arranged on a 40 by 40 grid and the coefficient estimates at each of these points were used to produce contour plots. For this example, emphasis will be placed on the inferential methods rather than the GWR maps themselves, although it is worth noting

Table 4.5 *Variables derived from 1990 US Census used in this example*

Variable Name	Description
PCTRURAL	Percentage of the county population classed as 'rural'
PCTBACH	Percentage of employable adults with a Bachelor's degree
PCTELD	Percentage of county population classed as elderly
PCTFB	Percentage of county population foreign born
POVPCT	Percentage of county population below the poverty level
PCTBLACK	Percentage of county population who are black

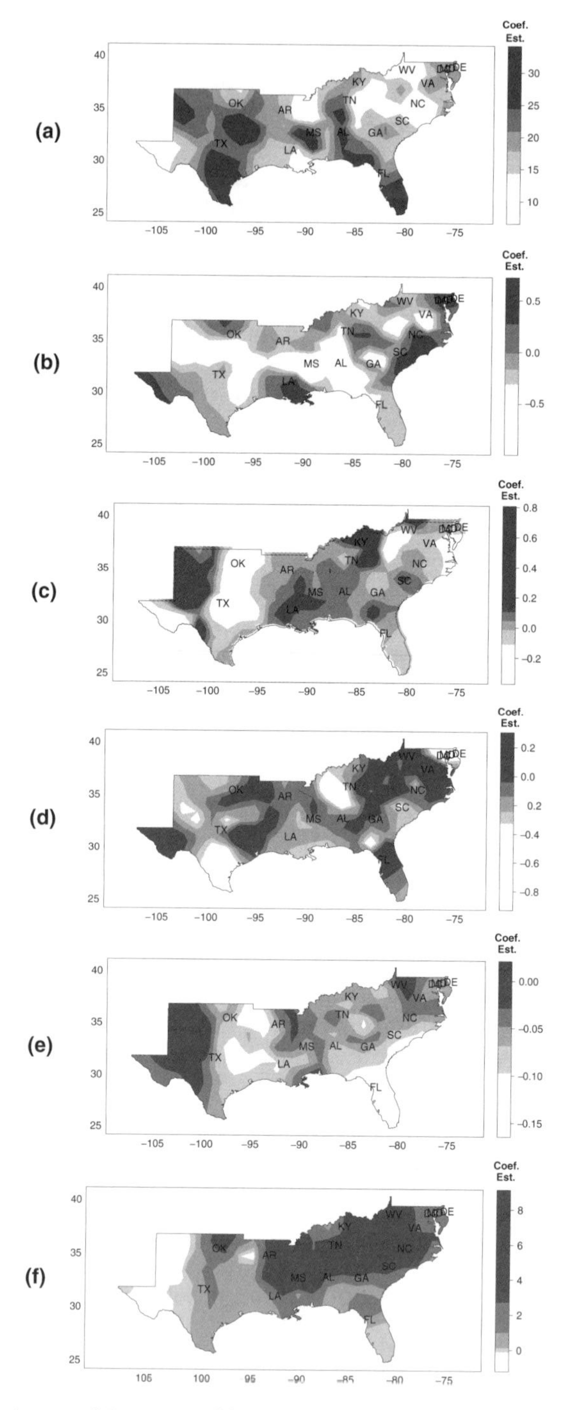

Figure 4.4 (a) Estimate of intercept; (b) estimate of PCTELD; (c) estimate of PCTBLACK; (d) estimate of POVPCT; (e) estimate of PCTRURAL; (f) estimate of PCTFB

that some interesting patterns occur. For example, the local parameter estimates for PCTELD are generally negative as expected but there is a cluster of positive estimates around the South Carolina coast and also centred on New Orleans. The negative effects of poverty on educational attainment seem particularly strong near the Tennessee–Alabama border and in southern Texas. The local parameter estimates for the percentage foreign born are generally strongly positive, particularly in a band from North Carolina to Mississippi, but are close to zero throughout most of Texas.

4.6.2 Estimates of Pointwise Standard Errors

Next, the levels of standard error are computed for each of the predictor variables, and for the intercept. As stated earlier, these are pointwise estimates. This implies that although it is possible to compute upper and lower 95% confidence limits for the coefficient estimates at each regression point, and to represent these as surfaces, the pair of surfaces does not represent a 95% confidence envelope for the regression surface as a whole. However, viewing the standard error surfaces does give a good idea of the relative degree to which sampling variability affects estimates at the regression points, which in itself can be a helpful tool.

For the southern US counties' model the standard error maps are shown in Figure 4.5a–f. These maps suggest that there is a great deal of variation in the patterns of standard error between the coefficients. For instance, the standard errors of the local parameter estimates for PCTELD are much lower in Texas and Florida, both states presumably having a relatively uneven distribution of elderly population. The standard errors of the PCTBLACK parameter estimates are highest for the central Appalachian area around Kentucky, West Virginia and Tennessee and lowest through Louisiana, Alabama, Georgia and into Virginia. The standard errors of POVPCT are lowest in Kentucky and West Virginia, where there is widespread rural poverty and also in southern Texas where there is a large immigrant population. Finally, it is worth noting the low standard errors on the local estimates of the percentage foreign born variable in southern Florida and southern Texas, two areas of very high recent immigration.

4.6.3 Working with the AIC

To decide which particular GWR model should be used in any given situation, a number of competing models should be evaluated and their AIC values compared. Note that these competing models may differ with respect to the calibration bandwidth, the choice of explanatory variables or both of these. If a single model is to be selected as the best, then this should be the one with the lowest AIC. Typically the AIC values can be presented in relative form, by subtracting the lowest AIC from each of the raw AIC values. In this case, the best model has a relative AIC of zero, and all other models have a higher AIC.

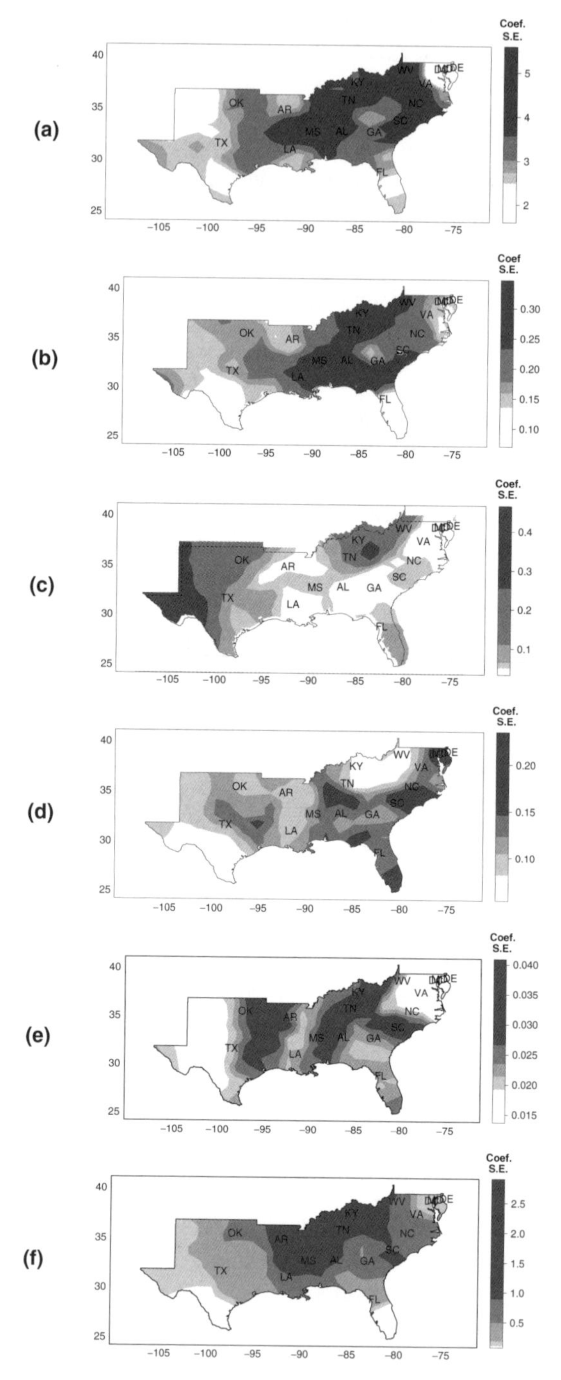

Figure 4.5 (a) Standard error of intercept; (b) standard error of PCTELD; (c) standard error of PCTBLACK; (d) standard error of POVPCT; (e) standard error of PCTRURAL; (f) standard error of PCTFB

As discussed earlier, the AIC provides not only a framework for variable choice, but also one for bandwidth selection. In the above computations, an adaptive bandwidth incorporating 5.5% of the data at each regression point was used. This may seem an arbitrary figure, but was in fact chosen as it minimises the AIC for all possible GWR models incorporating the variables used above. In Table 4.6 the AIC values are given for a global regression model, the 'best' GWR model as described above, and an alternative GWR model using an adaptive bandwidth incorporating 10% of the data at each regression point.

As a general rule, improvements in the AIC that are less than 3 in value could easily arise as a result of sampling error, whereas values greater than 3 are more likely to be due to a genuine difference in models. Table 4.6 indicates first that the best adaptive model is better than the global model. It also shows that the best model is a genuine improvement on the 10% model. The 10% model is also better than the global model; this is useful practical information. If one did not have the resources to select an optimal GWR bandwidth – for example, when working with a very large data set – one at least knows whether working with some *ad hoc* bandwidth is a notable improvement on working with a global model.

An alternative way of comparing the AIC values for a number of competing models is to consider the Akaike weights for each model. If the AIC for model i is denoted by AIC_i then the Akaike weight for model i is defined as

$$w_i = \frac{\exp(-\mathrm{AIC}_i/2)}{\sum_j \exp(-\mathrm{AIC}_j/2)} \tag{4.24}$$

The w_is sum to one and can be thought of as the 'weight of evidence' in favour of each model. Using this approach, we obtain some idea of the relative likelihood that each model is best, rather than choosing a single candidate. It has been noted that w_i has a Bayesian interpretation, see, for example, Akaike (1981; 1994). If we place equal prior probabilities on each of the competing models being the best, then the w_is are the posterior probabilities that each model is best, in the light of the data.

For the three models in Table 4.6 the Akaike weights are shown in Table 4.7. To three decimal places, the weights of evidence suggest that the 5.5% adaptive AIC model is certain to be the best of the three models presented here.

Table 4.6 *AIC values for a number of regression models*

Model	Relative AIC
Global	830
10% adaptive GWR	42
5.5% adapative GWR (Best)	0

Table 4.7 AIC weights for a number of regression models

Model	Akaike weights
Global	0.000
10% adaptive GWR	0.000
5.5% adaptive GWR (Best)	1.000

4.7 Summary

This chapter has covered the underlying ideas of two approaches to statistical inference for GWR models. Although this provides the basics, it is possible to extend these ideas. For example, in Chapter 8, we will consider AICs and confidence intervals for Binomial and Poisson-based GWR models. Also considered in Chapter 7 are exploratory approaches to inference for locally weighted summary statistics. These differ from classical approaches in the sense that they flag situations that *might* be interesting rather than seek strong evidence for the rejection of a null hypothesis.

One area we have not covered here, but that may also be of interest is the Bayesian approach, as considered in LeSage (1999b). This considers Monte Carlo Markov Chain approaches (see, for example, Gelman *et al.* 1995) to estimating posterior distributions for parameter estimates at the regression points, although the model used is somewhat different from that specified in equation (4.1). However, regardless of the actual model being used, there are two key points made in this chapter. First, it is possible to apply statistical inferential techniques to GWR models – the general ideas put forward here are linked with local likelihood and non-parametric regression techniques, although the Bayesian approach makes use of other ideas. Second, there are a number of possible inferential frameworks which may be used. In many cases, approaches other than the traditional significance testing techniques are overlooked, but they do exist, and may offer useful alternatives when assessing GWR-based statistical models.

5

GWR and Spatial Autocorrelation

5.1 Introduction

In this chapter we attempt to uncover links between two of the major challenges in spatial data analysis: spatial non-stationarity and spatial dependency. Three such links are examined:

1. the use of the GWR framework in providing local measures of spatial dependency;
2. the use of the GWR framework as an alternative to spatial regression modelling; and
3. the combination of both GWR and spatial regression models.

Spatial dependency or spatial association is often measured by spatial autocorrelation statistics which describe the similarity of nearby observations. If high values of an attribute tend to cluster together in some parts of a study area and low values tend to cluster together in other parts, the attribute is said to exhibit positive spatial autocorrelation. Conversely, if high values tend to be found in close proximity to low values and vice versa, the attribute is said to exhibit negative spatial autocorrelation. If the data are located in space so that no relationship exists between nearby values, the data are said to exhibit zero spatial autocorrelation (Cliff and Ord 1973; Miron 1984; Goodchild 1986; Odland 1988). Almost all spatial data exhibit some form of positive spatial autocorrelation and this has a variety of impacts on inference in spatial modelling, as described by, *inter alia*, Hordijk (1974); Ord (1975); Sen (1976); and Fotheringham *et al.* (2000).

Spatial autocorrelation is measured by a number of statistics having slightly different formulations but they all depend on some definition of spatial weighting which attempts to quantify the often subjective concept of proximity – what is meant by 'nearby'? For instance, two common types of weighting matrix that are often used are: (i) a discrete matrix where $w_{ij} = 1$ if polygons i and j are contiguous, in the case of areal data, or if j is the nearest neighbour to i in the case of point data, and 0 otherwise; and (ii) a continuous weighting matrix based on inverse distances where all locations have some impact on all others but the impact is largest for locations in closest proximity. Not only is the choice of either discrete or continuous spatial weighting often arbitrary, but there still remains a subjective decision regarding the particular method of spatial weighting within these broad classes. In the case of the discrete weighting function based on proximity, for example, the definition of contiguity is not always obvious (see, for example, Coombes 1978). In the case of a distance-based weighting function, there is no theory to guide the selection of a particular distance function. A detailed treatment of spatial weights is provided by Bavaud (1998).

This very brief description of spatial autocorrelation suggests that there might be similarities between measurements of spatial autocorrelation and GWR: both depend on the definition of spatial weighting matrices and both deal with spatial dependencies in data. This raises a number of questions that are explored both conceptually and empirically in this chapter. Some of the issues addressed become highly complex and we stress that the empirical results should be viewed as exploratory: in some cases, the statistical properties of the estimators remain to be discovered. We first describe the empirical context of this chapter and then explore a series of relationships between GWR and spatial autocorrelation.

5.2 The Empirical Setting

To investigate various aspects of the relationship between GWR and spatial autocorrelation, we have selected a subset of the London house price data set described in Chapter 2. The subset consists of 1406 houses sold in 1990 through the Nationwide Building Society within the neighbouring London boroughs of Brent and Ealing. The location of these houses plus some basic geography of Brent and Ealing are shown in Figure 5.1. Essentially, this is an area of mixed housing in north-west London outside the immediate area bordering the Thames but which lies well inside the Greater London boundary. The mean prices and numbers of various types of housing in the two boroughs are shown in Table 5.1. As would be expected from the location within a major city, there are relatively few detached houses and relatively large numbers of terraced properties and flats. While detached housing appears to be more expensive in Brent than in Ealing, the other forms of housing are of similar price.

5.3 Local Measures of Spatial Autocorrelation using GWR

As part of our review of local models in Chapter 1, we described a set of local measures of spatial dependency (Section 1.5.4). These include local measures of both the G_i statistic of spatial clustering (Getis and Ord 1992; Ord and Getis 1995) and Moran's I (Anselin 1995; 1998). However, it is perhaps simpler to derive local

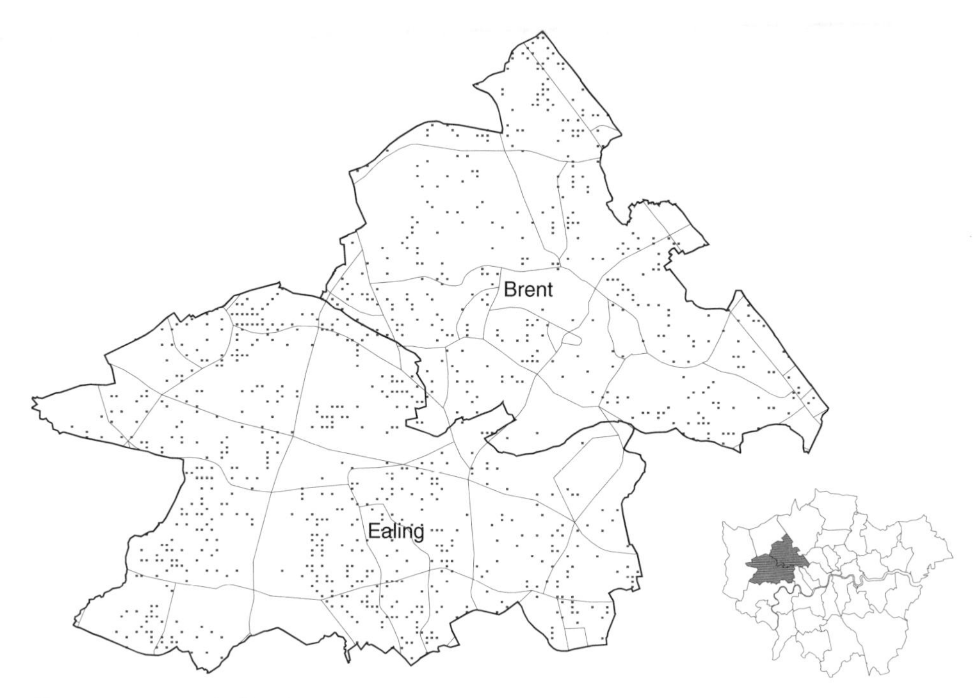

Figure 5.1 *Brent/Ealing locations*

Table 5.1 *Frequency and price in 1990 of properties in Brent and Ealing*

Borough	Type	Frequency	Mean price (£)
Brent	Detached	26	150 907
	Semi-detached	154	94 687
	Terrace	146	78 135
	Flat	209	58 057
Ealing	Detached	17	126 617
	Semi-detached	189	100 084
	Terrace	362	81 155
	Flat	290	60 533

measures of spatial dependency through the GWR framework. As we show, this also has the benefit of being able to produce local measures of conditional spatial dependency, a feature not possible with the above measures. In order to see how local unconditional and conditional measures of spatial dependency can be obtained relatively simply through the GWR framework, consider the following set of models.

Model 1a

$$P_i = \kappa + \rho P_i{}^* + \varepsilon_i \tag{5.1}$$

where P_i represents the price a house sold for at location i, κ and ρ are parameters to be estimated, ε_i is a random error term and $P_i{}^*$ is the average of the prices of the 10 nearest houses to i defined by,

$$P_i{}^* = \sum_{j=1}^{n} w_{ij} P_j / \sum_{j=1}^{n} w_{ij} \tag{5.2}$$

where

$$\begin{aligned} w_{ij} &= 1 \quad \text{if } j \text{ is one of the 10 nearest neighbours of } i \\ &= 0 \quad \text{otherwise} \end{aligned} \tag{5.3}$$

The model in (5.1) is known as a spatial autoregressive model and ρ is a measure of the degree of spatial autocorrelation, in this case in house prices (Ord, 1975; Brunsdon *et al.* 1998; Fotheringham *et al.* 2000). If the estimate of ρ is significantly positive, then positive spatial autocorrelation exists; if the estimate is significantly negative, then negative spatial autocorrelation exists; and if the estimate is not significant, then no spatial autocorrelation in house prices is present. Of course, ρ in (5.1) is a global parameter and describes the average degree of spatial autocorrelation across the region. Information on local spatial autocorrelation can be obtained from model 1b.

Model 1b

$$P_i = \kappa(u_i, v_i) + \rho(u_i, v_i) P_i{}^* + \varepsilon_i \tag{5.4}$$

This is a local version of model 1a which can be calibrated by geographically weighted regression and will produce a surface of local estimates of the spatial autocorrelation in house prices. This provides an alternative method of obtaining estimates of local spatial autocorrelation measures which can then be mapped to examine whether there are interesting spatial patterns in the degree to which spatial dependency exists in the data.

Model 2a

In models 1a and 1b the estimates of spatial autocorrelation that are obtained from the calibration of the models are unconditional – that is, they describe spatial dependency in one attribute, ignoring the effect of any other attributes. In some situations it may be useful to answer the question 'to what extent is an attribute spatially autocorrelated *given the distribution of another attribute*?'. For instance, suppose we were interested in the distribution of a certain insect type across a region. To examine this, suppose we have divided the region into grid cells and that we have data on insect counts in each of these grid cells. An unconditional autocorrelation statistic, global or local, would yield information on the extent to which the insects are clustered: the global statistic would present an average picture of this clustering; the local statistics would tell us whether there is any spatial variation in the degree of clustering of the insects. However, in undertaking such a measurement using models 1a or 1b, or any other statistical measure of spatial autocorrelation for that matter, we would be ignoring all other factors that might affect insect clustering. For instance, suppose we find that the insects exhibit strong positive spatial autocorrelation – what would this tell us? The insects might only cluster because they like a particular vegetation type that is clustered. If we accounted for the clustering of the vegetation on which the insects feed, the spatial distribution of the insects might exhibit negative spatial autocorrelation because of competition effects. Consequently, what is needed is a method of obtaining measures of spatial autocorrelation, both global and local, that are conditioned on other factors and that are hence independent of these other factors. This is easily achieved within the regression framework by adding other explanatory variables to models 1a and 1b. For instance, in model 2a we add the floor area of a house to the basic spatial autoregressive model of (5.1)

$$P_i = \kappa + \rho P_i^* + \alpha \mathrm{FLRAREA}_i + \varepsilon_i \tag{5.5}$$

Model 2b

Model 2b is the GWR equivalent of model 2a which produces local estimates of the conditional autocorrelation parameter:

$$P_i = \kappa(u_i, v_i) + \rho(u_i, v_i) P_i^* + \alpha(u_i, v_i) \mathrm{FLRAREA}_i + \varepsilon_i \tag{5.6}$$

An example of the calibration of such a model is provided by Brunsdon *et al.* (1998).

Model 3a

Model 3a develops this concept still further by adding explanatory variables for both the type of house and the age of the house, as defined in Chapter 2:

$$\begin{aligned} P_i = {} & \kappa + \rho P_i^* + \alpha_1 \mathrm{FLRAREA}_i + \alpha_2 \mathrm{BLDPWW1}_i + \alpha_3 \mathrm{BLDPOSTW}_i \\ & + \alpha_4 \mathrm{BLD60S}_i + \alpha_5 \mathrm{BLD70S}_i + \alpha_6 \mathrm{BLD80S}_i + \alpha_7 \mathrm{TYPDETCH}_i \\ & + \alpha_8 \mathrm{TYPTRRD}_i + \alpha_9 \mathrm{TYPFLAT}_i + \varepsilon_i \end{aligned} \tag{5.7}$$

Model 3b

This is the GWR equivalent of model 3a which yields local estimates of the parameters:

$$\begin{aligned} P_i = {} & \kappa(u_i,v_i) + \rho(u_i,v_i)P_i{}^* + \alpha_1(u_i,v_i)\mathrm{FLRAREA}_i \\ & + \alpha_2(u_i,v_i)\mathrm{BLDPWW1}_i + \alpha_3(u_i,v_i)\mathrm{BLDPOSTW}_i \\ & + \alpha_4(u_i,v_i)\mathrm{BLD60S}_i + \alpha_5(u_i,v_i)\mathrm{BLD70S}_i + \alpha_6(u_i,v_i)\mathrm{BLD80S}_i \\ & + \alpha_7(u_i,v_i)\mathrm{TYPDETCH}_i + \alpha_8(u_i,v_i)\mathrm{TYPTRRD}_i \\ & + \alpha_9(u_i,v_i)\mathrm{TYPFLAT}_i + \varepsilon_i \end{aligned} \tag{5.8}$$

The six models described above are calibrated using the housing data set described in Section 5.2. The global models are calibrated by OLS regression while the local models are calibrated by GWR even though, as Ord (1975) notes, OLS will not yield the correct maximum likelihood estimates of the autoregressive models. Our use of OLS here is supported by four arguments:

1. The likelihood functions for the local autoregressive models are not currently known.
2. Even if they were known, they would be computationally complex. It would be useful to find some means of reducing the computational complexity of this estimation procedure and the work of Pace and Barry (1997) is perhaps worth noting here.
3. In any model that attempts to account for spatial dependency amongst the error terms, there needs to be some definition of a spatial weights matrix that specifies the nature of the spatial relationships between zones or points. The results of spatial regression modelling are probably more sensitive to the often arbitrary definition of this weights matrix than they are to the method used to calibrate the models. Evidence for this can be seen from Table 5.2 which contains the results of three calibrations of model 3a (equation 5.7). The first calibration uses OLS and a spatial weights matrix defined in terms of the 10 nearest neighbours. The second calibration uses OLS but with a broader definition of the spatial weights matrix which takes an average of all the house prices within a radius of 1 km. The third calibration uses the same weights matrix as the second but uses maximum likelihood (ML) rather than OLS. The calibration results for OLS and ML are more similar than are the OLS estimates obtained with the different definitions of the spatial weights.
4. From Table 5.2 the results of the ML and OLS calibrations are very similar both in terms of the parameter estimates and the standard errors. The one exception is the standard error of the autoregressive parameter which is much higher for ML than for OLS, suggesting we need to be careful about any inference we may draw from the OLS estimate. Although there is no guarantee that such a degree of similarity would be found for the local model results, the results are nevertheless encouraging.

In terms of presenting the results of calibrating the above models, we first focus on the estimates of the spatial autocorrelation coefficient from each model. The

Table 5.2 *Three calibrations of a spatial autoregressive model*

	OLS (10 nn)	OLS (1 km)	ML (1 km)
Variable	β	β	β
Intercept	2 137 (2 858)	−10 551 (3 466)	−6 260 (3 251)
Autoregressive term	0.517 (0.029)	0.685 (0.039)	0.622 (0.093)
FLRAREA	459 (18.3)	466 (18.4)	473 (18.4)
BLDPWW1	4 380 (1 112)	2 723 (1 137)	3 163 (1 133)
BLDPOSTW	−2 122 (1 959)	−1 068 (1 975)	−1 343 (1 992)
BLD60s	−3 022 (2 484)	−3 194 (2 497)	−3 040 (2 487)
BLD70s	−3 755 (2 680)	−2 269 (2 696)	−2 474 (2 699)
BLD80s	10 980 (2 136)	12 600 (2 146)	12 619 (2 148)
TYPDETCH	18 953 (2 813)	20 039 (2 827)	20 104 (2 868)
TYPTRRD	−9 735 (1 217)	−10 474 (1 234)	−10 705 (1 225)
TYPFLAT	−14 433 (1 484)	−17 092 (1 514)	−16 931 (1 505)

Note: Figures in parentheses are standard errors.

results of the three global models (1a, 2a, 3a) are presented in Table 5.3.[1] The raw house prices (model 1a) exhibit a large amount of positive spatial autocorrelation (the global estimate of ρ is 0.91). The degree of autocorrelation falls considerably once the spatial distribution of other factors is considered (the global estimates of ρ from models 2a and 3a are 0.56 and 0.52, respectively). However, these global values are relatively uninformative compared to the results of the local spatial autocorrelation estimates, obtained by GWR on models 1b, 2b and 3b, which are shown in Figures 5.2, 5.3 and 5.4, respectively.

[1] Given the earlier comments on the standard errors of the autoregressive parameter estimates, we need to be very careful about interpreting the t statistics in Table 5.3. However, given the t values are so large, it would seem highly unlikely that the substantive interpretation of the results would change if the estimates were obtained from ML instead of OLS.

Table 5.3 Global estimates of spatial autocorrelation

Model	Estimate	SE	t
1a	0.91	0.041	22.5
2a	0.56	0.030	18.9
3a	0.52	0.029	18.1

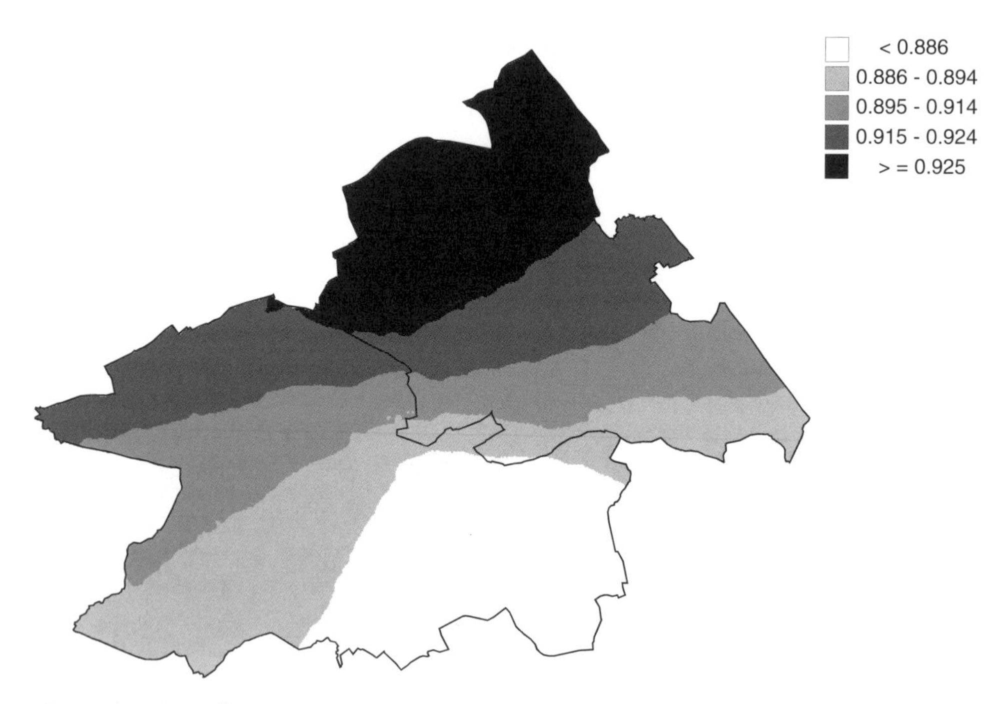

Figure 5.2 *Spatially varying autoregressive parameter estimates from the model in (5.4)*

Figure 5.2 shows the local estimates of the spatial autocorrelation exhibited by the raw house prices. These local estimates do not vary greatly because all the estimates are very high (high-priced houses tend to cluster together as do low-priced houses throughout the region). The spatial pattern is a simple one with highest local spatial autocorrelation found towards the north and the lowest found towards the south. This suggests there is slightly more mixed-price housing in the south but the autocorrelation levels are very high throughout the region.

In Figures 5.3 and 5.4, we observe the local nature of spatial autocorrelation of the conditional house prices. In Figure 5.3 prices are conditioned solely on the floor area of each house; in Figure 5.4 the prices are conditioned on floor area, house type and the age of the property. The spatial patterns of local autocorrelation are much more complex in both cases but show broadly similar patterns. There are now large spatial variations in the level of the conditional autocorrelation with even some local negative autocorrelation appearing in Figure 5.3. The local autocorrelation is generally stronger in Brent than it is in Ealing suggesting

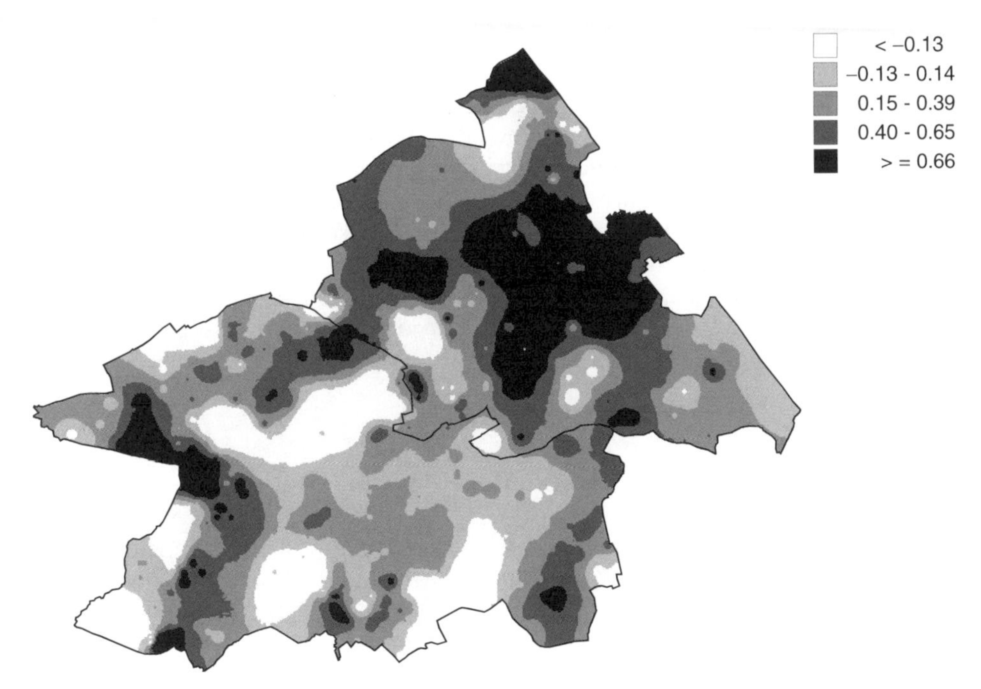

Figure 5.3 *Spatially varying autoregressive parameter estimates from the model in (5.6)*

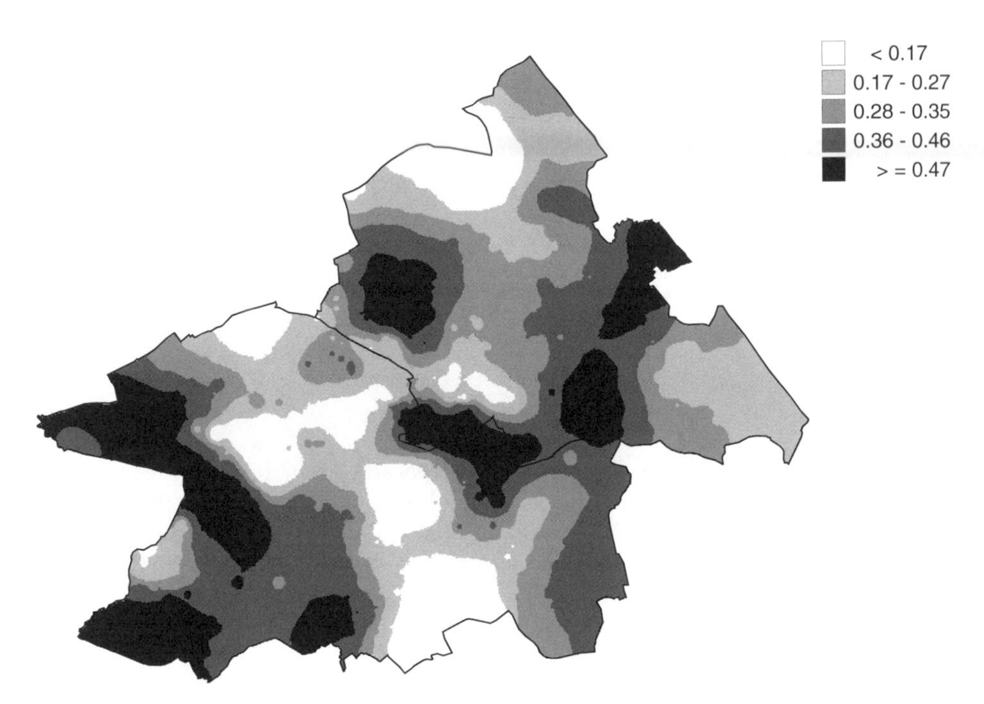

Figure 5.4 *Spatially varying autoregressive parameter estimates from the model in (5.8)*

that there are some interesting differences in the way house prices are distributed in the two boroughs. For example in Figure 5.3, where variations in floor area are accounted for, house prices in Ealing exhibit much greater local variation (the degree of local autocorrelation is very low) than in Brent where there appears to be much less variation in house prices. An investigation into the local geography of the two places might reveal why this is so. It could be, for example, that Ealing contains some features that promote differential house prices but which are not contained in our model (such as proximity to underground stations for example). Whatever the explanation, it is clear that the surfaces of local spatial autocorrelation in Figures 5.2–5.5 derived from GWR provide much more information on the spatial distribution of house prices than do the single global estimates in Table 5.2.

5.4 Residuals in Global Regression Models and in GWR

A diagnostic statistic indicating problems in regression modelling with spatial data is the degree of spatial autocorrelation exhibited by the residuals from the model (Cliff and Ord 1973; 1981). If the residuals from the model exhibit significant positive autocorrelation, the standard errors of the parameter estimates will be underestimated leading to potential problems with inference. Until now the solution to this problem was to apply spatial regression techniques (see below and also Fotheringham *et al.* 2000 for a description of such techniques). However, GWR has the potential to provide a more logical solution to this problem. To see this, consider Figure 5.5.

Assume we have a non-stationary process that can be modelled by the following local regression model in which the slope parameter is allowed to vary spatially but the intercept is fixed:

$$y_i = \alpha + \beta(u_i, v_i)x_i + \varepsilon_i \tag{5.9}$$

where $\beta(u_i, v_i)$ represents the locally varying slope parameter that can be estimated by GWR. Suppose that the real (unobserved) values of $\beta(u_i, v_i)$ are those shown in Figure 5.5a. Suppose further that we model this process incorrectly with a global model of the form:

$$y_i = \alpha + \beta x_i + \varepsilon_i \tag{5.10}$$

which yields an estimate of the global slope parameter of 0.5. This situation is depicted in Figure 5.5b. It is then clear that the spatial pattern of the *residuals* from the global model in (5.10) would be that shown in Figure 5.5c where the residuals exhibit positive spatial autocorrelation. Whenever the local parameter is greater than 0.5, the global model will underestimate the value of y_i; whenever the local parameter estimate is less than 0.5, the global model will overestimate y_i.

Consequently, it is possible that much of the observed spatial autocorrelation in residuals that is frequently observed in the calibration of global models applied to

(a) Real values of $\beta(u_i, v_i)$

0.9	0.8	0.8	0.7	0.5
0.8	0.7	0.6	0.5	0.4
0.7	0.6	0.5	0.4	0.4
0.6	0.5	0.4	0.3	0.2
0.5	0.4	0.3	0.2	0.1

(b) Estimated value of β from global model

0.5	0.5	0.5	0.5	0.5
0.5	0.5	0.5	0.5	0.5
0.5	0.5	0.5	0.5	0.5
0.5	0.5	0.5	0.5	0.5
0.5	0.5	0.5	0.5	0.5

(c) Residuals $(y_i - \hat{y}_i)$

+	+	+	+	0
+	+	+	0	–
+	+	0	–	–
+	0	–	–	–
0	–	–	–	–

Figure 5.5

spatial data results from applying a global model to a non-stationary process. Rather than allowing the non-stationarity to be reflected through the error terms in the model, it would seem better to model this non-stationarity directly through GWR. Thus, GWR could provide a more directly interpretable solution to the problem of spatially autocorrelated error terms in regression models applied to spatial data. This line of reasoning is similar to that of Miron who states: 'Spatial autocorrelation is neither a magical phenomenon nor a statistical nuisance... It is a specific consequence of the failure to accurately measure or specify variables or relationships' (1984: 206).

To examine how calibrating local, rather than global, models of spatial relationships can help solve the problem of spatial autocorrelation, we now return to the empirical example described above and calibrate the following models:

Model 1a

$$P_i = \kappa + \alpha \text{FLRAREA}_i + \varepsilon_i \tag{5.11}$$

Model 1b

$$P_i = \kappa(u_i, v_i) + \alpha(u_i, v_i)\text{FLRAREA}_i + \varepsilon_i \tag{5.12}$$

Model 2a

$$P_i = \kappa + \rho P_i{}^* + \alpha \text{FLRAREA}_i + \varepsilon_i \tag{5.13}$$

Model 2b

$$P_i = \kappa(u_i, v_i) + \rho(u_i, v_i) P_i{}^* + \alpha(u_i, v_i)\text{FLRAREA}_i + \varepsilon_i \tag{5.14}$$

Model 3a

$$\begin{aligned} P_i = {} & \kappa + \alpha_1 \text{FLRAREA}_i + \alpha_2 \text{BLDPWW1}_i + \alpha_3 \text{BLDPOSTW}_i \\ & + \alpha_4 \text{BLD60S}_i + \alpha_5 \text{BLD70S}_i + \alpha_6 \text{BLD80S}_i + \alpha_7 \text{TYPDETCH}_i \\ & + \alpha_8 \text{TYPTRRD}_i + \alpha_9 \text{TYPFLAT}_i + \varepsilon_\text{i} \end{aligned} \tag{5.15}$$

Model 3b

$$\begin{aligned} P_i = {} & \kappa(u_i, v_i) + \alpha_1(u_i, v_i)\text{FLRAREA}_i + \alpha_2(u_i, v_i)\text{BLDPWW1}_i \\ & + \alpha_3(u_i, v_i)\text{BLDPOSTW}_i + \alpha_4(u_i, v_i)\text{BLD60S}_i + \alpha_5(u_i, v_i)\text{BLD70S}_i \\ & + \alpha_6(u_i, v_i)\text{BLD80S}_i + \alpha_7(u_i, v_i)\text{TYPDETCH}_i \\ & + \alpha_8(u_i, v_i)\text{TYPTRRD}_i + \alpha_9(u_i, v_i)\text{TYPFLAT}_i + \varepsilon_i \end{aligned} \tag{5.16}$$

Model 4a

$$\begin{aligned} P_i = {} & \kappa + \rho P_i{}^* + \alpha_1 \text{FLRAREA}_i + \alpha_2 \text{BLDPWW1}_i + \alpha_3 \text{BLDPOSTW}_i \\ & + \alpha_4 \text{BLD60S}_i + \alpha_5 \text{BLD70S}_i + \alpha_6 \text{BLD80S}_i + \alpha_7 \text{TYPDETCH}_i \\ & + \alpha_8 \text{TYPTRRD}_i + \alpha_9 \text{TYPFLAT}_i + \varepsilon_i \end{aligned} \tag{5.17}$$

Model 4b

$$\begin{aligned} P_i = {} & \kappa(u_i, v_i) + \rho(u_i, v_i) P_i{}^* + \alpha_1(u_i, v_i)\text{FLRAREA}_i \\ & + \alpha_2(u_i, v_i)\text{BLDPWW1}_i + \alpha_3(u_i, v_i)\text{BLDPOSTW}_i \\ & + \alpha_4(u_i, v_i)\text{BLD60S}_i + \alpha_5(u_i, v_i)\text{BLD70S}_i + \alpha_6(u_i, v_i)\text{BLD80S}_i \\ & + \alpha_7(u_i, v_i)\text{TYPDETCH}_i + \alpha_8(u_i, v_i)\text{TYPTRRD}_i \\ & + \alpha_9(u_i, v_i)\text{TYPFLAT}_i + \varepsilon_i \end{aligned} \tag{5.18}$$

For each model, we calculate the Moran's I of the residuals to examine the effect of calibrating the models locally (models X.b) by GWR rather than globally (models X.a). It is our supposition that the local calibration removes much of the problem of spatially autocorrelated error terms encountered in traditional global models. Consequently, we are interested in the comparison of the results from models 1a and 1b, 2a and 2b, 3a and 3b, and 4a and 4b. These are given in Table 5.4.

The results suggest that the calibration of local rather than global models reduces the problem of spatially autocorrelated error terms by allowing geographically

Table 5.4 Moran's *I* of the residuals from 8 models

Model	Global version (OLS)	Local version (GWR)
1	0.5433	0.2503
2	0.3561	0.2026
3	0.5591	0.3387
4	0.3942	0.2934

varying relationships to be modelled through spatially varying parameter estimates rather than through the error term. As expected, the error terms are most strongly autocorrelated for the models that do not contain the autoregressive term (0.5433 in the case of model 1 and 0.5591 in the case of model 3). For both these models the reduction in autocorrelation produced by GWR is large: the spatial autocorrelation parameter is 0.2503 for model 1 and 0.3387 for model 3. Even where there is an autoregressive term in the global model, however, local calibration can still reduce the problem of spatially autocorrelated error terms: Moran's I is 0.3561 for model 2a and only 0.2026 for model 2b and is 0.3942 for model 4a and 0.2942 for model 4b. These are encouraging results.

The reduction in the degree of spatial autocorrelation through GWR can be seen if we compare maps of the residuals from the above models. However, to avoid repetition, we present maps only for models 3a, 3b and 4a which are shown in Figures 5.6, 5.7 and 5.8, respectively. In Figure 5.6, the global residuals from the

Figure 5.6 Residuals from the model in (5.15)

Figure 5.7 *Residuals from the model in (5.16)*

Figure 5.8 *Residuals from the model in (5.17)*

model of house prices on floor area, house type and age clearly exhibit positive spatial clustering with large areas of positive residuals grouped together and large areas of negative residuals grouped together. For instance, the model underpredicts house prices in large areas of Brent and the western edge of Ealing but overpredicts prices in a large area of central and eastern Ealing. The spatial autocorrelation in the residuals from the equivalent GWR model, shown in Figure 5.7, is no longer evident: there are no obvious patterns to the residuals which appear random across the region. It is interesting that spatial patterns in the residuals are still present in Figure 5.8 even though an autoregressive term has been added to the global model. This does not remove the autocorrelation problem nearly as well as GWR.

These results suggest that GWR provides an alternative, and perhaps more logical, solution to the problem of spatially autocorrelated error terms in spatial modelling compared with the various forms of spatial regression modelling. However, the two approaches can be combined as shown later in Section 5.6.

5.5 Local Parameter Estimates from Autoregressive and Non-Autoregressive Models

In the above section we provide evidence that the addition of a spatial autoregressive term to local models has relatively little impact on the spatial distribution of the error terms. In this section we examine the impact of autoregressive terms on the spatial distribution of local parameter estimates. To do this, we again continue with the empirical example of house prices in Brent and Ealing described above. In this case, however, we concentrate on the surfaces of the local parameter estimates obtained from the calibration of the local models given in equations (5.16), model 3b, and (5.18), model 4b. These models have the same sets of explanatory variables but model 4b has an additional autoregressive term.

Rather than display the surfaces associated with all the parameters in the two models, we first concentrate only on those surfaces exhibiting a statistically significant degree of spatial non-stationarity in the non-autoregressive model. The results of a Monte Carlo test on the significance of the spatial variation in each local estimate from model 3b (see Chapters 4 and 9 for more details of this test) are presented in Table 5.5.[2] The results suggest that the local parameter surfaces for the variables FLRAREA, BLDPWW1, BLDPOSTW, BLD80s and TYPTRRD exhibit significant spatial non-stationarity. For the other variables, the degree of spatial variation is not sufficient to be able to reject the null hypothesis that the surface has arisen from a stationary process and that the observed variation in the local estimates results solely from sampling variation.

The comparison we wish to make is between the two models so we describe the maps of the local parameter estimates in pairs. However, in each case, the spatial

[2] Unfortunately, p values associated with the spatial variation of the parameters from the local autoregressive model cannot be computed with the current version of the GWR software (see Chapter 9 for details of the software). This is because the Monte Carlo test employed uses a reallocation process of each row of data across the locations for which data are recorded. However, this does not make sense in terms of the spatial autoregressive variable which would not remain constant.

Table 5.5 Testing the statistical significance of the spatial variation in local parameter estimates from model 3b in (5.16)

Parameter	*p* value
Intercept	0.140
FLRAREA	0.000
BLDPWW1	0.000
BLDPOSTW	0.000
BLD60s	0.600
BLD70s	0.170
BLD80s	0.010
TYPDETCH	0.300
TYPTRRD	0.020
TYPFLAT	0.150

Note: The *p* value denotes the probability that the estimated local parameter surface could have resulted solely from sampling variation. We have taken probabilities less than 0.05 to indicate significant spatial non-stationarity.

distributions of the local parameter estimates from the two models (with and without the autoregressive term) are virtually identical. Consequently, rather than repeat the evidence we simply demonstrate the results for the first two variables, FLRAREA and BLDPWW1. Figure 5.9(a), for example, describes the spatial distribution of the local floor area parameter obtained from the non-autoregressive model; Figure 5.9(b) describes the same distribution obtained from the autoregressive model. It is clear that the two distributions are virtually identical and that the same pattern of local parameter estimates is found even in the autoregressive model where spatial dependency has been partially accounted for. The pattern suggests there are two very different housing sub-markets within the study area. In the central part of Brent, there is an area of housing where, accounting for variations in size, age and house type, house prices per square metre are much lower than average. This is flanked on both sides by areas of Brent in which housing is more expensive than average, *ceteris paribus*. Conversely, Ealing is dominated by an area of much higher conditional house prices per square metre although to the far west there is an area of housing with lower prices than average. By an interesting coincidence, a recent paper (Harris, 2001) contains maps of socio-economic data within Brent and the areas identified through GWR as having low house prices per square metre (accounting for the other factors in the model) correspond very closely to the most deprived parts of the borough.

The distribution of the relative attractiveness of pre-World War I housing in Figure 5.10 suggests there are larger pockets of relatively attractive pre-WWI housing in Ealing than in Brent. In both boroughs there are also large areas where pre-WWI housing is seen as less attractive than inter-war housing. There is a noticeable difference between east and west Ealing where the local parameter estimates exhibit a huge swing from highly negative values in the west to highly positive values in the east. The results suggest that whatever causes the local variation in parameter estimates from the hedonic price model in equation (5.16) is *not* accounted for by the addition of an autoregressive term to the local modelling framework.

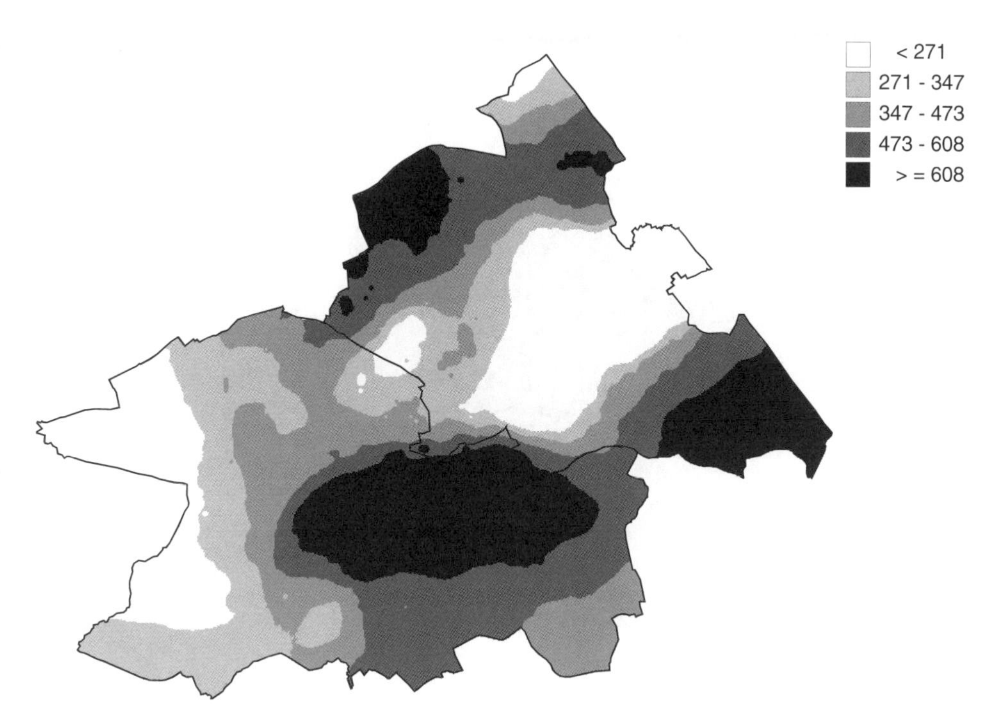

Figure 5.9 *(a) Local estimates of the floor space parameter from the model in (5.16)*

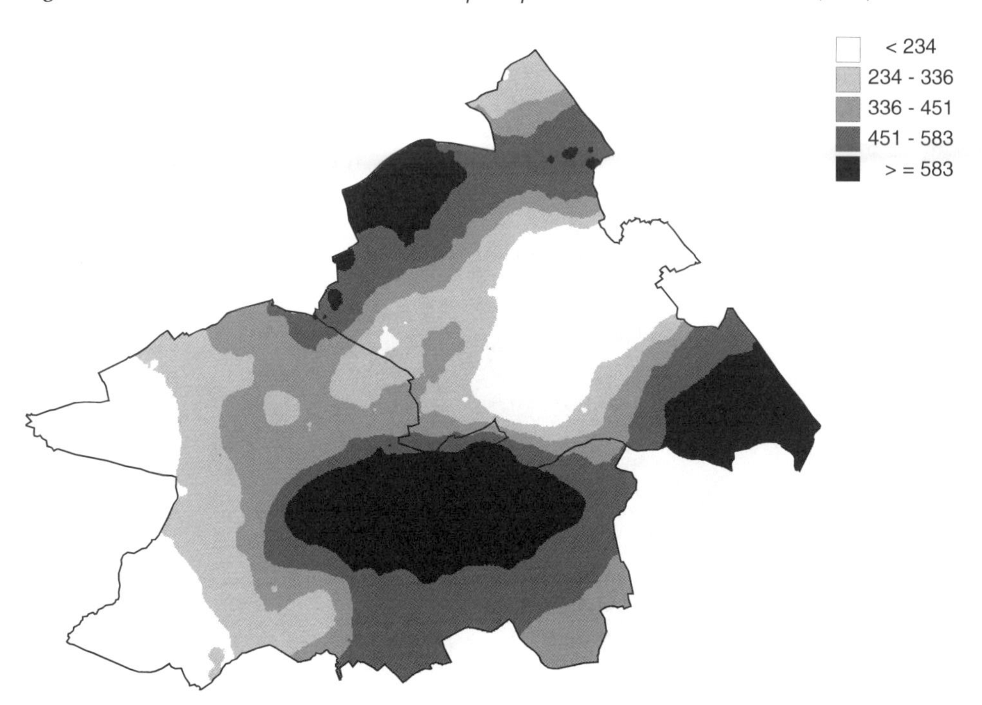

Figure 5.9 *(b) Local estimates of the floor space parameter from the model in (5.18)*

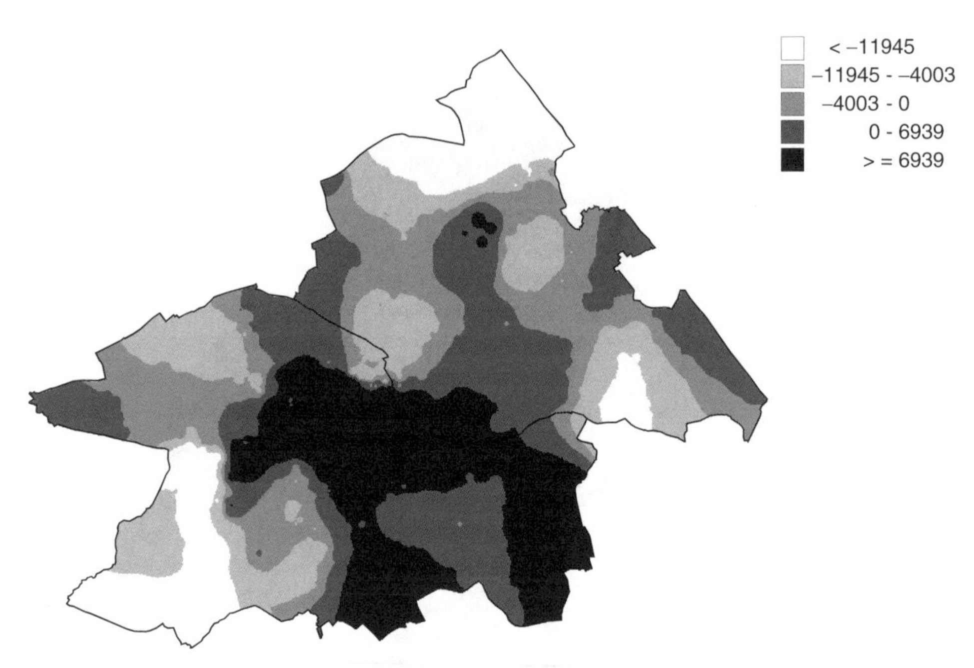

Figure 5.10 *(a) Local estimates of the Pre WW1 parameter from the model in (5.16)*

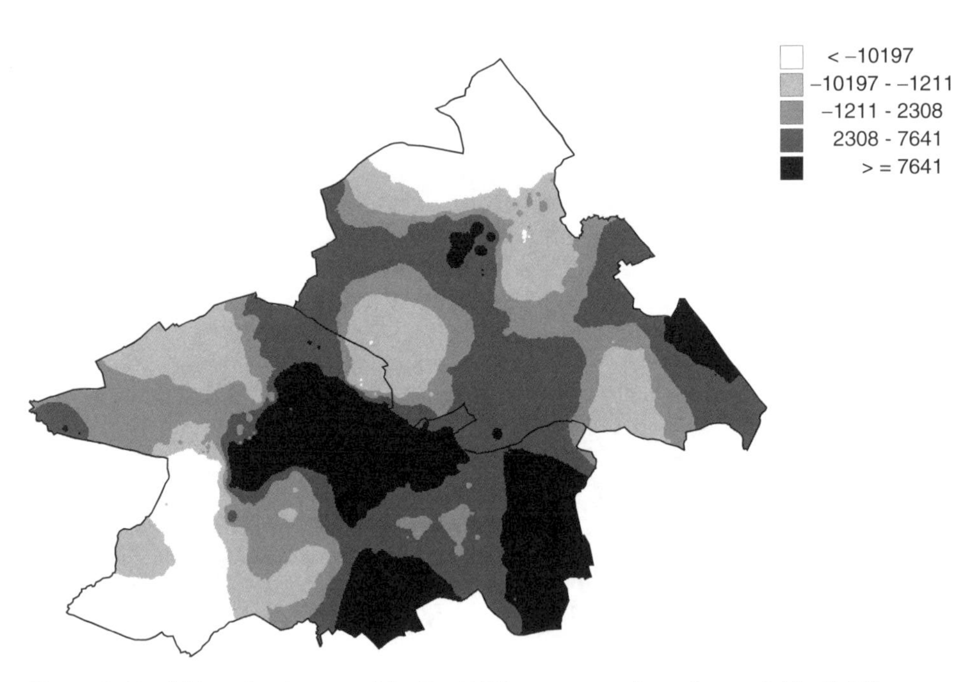

Figure 5.10 *(b) Local estimates of the Pre WW1 parameter from the model in (5.18)*

5.6 Spatial Regression Models and GWR

Until now, we have concentrated on highlighting how the GWR framework can be seen as an alternative to spatial regression modelling. In this last major section, we explore how the two frameworks can be combined. Because the models we propose as hybrids between the GWR and spatial regression frameworks are highly complex, some of their properties are unknown. Consequently, no empirical work is included in this section; instead, we concentrate on a conceptual and theoretical discussion.

5.6.1 Overview

Over the past three decades, the most usual way to take account of spatial processes in regression models has been to allow for spatial autocorrelation in the regression model. The basic regression model can be written as

$$y_i = \alpha_0 + \sum_{j=1}^{m} \alpha_j x_{ij} + \varepsilon_i \qquad (5.19)$$

where the ε_is are independent Gaussian random variables with mean zero and variance σ^2. In vector/matrix form, we can rewrite (5.19) as

$$\boldsymbol{y} = \boldsymbol{X\beta} + \boldsymbol{\varepsilon} \qquad (5.20)$$

where $\boldsymbol{\varepsilon} \sim \boldsymbol{N}(0,\sigma^2\boldsymbol{I})$. More simply, we can write this as:

$$\boldsymbol{y} \sim \boldsymbol{N}(\boldsymbol{X\beta},\sigma^2\boldsymbol{I}) \qquad (5.21)$$

With spatial data it is often difficult to make the assumption that the error terms are independent, in which case it is better to adopt the general model,

$$\boldsymbol{y} \sim \boldsymbol{N}(\boldsymbol{X\beta},\boldsymbol{A}) \qquad (5.22)$$

where $\boldsymbol{A}$ is a positive definite symmetric covariance matrix which allows non-zero covariances amongst the error terms. Typically, $\boldsymbol{A}$ is chosen so that elements of $\boldsymbol{y}$ that are closer to each other in space also have higher covariance. For brevity, we denote $\boldsymbol{X\beta}$ as $\boldsymbol{\mu}$, as this will prove to be useful below. Then, we can write the model as:

$$\boldsymbol{y} \sim \boldsymbol{N}(\boldsymbol{\mu},\boldsymbol{A}) \qquad (5.23)$$

This represents a generic assumption for spatial regression models. Before considering the relationship between such models and GWR it is helpful to review the ideas underlying the specification of autocorrelated models. There are two ways in which such models are specified – usually referred to as the conditional autoregressive (CA) and simultaneous autoregressive (SA) models.

5.6.2 Conditional Autoregressive (CA) Models

In CA models, we assume that the y-variable is dependent not only on a number of explanatory x-variables but also on other nearby y-variables. The model is specified as

$$y_i|\{y_j\colon j \neq i\} \sim N(\mu_i + \sum_{j=1}^{n} c_{ij}(y_j - \mu_j), \tau^2). \tag{5.24}$$

That is, the distribution of y_i conditional on all the other y-values is normal. Note that the distribution is expressed in terms of $y_i - \mu_i$, the difference between the observed y_i and the expected value of y_i obtained when considering the x-variables. Some restrictions are imposed on the spatial weight values, the c_{ij} values. First, $c_{ij} = c_{ji}$ and, second, $c_{ii} = 0$. The former restriction means the weights must be symmetrical; the latter restriction simply means that the conditional distribution of y_i cannot depend on y_i itself – only on other y-values. Typically, the c_{ij}s are chosen to reflect the spatial structure of the data. If the data are associated with a set of zones, then c_{ij} might be defined as 1 if zones i and j are contiguous, and 0 otherwise. For point data, c_{ij} might be defined as a continuous function of distance such as $kd_{ij}^{-\alpha}$ where $\alpha = 1$ or 2 and d_{ij} is the distance between points i and j (assuming that there are no coincident points). This latter scheme could also be applied to zonal data using distances between zone centroids.

This type of model was applied in Sections 5.3 and 5.4 above to the analysis of house prices. As well as considering the attributes of a house such as its floor area as determinants of a house's value, the price a given house sells for might also be explained by the values of neighbouring houses. It is not unusual, for example, for home-owners and real estate sellers to base their opinions on the price of a house on the prices of neighbouring houses.

We can rearrange the format of equation (5.24) to that of equation (5.23) so that it is rewritten as:

$$\boldsymbol{y} \sim \boldsymbol{N}(\boldsymbol{\mu}, (\boldsymbol{I} - \boldsymbol{C})^{-1}\tau^2) \tag{5.25}$$

from which it can be seen that the reason for the restriction $c_{ij} = c_{ji}$ is to stop the matrix $\boldsymbol{I} - \boldsymbol{C}$ becoming ill-defined.

5.6.3 Simultaneous Autoregressive (SA) Models

A simultaneous autoregressive model can be defined as

$$y_i \sim N(\mu_i + \sum_{j=1}^{n} b_{ij}(y_i - \mu_j), \lambda^2) \tag{5.26}$$

This differs from the CA model because the distribution of y_i is not conditional. In this case the marginal distributions for all the y_is are specified as a system of simultaneous equations. Again, the restriction on the spatial weights that $b_{ii} = 0$ is imposed but there is no longer a symmetry constraint on the weight matrix. Equation (5.26) can also be re-arranged into the form of equation (5.23)

$$\boldsymbol{y} \sim N(\boldsymbol{\mu}, \lambda^2(\boldsymbol{I} - \boldsymbol{B})^{-1}(\boldsymbol{I} - \boldsymbol{B}^{\mathrm{T}})^{-1}) \tag{5.27}$$

5.6.4 GWR, Conditional Autoregressive Models and Simultaneous Autoregressive Models

Both CA and SA models have been proposed in the literature as useful alternatives to OLS when one suspects that there are non-zero covariances between the error terms in the model. However, as Miron (1984: 207) notes: 'these are relief from a symptom of a problem not a cure for the problem itself'. It is also worth remembering that if the non-zero covariances result from model misspecification, the estimates of the parameters of the CA and SA models might still be biased. We suggest above that GWR might form an alternative, and preferable, solution when spatial dependency results from spatial non-stationarity.

However, in some instances, it might be useful to combine the two approaches. To see how this might be achieved, note that in (5.23) the exact form of $\boldsymbol{\mu}$ does not matter for the derivation of the CA and SA models. Thus, we can make $\boldsymbol{\mu}$ a local function of attributes as in GWR and allow $\boldsymbol{\mu}$ to vary with location i:

$$\mu_i = \beta_0(u_i, v_i) + \sum_{j=1}^{m} \beta_j(u_i, v_i)x_{ij} + \varepsilon_i \tag{5.28}$$

This appears to be a standard GWR model but here we specify the error term vector to be distributed as $N(\mathbf{0}, \boldsymbol{A})$, where $\boldsymbol{A}$ is a non-diagonal (but symmetrical) matrix, instead of as $N(\mathbf{0}, \sigma^2\boldsymbol{I})$. This could give a means of modelling both spatial non-stationarity and spatial dependency. However, in practice, it would still remain difficult to disentangle fully the effects of spatial non-stationarity from spatial dependency, a point made by Bailey and Gatrell (1995). Also, calibration of such a model would prove difficult and the inferential aspects of the model have still to be established. One line of reasoning that could be exploited here is that we could model autocorrelation in the error terms using a Kriging-based approach by calibrating a relationship between the error dependency and a continuous function of distance (as in the estimation of the elements of a variogram, for example).

Note that although the combination of equations (5.28) and (5.23) yield a GWR version of a spatial regression model, the degree of autocorrelation, as specified by $\boldsymbol{A}$, could still be stationary. For example, the correlation between two locations i and j – the ijth element of the matrix $\boldsymbol{A}$, could depend *only* on the distance d_{ij} between these locations – that is, $a_{ij} = f(d_{ij})$. In such an instance, calibration of the models could be carried out using a technique based on Kriging. Methods of dealing with correlated errors in kernel regression, such as those proposed by Altman (1990) or Herrmann *et al.* (1992) could also be applied.

An alternative approach is to assume that the degree of correlation between residuals could depend on their absolute location as well as the distance between their locations. In this case $a_{ij} = f(u_i, v_i, u_j, v_j)$ so that the variogram approach breaks down. Work to date considering special cases of this kind of model has been largely exploratory, but has involved using local calibrations of models such as the Ord autoregressive model used earlier in this chapter, where the autoregressive coefficient is non-stationary. However, it should be noted that again the inferential theory for such a model has yet to be developed.

5.7 Summary

In this chapter we have considered the relationship between GWR and spatial autocorrelation in two ways. In the first, an empirical study of house prices, where it is strongly suspected that autoregression processes cause autocorrelation, we demonstrate how GWR can be used to obtain local measures of both unconditional and conditional spatial autocorrelation. In the second, we consider the link between GWR and spatial autocorrelation from a theoretical viewpoint. In both situations there are a number of challenges.

First, one has to address the theoretical issues of defining and handling spatially autocorrelated GWR models. In particular, it is important to gain some understanding of how these models may be calibrated and how inference may be carried out. The latter would enable us to answer such questions as 'does autocorrelated GWR provide a better model for the data than an ordinary GWR model?' and 'does ordinary GWR provide a better model for the data than a global autoregressive model?'

Second, there are practical computational issues that need addressing. The maximum likelihood (or possibly local likelihood) approach to calibrating autocorrelated models is computationally intensive. It involves the computation of the eigenvalues of an $n \times n$ matrix simply to compute regression coefficient estimates. If standard errors of these are also to be computed, a matrix of the same size needs to be inverted in addition to the previous computation. Furthermore, if the GWR coefficients are to be computed at m regression points, then the above procedures must be repeated m times! Thus, in this investigation we have worked with OLS methods, rather than maximum likelihood. Given the computational load of working with the maximum likelihood approach, it would seem that OLS is currently the only viable option for large data sets. There is a need, therefore, for investigation into the cost of this computational convenience in terms of bias in the estimation of coefficients. Such investigation needs to be tempered with the recognition that the results of calibrating spatial regression models are highly sensitive to the definition of a spatial weights matrix. This sensitivity will often lead to much greater variation in parameter estimates than that produced through the use of different calibration methods.

Despite these comments, there is evidence in this chapter to suggest some intriguing possibilities. It would seem that allowing for non-stationarity in the regression parameters can account for at least some, and possibly a large part, of the autocorrelation in error terms in a global model calibrated with spatial data. It also appears that a GWR approach to spatial autoregressive modelling provides a relatively easy method of calculating both unconditional and conditional measures of local spatial autocorrelation. Mapping these measures has the potential to reveal local patterns in the spatial distribution of a variable that would otherwise go unseen.

6

Scale Issues and Geographically Weighted Regression

6.1 Introduction

Scale has a pervasive and potentially complex influence in the analysis of spatial data. On the one hand, the results of many statistical techniques appear to be sensitive to the scale of the spatial units for which data are available, even when the underlying relationships being measured are scale independent. On the other hand, in some situations different spatial processes might operate at different spatial scales and, consequently, measurements of these processes should vary across different scales. While all types of spatial analysis and spatial modelling are affected by scale to some degree, certain types of analysis might be more robust than others to changes in scale. Here we consider two main issues connected with scale and GWR. One is the extent to which GWR can provide information on the appropriate scale of analysis; the other is the sensitivity of GWR results to variations in the scale of analysis. First, we make a few introductory comments about scale and GWR.

Suppose we are interested in the relationship between $\boldsymbol{Y}$, a vector of observed values of the dependent variable, and $\boldsymbol{X}$, a matrix of observations on a set of independent variables. For example, a linear relationship between $\boldsymbol{Y}$ and $\boldsymbol{X}$ would be written as:

$$y_i = \alpha_0 + \alpha_1 \, x_{1i} + \alpha_2 \, x_{2i} + \ldots \alpha_k \, x_{ki} + \varepsilon_i \tag{6.1}$$

where y_i represents an observed value of y at place i, x_{ki} represents the observed value of the kth independent variable at place i, $\alpha_1 \ldots \alpha_k$ represent parameters to be estimated in the calibration of the model and ε_i is a random error term.

Suppose further that the data are measured at each of a set of N locations distributed in geographic space as shown in Figure 6.1a. A question that could be

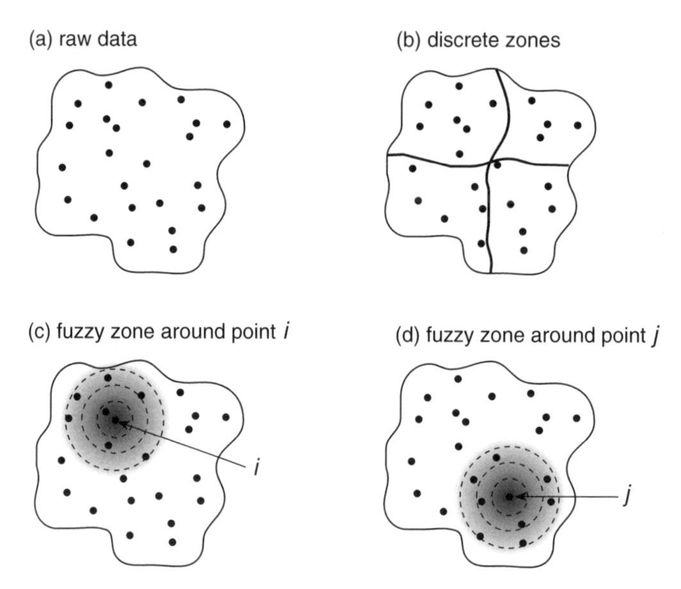

Figure 6.1 *Zoning systems and GWR*

asked is: What is the appropriate scale at which to analyse these data? One possibility that might spring immediately to mind is to analyse the relationship between $\boldsymbol{Y}$ and $\boldsymbol{X}$ separately at each of the N points where data are reported. However, the problem with such a procedure is that the model in (6.1) cannot be calibrated with observations at only one place. Another possibility therefore is to use all the data in a single calibration of the model, which is what usually happens, although clearly one has to hope that the relationships being examined do not vary over space. If the relationships do vary over space, the parameter estimates yielded in the calibration of equation (6.1) then become 'average' values which will hide the potentially interesting spatial variations in relationships as well as producing poorer fits to the data. This then raises the following two questions:

> What is the appropriate partitioning of space in which to study the relationships between y_i and $x_{li} \ldots x_{ki}$?

and

> Can discrete partitions of this space be defined in which the relationships being examined are relatively homogeneous (Figure 6.1b) or should the partitions be 'fuzzy' (Figures 6.1c and 6.1d)?

Figure 6.1b shows the study area divided into discrete zones in which the model can be calibrated separately. While this is a common way of proceeding in spatial analysis, it is clearly inappropriate in many instances because spatial processes are generally continuous and are not contained within discrete, and often arbitrary, boundaries. The imposition of discrete boundaries implies that a spatial process within one zone might be quite different from that in the neighbouring zone.

An alternative way of thinking about this issue is to consider a set of undefined or 'fuzzy' zones centred on locations in the study region. These locations might be those at which data are recorded or they might be independently selected from within the region as a basis for further mapping, or they may be a mixture. Whatever the definition of the centroids of the zones, for each zone we can think of a likelihood of a data point belonging to each zone. We can base our estimate of this likelihood on the distance between the data point and the zone centroid: the closer a data point is to a zone centroid, the greater is the likelihood that the data point belongs to that zone. Consequently, all data points potentially belong to all zones but the likelihood of belonging to a nearby zone is greater than that of belonging to a distant one. This is the situation described in Figures 6.1c and 6.1d where we attempt to describe these fuzzy zones around just two centroids. We could construct similar zones around any location in the region and within any zone we can answer the question: What are the relationships around this location? The answer would not be a set of discrete model calibrations but would be a continuous surface of parameter estimates for any relationship.

The difference in the two ways of defining subdivisions of the data – either in discrete or fuzzy zones – has important ramifications for the way in which the model in equation (6.1) is calibrated. In the case where the study region can be subdivided into discrete zones, the equation becomes

$$y_{i \in M} = \alpha_{0M} + \alpha_{1M}\, x_{1i \in M} + \alpha_{2M} x_{2i \in M} + \ldots \alpha_{kM} x_{ki \in M} + \varepsilon_{i \in M} \qquad \text{for all } M \quad (6.2)$$

where M represents one of the discrete zones and the notation $i \in M$ denotes a data point i in zone M. The result of calibrating equation (6.2) will be a set of parameters labelled M for each of the discrete zones. Unfortunately, this implies that a relationship, say, that described by α_{1M}, experiences a discontinuity from one zone to another, which is not a realistic representation of most spatial processes.[1]

In the case of fuzzy zones, the model to be calibrated is:

$$y_i = \alpha_0(x,y) + \alpha_1(x,y)x_{1i} + \alpha_2(x,y)x_{2i} + \ldots \alpha_k(x,y)x_{ki} + \varepsilon_i \qquad (6.3)$$

where the parameters to be estimated are specific to each regression point (x,y) which is the centroid of a fuzzy zone. Equation (6.3) represents a GWR model and so GWR can be seen as a technique for allowing fuzzy zones to be placed around each regression point. The data within each zone are weighted according to the probability of group membership which, in turn, is related to distance from each

[1] Multi-Level Modelling, a local modelling technique, also assumes that the zones bounding spatial processes are discrete. This is necessary in order to specify precisely a spatial hierarchy prior to the application of the model. This might be tenable in some instances where the application of a particular policy or set of procedures might occur within political boundaries but in most other instances the assumption is untenable.

regression point as defined by an appropriate weighting function (see Chapter 2 for examples). Clearly, the nature of the weighting function in terms of its bandwidth or distance-decay reflects the scale of the zone around each regression point. When the bandwidth is small, only data points in close proximity to the regression point can have a substantial impact on the estimated parameters for the regression point. This situation is one where GWR is applied at a very local scale. When the bandwidth is large, data points from further away can have an impact on the estimated parameters for the regression point. This situation is one where GWR is applied at a broader, regional scale. We now explore this further with an empirical example.

6.2 Bandwidth and Scale: The Example of School Performance Analysis

6.2.1 Introduction

Within the United Kingdom, there was a radical shift in parental power in education during the 1990s with parents being encouraged to 'shop around' for the best schools for their children. It is now possible in theory for parents to select the state school which they would like their children to attend and the Government encourages this selection process by publishing league tables of scholastic achievement by pupils at each state-run school.[2] The data on school performance exhibit some interesting geographical variations and it is a politically 'hot' issue as to what factors might cause such variations. For instance, does school size affect performance? To what extent is scholastic achievement a product of environment? Are there areas where school performance is consistently below average, and if so, what socio-economic characteristics do such areas have? Is it possible to identify schools that are performing well relative to their intake of pupils? Clearly these are all very sensitive issues for a government whose stated aim is the elimination of inequalities (see, for example, *The Times*, 7 December 1998). The issue is particularly sensitive if it appears that children aged between 4 and 11 are being disadvantaged so early in their scholastic careers because of their environment.

For the above reasons, it is useful to examine the relationship between school performance and the socio-economic characteristics of school catchment areas. Such an analysis is not new and examples of attempts to discover relationships between school performance and catchment area characteristics can be found in, *inter alia*, Conduit *et al.* (1996); McCallum (1996); Coombes and Raybould (1997);

[2] In practice, however, this inevitably collapses to parents being able to express a preference for a particular school with no guarantee that the preferred school will be able to provide a place because the more popular schools quickly reach their capacity. In fact most schools, particularly primary ones, faced with capacity limitations, give preference to children living close to the school. The net result is that school catchment areas are still strongly geographically based around each school. However, as a result of the theoretical situation in which parents might be able to select a state primary school for their children, as opposed to the children just going to the nearest one, state schools are evaluated on a standard set of attainment criteria across the country. These attainment levels are then employed to produce 'league tables' of school performance, which can be used by parents as a basis for comparing schools.

Brown *et al.* (1998). However, using GWR allows us not just to determine whether or not relationships between school performance and catchment area characteristics exist, but also to determine if there are any interesting spatial variations in these relationships. That is, perhaps some attributes of school catchment areas have an effect on school performance in some regions and not in others; such variations would be hidden in global modelling results.

6.2.2 The School Performance Data

Figure 6.2a shows the spatial distribution of the 3 687 primary schools in northern England for which data on pupils' attainment levels were made available.[3] County boundaries and names are also displayed. Figure 6.2b displays the major urban areas within this region. The varying density of the schools obviously reflects the spatial distribution of population with heavy concentrations of schools in the metropolitan areas of Tyneside and Cleveland in the north-east, Leeds in the south, and Manchester in the south-west.

For each school we have the percentage of students who reached or exceeded a pre-defined level of attainment in maths in 1997.[4] This may be reasonably modelled by a binomial distribution (students either attain the desired level or they do not). Consequently, given the variance of the binomial distribution is $p(1-p)/n$, with p being the proportion of students at a school who achieve the prescribed standard, the variance of the attainment score will decrease as it approaches 0 or 1. This makes the use of regression analysis, which assumes the error terms to have constant variance, highly questionable. To remove this problem, the attainment scores can be transformed in the following way to produce a variance-stabilised maths score for each school which can be regressed on a set of independent attributes:

$$TM_i = \arcsin\,[\mathrm{sqrt}(M_i)] \tag{6.4}$$

where TM_i is the transformed maths score and M_i is the raw score (Bartlett, 1936).

The data on school performance in maths were then regressed on the following attributes drawn from the school catchment areas (see Fotheringham *et al.* 2001 for more details of how these explanatory variables were obtained):

---

[3] The authors would like to thank Dr Robin Flowerdew of the School of Geography and Geosciences at the University of St Andrews for making these data available to us.

[4] This is not an ideal indicator of school performance for obvious reasons. It fails to differentiate pupils who are able from those who are very bright and there could, in theory, be quite big differences in pupils' abilities between two schools having the same percentage of students achieving the required level of attainment. Also, it should be noted that pupils who are absent on the day of the test are counted as having failed and that 'statemented' pupils, those with special educational needs, are also counted in the denominator. While these latter two issues cause some undesirable variability in the scores, they do ensure a certain degree of reliability in the testing procedure with schools unable to 'hide' their poorer students. However, despite these caveats on the school performance data, the data do separate schools where children are performing well from those where children are performing poorly. In addition, the data are available for every state-supported school in the country.

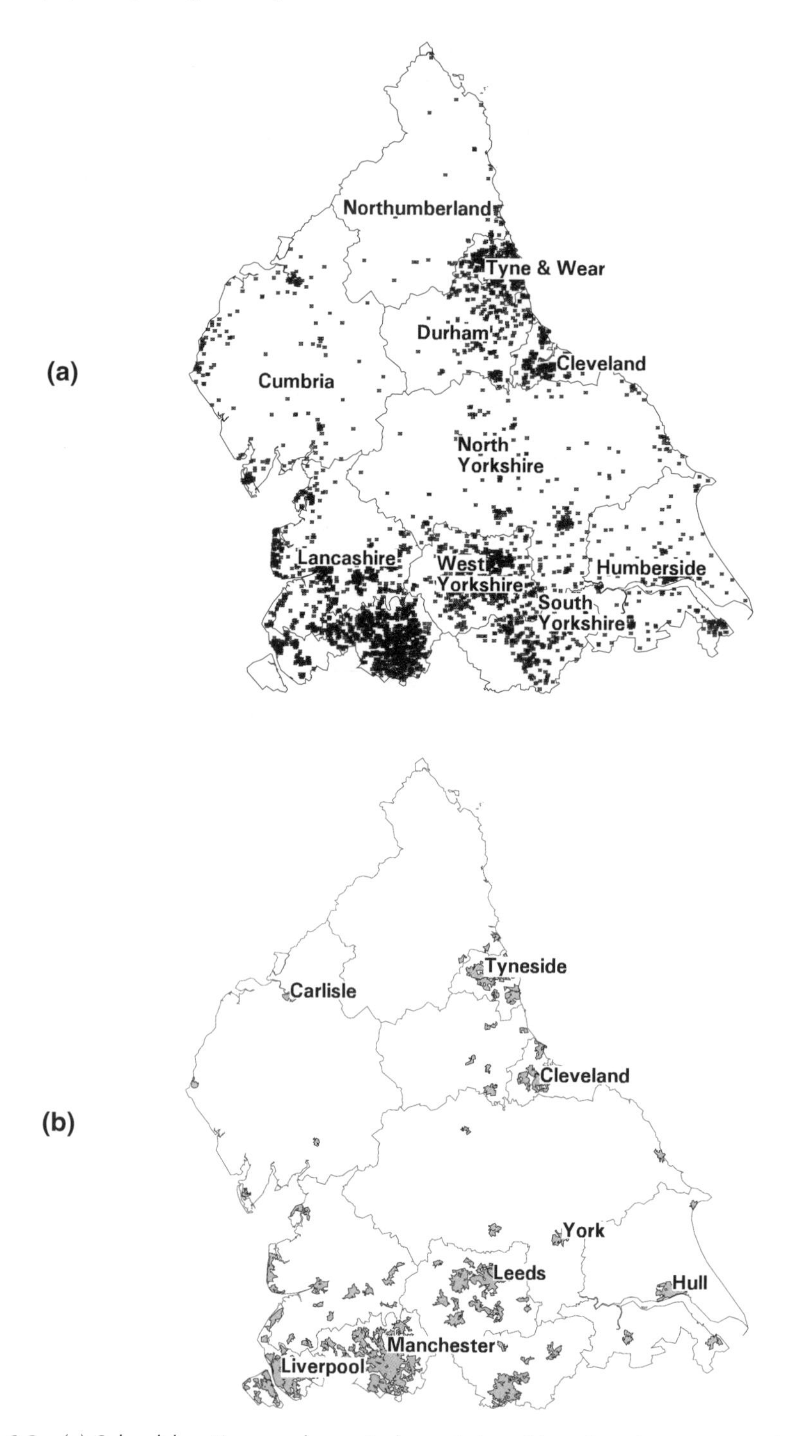

Figure 6.2 *(a) School locations and county boundaries; (b) main urban areas and county boundaries*

- SC1: the percentage of families where the head of household has a professional or managerial occupation. Given that the 1991 UK Census of Population did not ask any income information, this variable is often used as a proxy for income. High levels of SC1 would typically indicate a relatively rich catchment area.
- CH: the percentage of households living in state-provided council housing.
- UN: the percentage of unemployment.
- LP: the percentage of families headed by a single parent.
- B: the percentage of black residents
- I: the percentage of Indian residents
- C: the percentage of Chinese residents.

In addition, a variable denoting school size was also used as an explanatory variable in order to get some idea of any relationship between school size and performance. This variable, termed the school roll (SR), is supplied with the school performance data and is the total number of pupils from each school eligible to take the standardised test.[5]

There is an issue that the school performance data were recorded in 1997 and we are using 1991 census data to describe the schools' catchment areas. However, given the general stability in socio-economic patterns of population in the UK, this is probably of little consequence: what were relatively deprived areas in 1991 will almost certainly have remained relatively deprived areas in 1997.

6.2.3 Global Regression Results

In order to investigate whether there are any environmental factors which might explain, in part, the spatial variation in school maths performance, the following regression model was calibrated by weighted least squares regression with data on all 3 687 schools.

$$TM_i = \alpha_0 + \alpha_1 \mathrm{SR}_i + \alpha_2 \mathrm{SC1}_i + \alpha_3 \mathrm{CH}_i + \alpha_4 \mathrm{UN}_i + \alpha_5 \mathrm{LP}_i + \alpha_6 \mathrm{B}_i + \alpha_7 \mathrm{I}_i + \alpha_8 \mathrm{C}_i + \varepsilon_i \quad (6.5)$$

where the αs are parameters to be estimated and the variables are defined above. Weighted least squares regression is used to calibrate the global model with weights equal to $1/SR_i$ because the variance of the maths scores is likely to be an inverse function of the school roll: the scores for smaller schools will be more sensitive to absentees and special needs pupils. As mentioned earlier, the denominator of the school performance indicator includes both pupils with special needs and those who were absent on the day of the test.

[5] Although this gives a good idea of the size of the school, it would have been preferable to have the number of children *per class* eligible to take the test rather than the total number of a certain age. Class size is more likely to have an effect on pupils' performance (smaller classes in theory being better for educating pupils) than school size, for which the school roll variable is a surrogate. It is impossible to know, for example, whether a school roll of 60 represents two classes of 30 or three classes of 20 in a particular school.

Table 6.1 *Global regression results of the school performance data*

Variable	Parameter Estimate	*t*
Intercept	1.11	94.7
SR	−0.0013	−6.5
SC1	0.0060	6.9
CH	−0.0029	−7.4
UN	−0.0035	−2.9
LP	−0.0234	−4.5
B	0.0013	0.4
I	−0.0074	−7.7
C	0.0133	1.3

Calibration of this model by weighted least squares yields the results shown in Table 6.1. It appears that schools with good performances in maths are characterised by low numbers of pupils and catchment areas with high percentages of people in professional and managerial occupations, low percentages in council housing, low percentages of Indian residents, low unemployment rates, and low percentages of lone parent households.

However, the model explains only 24% of the variance in the transformed maths scores and there are clearly many other determinants of school performance not accounted for in the model. The more obvious variables missing from the model are those related to the school such as the average class size, resources such as library and computing facilities, the number of special needs pupils, and the level of truancy. Also missing are attributes which are difficult to quantify such as the degree of parental involvement and the quality of the teaching. To some extent, the absence of these variables from the model is exacerbated by the fact that the model being calibrated is a global one that is assumed to apply equally to *all* parts of the region. In fact, it may not apply to *any* part of the region. The estimated parameters represent global averages of processes that might exhibit a substantial degree of spatial variation. We now examine the calibration of the model by GWR which allows us to investigate the nature of any spatial non-stationarity in the relationships depicted in equation (6.5).

6.2.4 Local Regression Results

As we have seen in earlier chapters, instead of producing a single global average parameter estimate for each relationship, GWR produces a set of local parameter estimates, the spatial distribution of which can then be mapped. In this case, we have divided each local parameter estimate by the estimate of its standard error (see Chapters 2 and 4 on how to obtain the best estimate of the standard error). This is useful because it accounts for greater uncertainty in the estimated parameters where the data are sparse. These *t*-surfaces are used, not in a formal sense, but in a purely exploratory role, to highlight parts of the map where interesting relationships appear to be occurring.[6] The t surfaces for the school roll and social class 1 parameters in equation (6.5) are shown in Figures 6.3a and 6.4a, respectively.[7] These figures display absolute t values greater than 1.96 and 2.58; these being frequently used in global modelling where they correspond to the 95% and 99% significance levels, respectively. In Figures 6.3b and 6.4b we display the same t values but use a different shading scheme based on 4.35 and 4.69, the Bonferroni-adjusted critical values of the t distribution at the 95% and 99% significance levels, respectively.[8]

The global relationship between school size and performance is significantly negative with a t value of −6.5 (Table 6.1) suggesting that in general pupils in larger schools do less well than pupils in smaller schools. This would appear to be a strong argument in favour of reducing school sizes. However, in Figure 6.3a we see that the majority of the local estimates do not exhibit any strong relationship between school size and performance. In fact there appear to be only two areas of the region where there is a strong negative relationship between school size and performance. These are centred on schools in Tyneside and in Leeds. There is even an area in the north-west, albeit with not many schools, in which locally larger schools appear to be performing better than smaller ones, *ceteris paribus*. To some extent the local parameter estimates exhibit a rural–urban dichotomy with the primarily rural areas having only weakly negative or positive parameter estimates and the predominantly urban areas having more negative estimates. Perhaps pupils in larger rural schools tend to perform better than pupils in smaller rural schools but the situation is reversed in urban areas. One possible explanation for this difference is that perhaps the smaller rural schools lack the resources of larger rural schools. Smaller rural schools often also have mixed age classes because of the small number of students in each year. Larger urban schools, on the other hand, perhaps are too large and would benefit from smaller class sizes where teachers could have a greater influence. These variations in relationships tend to exist still in Figure 6.3b with the much higher, Bonferroni-adjusted critical t values. However, no local parameter estimate now exceeds the 95% critical value. Overall, the local results indicate that the relationship between school size and performance is perhaps more complex than is suggested by the global modelling results.

---

[6] Bowman and Azzalini (1997) undertake similar multiple tests with kernel density surfaces to good effect and also provide a general discussion about this type of approach.

[7] Similar t surfaces were reported in an earlier paper by Fotheringham *et al.* (2001) but these t surfaces used estimates of the local standard errors obtained in the local regressions. In Chapter 2 we describe a better method of obtaining values for the standard error and this method is used above. However, in this instance, it appears that there is little difference in the two ways of estimating the local standard errors.

[8] To avoid potential problems caused by multiple hypothesis tests, the Bonferroni correction adjusts the critical value of the test upwards by setting a new significance level equal to the original significance level (e.g. 0.05) divided by the effective degrees of freedom in the GWR model (see Chapters 2 and 4). In this case, the critical value of t equated with an original significance level of 0.05 is 4.35 and that with a significance level of 0.01 is 4.69. It should be noted that this is a very conservative correction to the multiple hypothesis testing problem and some claim it to be too conservative.

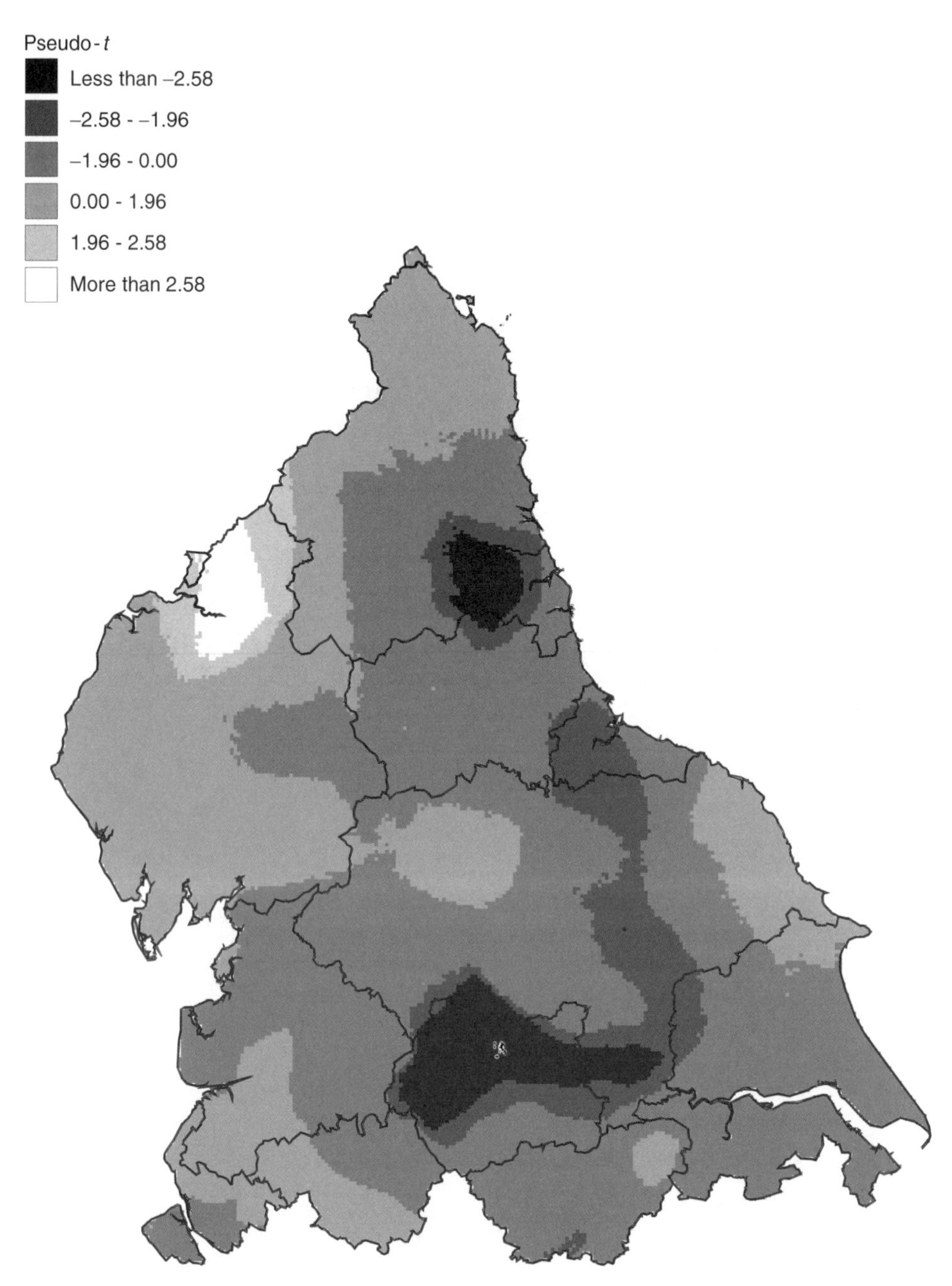

Figure 6.3 *(a) School roll t*

Similarly, the local t-surfaces for the relationship between school performance and the percentage of people in the school catchment area employed in professional and managerial occupations (SC1), shown in Figures 6.4a and 6.4b, exhibit

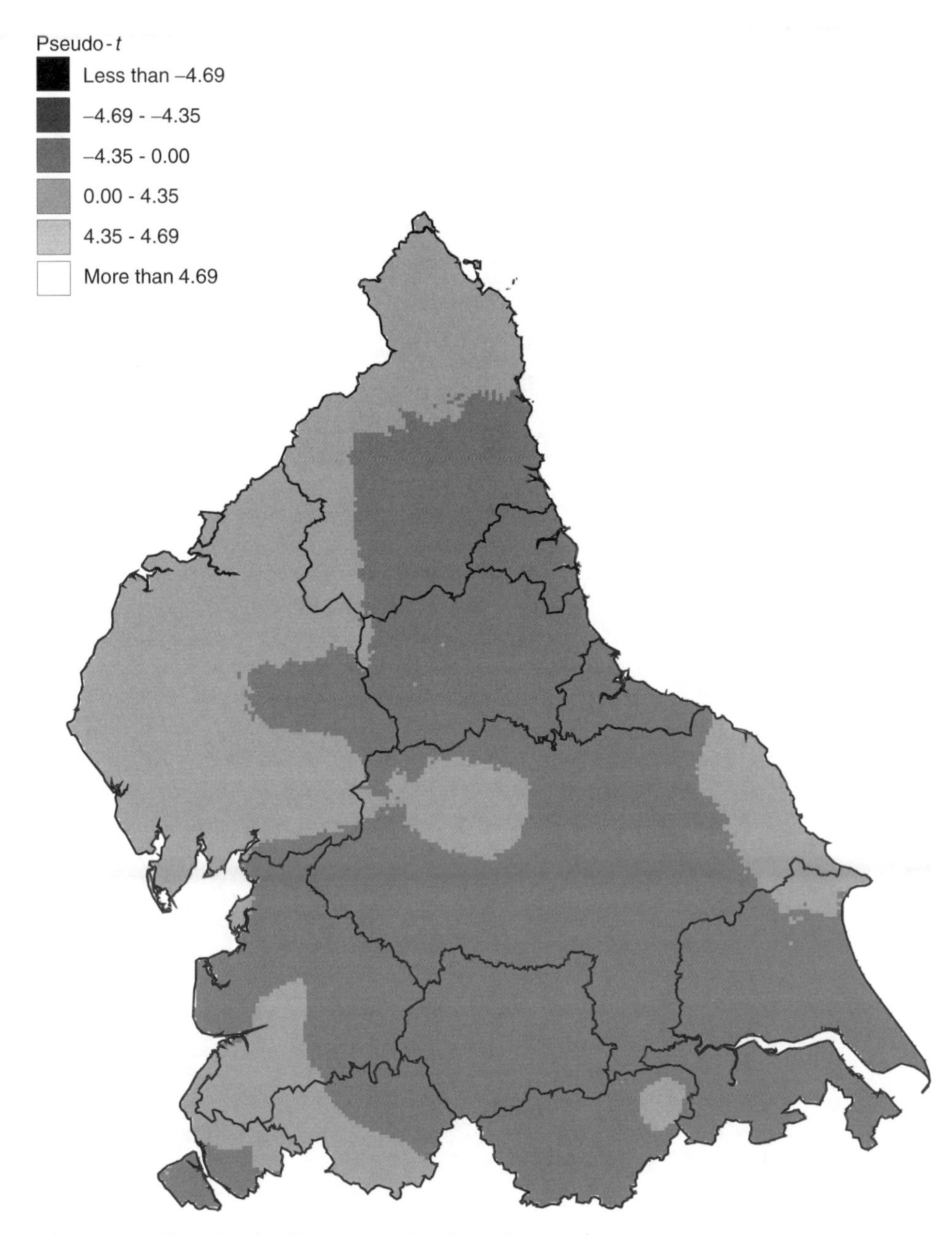

Figure 6.3 *(b) School roll parameter Bonferroni intervals*

more complexity than is suggested by the single global parameter estimate. The global relationship is significantly positive with a t value of 6.9 and most of the local parameter estimates in Figure 6.4a are strongly positive. However, the intensity of the relationship is not constant across northern England and there appear to

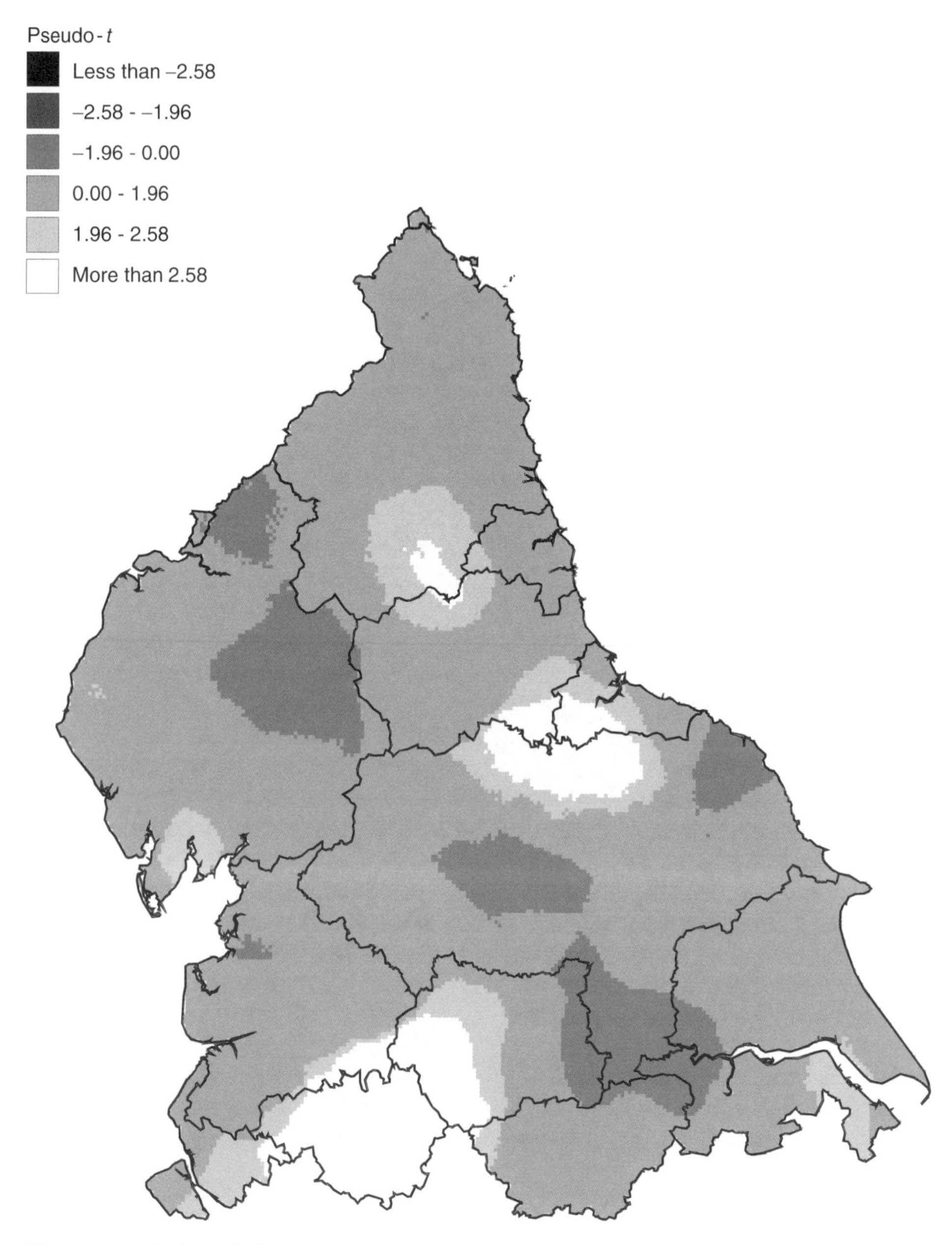

Figure 6.4 *(a) Social class I t*

be three areas in which the relationship between school performance and social class is particularly strong. One of these areas is just west of Tyneside, the second is south-west of Cleveland and the third is a large area centred on Manchester. Interestingly, using the Bonferroni-corrected critical values in Figure 6.4b, only

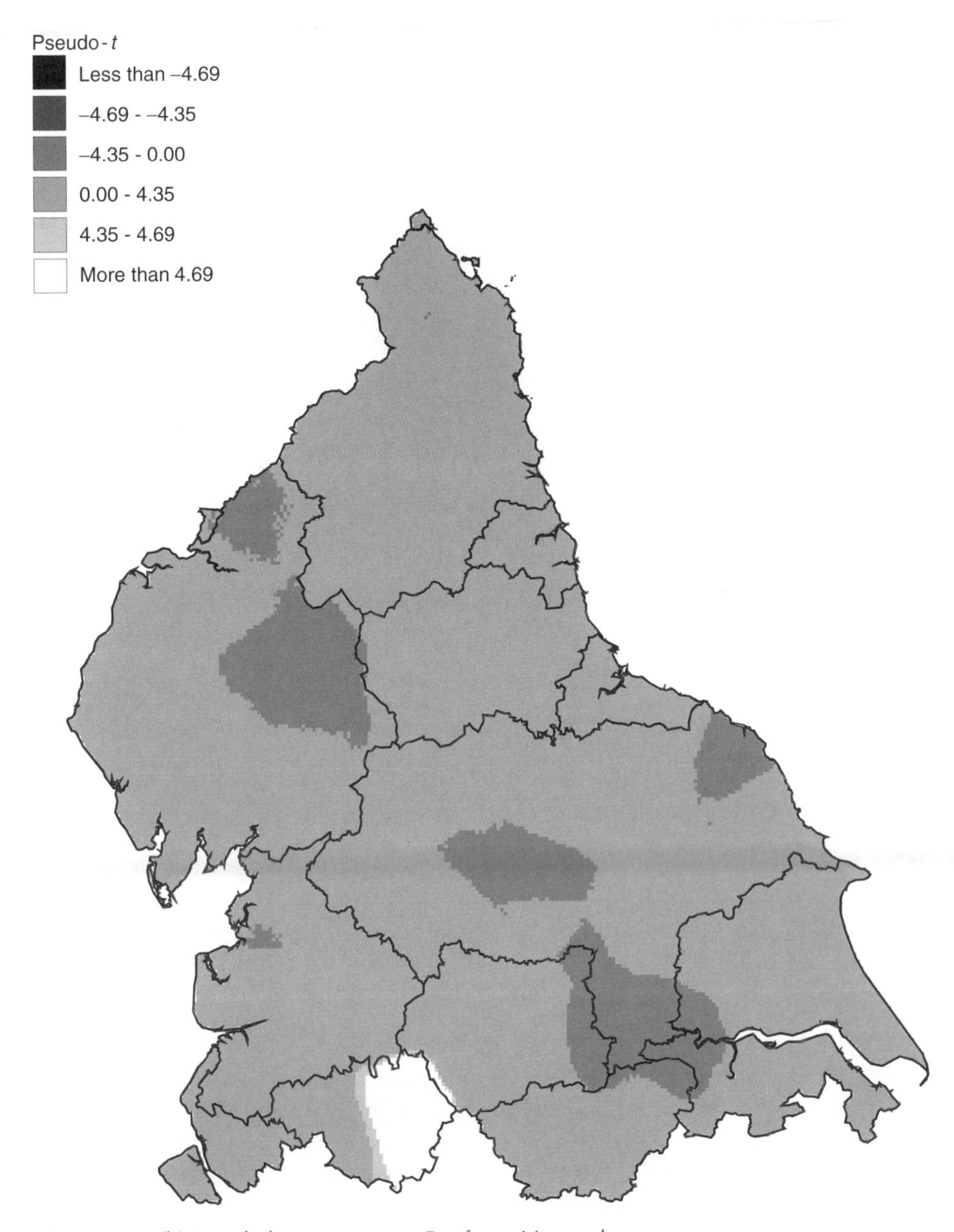

Figure 6.4 *(b) Social class I parameter Bonferroni intervals*

the Manchester region still remains as an area with a notable relationship between school performance and social class. All three highlighted areas have very diversified populations and it would appear that here school performance is strongly determined by the characteristics of the school catchment areas. Schools with high concentrations of parents in professional and managerial occupations perform considerably better than those with low concentrations. Further insight must be left to a more detailed local investigation into the school systems. One of the values of producing maps of local relationships such as those in Figures 6.3 and 6.4 is that they identify areas for more detailed investigation.

The above results are obtained from a GWR model in which the bandwidth is optimised as a part of the GWR calibration. In this case, the optimal bandwidth for a fixed spatial kernel with a bi-square weighting function (see Chapter 2) is around 15 kms. However, we can use GWR in another way by allowing the bandwidth to vary across a range of values and examining the results of the local parameter surfaces at different bandwidths. In such instances, we do not need any calibration routine – we simply input the bandwidth into the GWR routine and repeat the procedure with different values of the bandwidth. The computer program described in Chapter 9 allows the user this option.

One might ask, 'why would we want to input a value for the bandwidth rather than finding an optimal value for it?' As we have already mentioned in Chapter 2, the results of any GWR analysis will be sensitive to the value of the bandwidth; hence the reason that an optimal value for the bandwidth should normally be found. However, by pre-setting the value of the bandwidth at different values, we can describe the sensitivity of the GWR results to variations in the bandwidth and, consequently, to variations in the spatial scale of the analysis.

Figures 6.5–6.7 demonstrate the results of such sensitivity to spatial scale. In each figure the local values of the intercept parameter from equation (6.5) are displayed for three different bandwidths of 55 kms, 35 kms, and 15 kms, respectively. An interpretation of the local intercept is that each local estimate describes the level of school performance holding all the independent variables in the model constant. In this particular example it usefully indicates where school performance is locally much better or worse than would be expected given school size and the various characteristics of the catchment areas. The results at the three different bandwidths allow us to make judgements about school performance at different spatial scales. For instance at a wider regional scale where the bandwidth is 55 kms (Figure 6.5), the spatial distribution of the intercept parameter suggests a broad east–west division with schools in the eastern half of the region generally appearing to perform better, given their size and catchment characteristics, than schools in the west. At a medium scale where the bandwidth is 35 kms (Figure 6.6), more detail reveals a more complex picture. Schools in the far north and in the south-west centred on Manchester appear to be under-performing whereas schools in the north-west and in a ridge running down the south-east appear to be performing better than local conditions would suggest. Finally, at a much smaller scale of analysis where the bandwidth is only 15 kms (Figure 6.7), even more detail can be seen on school performance, with the influence of individual schools becoming apparent in some areas. The north-west appears to have some marked variations in the determinants of school performance with schools in one part of the region doing much better than conditions would suggest and schools in an adjacent area performing much worse than expected. Similarly, the eastern side of the region now appears as a more complicated mosaic of better-than-expected and worse-than-expected school performances. Interestingly,

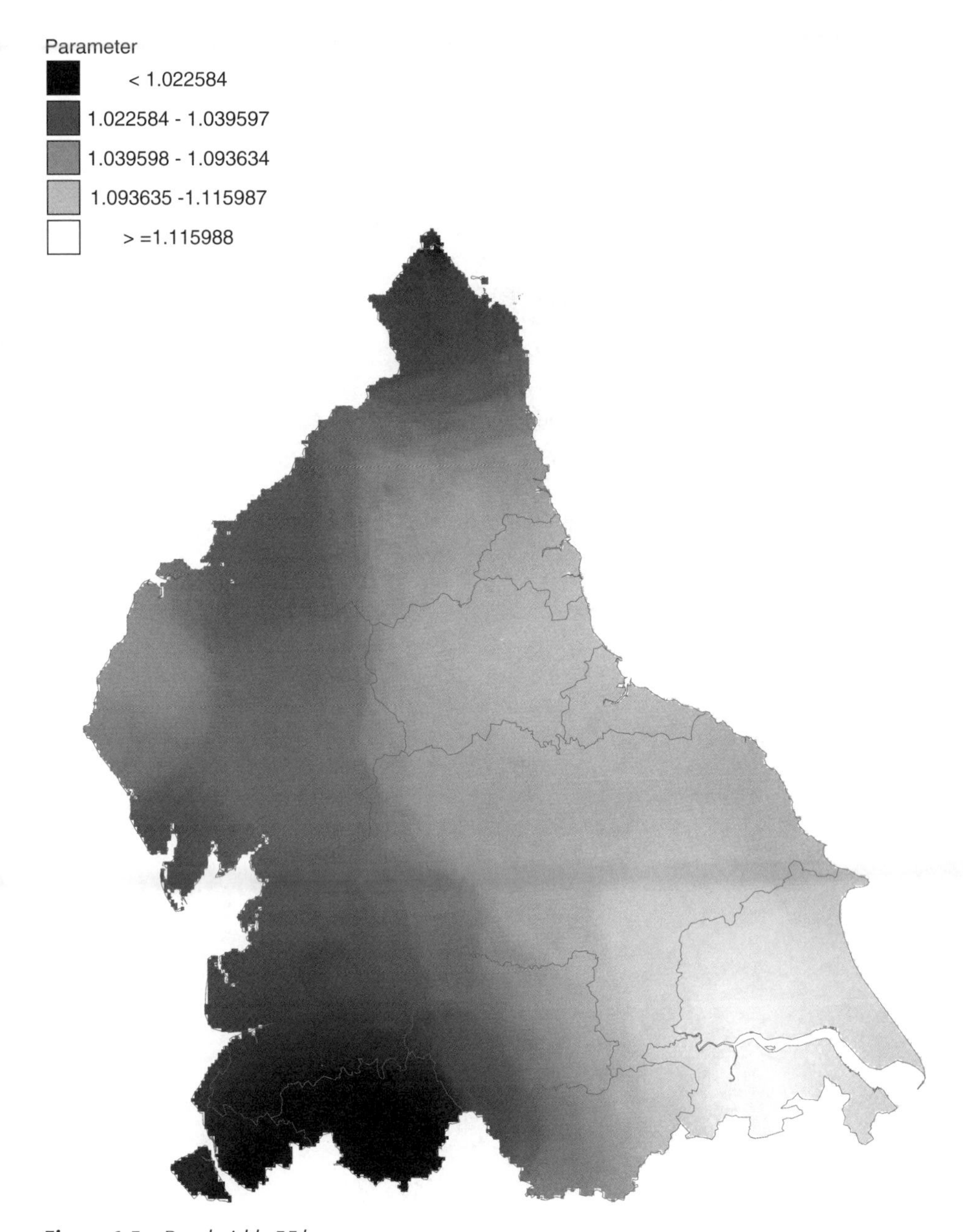

Figure 6.5 *Bandwidth 55 km*

the area of relatively low intercept values in the far north of the region has now been replaced by relatively high intercept values. This is probably a quirk of the calibration results in this part of the study area which has a very low population density and in which there are very few schools (see Figure 6.2a).

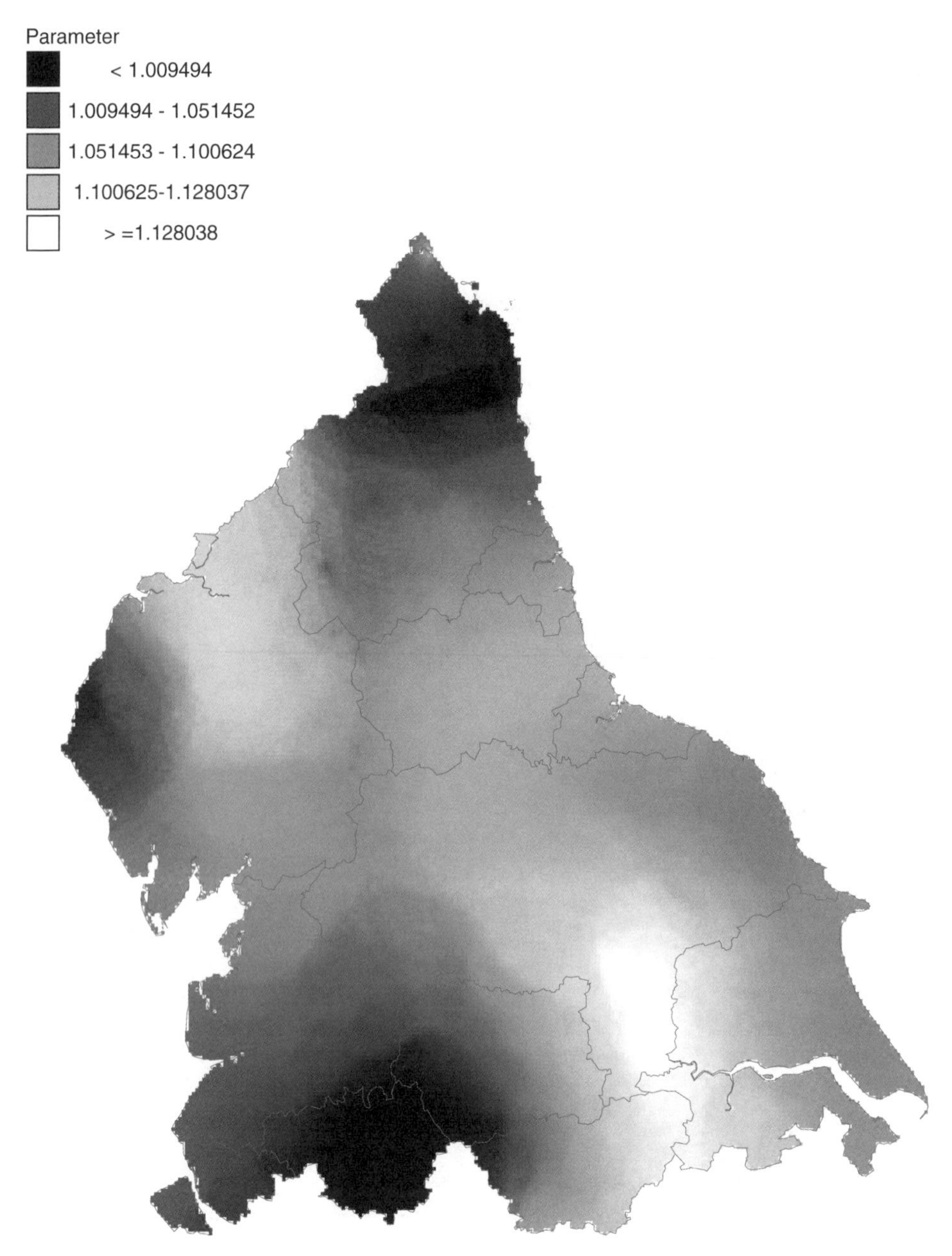

Figure 6.6 *Bandwidth 35 km*

These results suggest that complex 'movies' could be constructed from GWR results by using a series of different bandwidths. These movies would be the equivalent of a spatial microscope. At large bandwidths, only broad detail in the spatial variations of relationships could be identified and decreasing the size of

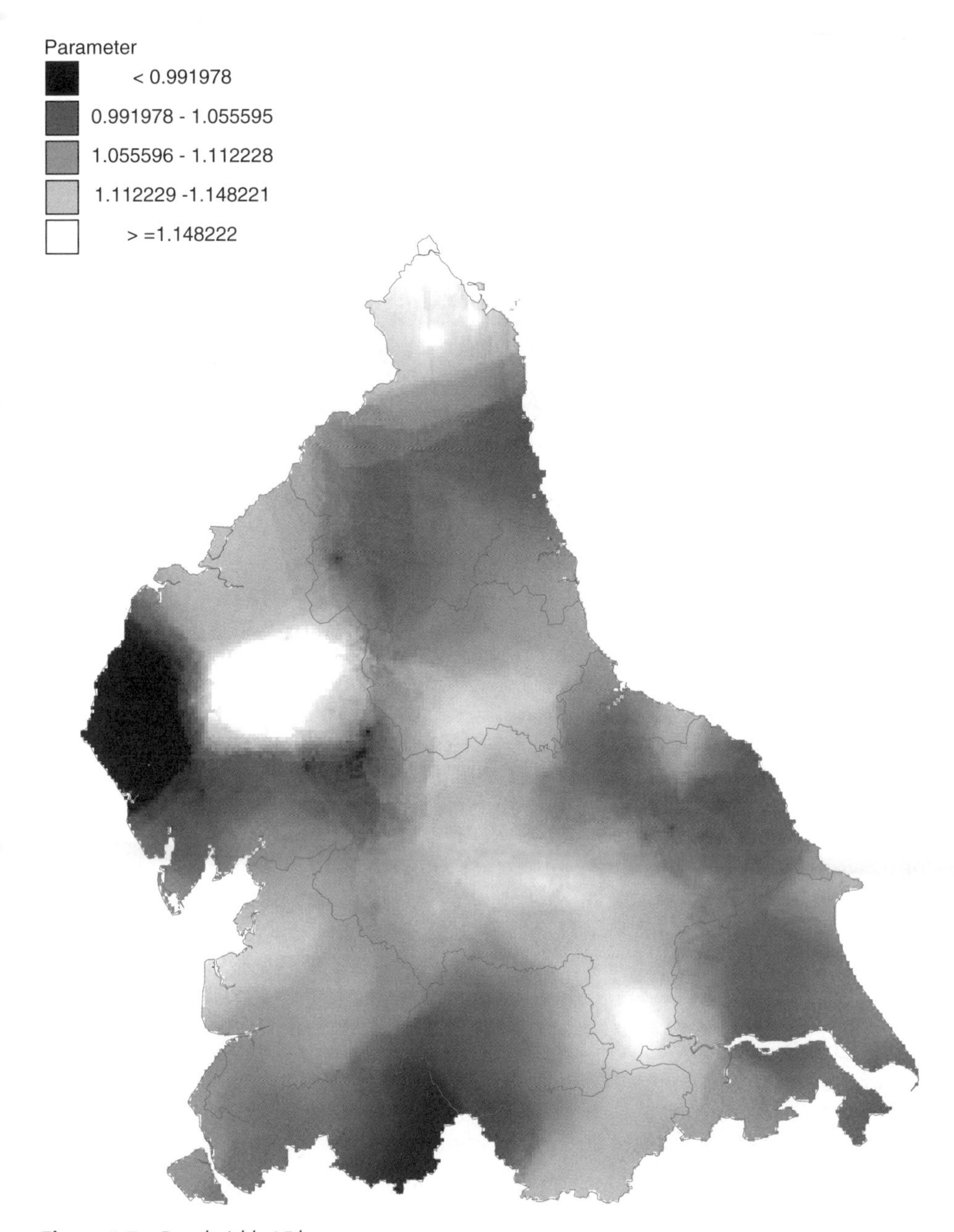

Figure 6.7 *Bandwidth 15 km*

the bandwidth would be akin to focussing a microscope and being able to see much more detail in the pattern. In most cases, greater detail would be desirable but there may be instances where some broad regional understanding of patterns might be useful. Also it should be borne in mind that there is a limit to the level of detail possible because as the bandwidth gets smaller, the degrees of freedom in each local model calibration will decrease and will, at some point, lead to unstable regression results.

6.3 GWR and the MAUP

6.3.1 Introduction

A somewhat different scale-related problem that has long been identified in the analysis of spatially aggregated data is that the results of such an analysis often depend on the definition of the areal units for which data are reported. This has become known as the modifiable areal unit problem (Openshaw 1984) or, in some cases, the zone definition problem (Fotheringham *et al.* 1995). Gehlke and Biehl (1934), for example, noted that aggregation from census tracts to larger areal units increased the value of a correlation coefficient, and posed the question as to whether the correlation coefficient was an appropriate statistic for use with aggregate spatial data. Yule and Kendall (1950: 310–14) observed the same phenomenon in an analysis of wheat and potato yields among the counties of England. In the same year Robinson (1950) reported a similar trend when comparing correlations based on individual measurements to those aggregated to counties within states. Robinson (1956) suggested that areal weighting would ameliorate the problems in calculating correlation coefficients for areal units of different sizes. However, Thomas and Anderson (1965) re-examined the findings of Gehlke and Biehl, and Yule and Kendall, concluding that the differences in the values of the correlation coefficients in both studies could have resulted from random variation.

Openshaw and Taylor (1979) discovered that they could obtain almost any value of the correlation between voting behaviour and age in Iowa merely by aggregating counties in different ways, although Tobler (1989) pointed out that this result could have been anticipated theoretically. More recently, Fotheringham and Wong (1991) described how the stability or instability of parameter estimates obtained from a multivariate model under different levels of aggregation could be visualised. They demonstrated that some relationships can be relatively stable to data aggregation, while others appear to be very sensitive. Fotheringham *et al.* (1995) described the same problem in the context of location-allocation modelling where sensitive locational decisions are shown to be affected by the definition of the zones for which data are reported. Both these latter two studies also demonstrate the two components of the modifiable areal unit problem:

1. *The scale effect*: different results can be obtained from the same statistical analysis at different levels of spatial resolution.
2. *The zoning effect*: different results can be obtained due to the re-grouping of zones at a given scale.

However, despite the fact that the modifiable areal unit problem has been observed for decades, we appear to be little closer to dealing with the problem effectively. Of course, one solution is to use spatially disaggregated data, although such data are frequently unavailable for reasons of confidentiality. For instance, much of the Census of Population data in both the USA and the UK are only available for aggregate spatial units and even though samples of microdata are available, they are only available for extremely coarse spatial units and so are of limited use in spatial analysis. Another 'solution' is to report the results of the spatial analysis at the most spatially disaggregated level possible and then to demonstrate visually the sensitivity of the results to both the scale and zoning effects (Fotheringham and Wong, 1991). In this way, if some results can be shown to be relatively stable over a wide range of zoning systems, this can induce greater confidence that the results at the most disaggregated level have some meaning and are not simply artefacts of the way the data are arranged.

For Openshaw and Rao (1995), the answer lies in the construction of zoning systems that are in some sense 'optimal'. It is not always clear whether this construction is intended to influence those releasing the data. It might be regarded as somewhat eccentric to create a zoning system for which the fit of a particular spatial model is 'optimal' if this zoning system is to be used for some completely different purpose. Equally, if the re-aggregation is to take place after the data are released, we know that we are stuck with the areal units that we have; if we re-aggregate them to some other units, we are essentially creating arbitrary results. If we create a zoning system for one regression model, then include another variable and re-zone to get the 'best fit', it is not immediately clear how we compare the two models, apart from the rather trivial observation that the goodness-of-fit has changed.

For Steel and Holt (1996), Holt *et al.* (1996) and Tranmer and Steel (1998) the answer lies in circumventing the modifiable areal unit problem with data on additional relationships. They propose a model structure in which an extra set of *grouping variables* is included. These grouping variables, z, can be measured at the individual level and are in some way related to the processes being measured at the aggregate level. They are used to adjust the aggregate-level variance–covariance matrix for the model so that it more closely approximates the unknown, individual-level variance–covariance matrix. The details of this process can be found in Chapter 10 of Fotheringham *et al.* (2000). The problem, of course, is to find an estimate of the individual-level covariance matrix for suitable z variables. In some circumstances it may be possible to obtain individual-level data. Holt *et al.* (1996), for example, use data from the UK 1991 Census of Population for 371 enumeration districts in South London and they make use of the 'Sample of Anonymised Records' (Dale and Marsh, 1993) to provide individual level records for a 2% sample of the district's population to estimate the individual-level covariance matrix of the grouping variables. The technique could be applied to US census data using the Public Use Microdata Sample (PUMS) to estimate the individual-level variance–covariance matrix. The technique obviously depends on the availability of individual-level data although Holt *et al.* (1996) point out that these need not be for the same area. If suitable individual-level data are not available, the technique is of little use. It also depends on finding very strong relationships between sets of variables. As Voas and Williamson (2001: 67) point out in a study of the inter-relationships between sets of census variables:

> even complete knowledge concerning nearly half the variables would not let us predict the other values with any degree of accuracy. These findings have discouraging implications for the approach to ecological analysis developed in Steel and Holt (1996).

It should also be noted that in Steel and Holt's technique the measurement of both the individual-level covariance matrix and the relationships between the grouping and model variables depends on some definition of the areal extent of the individual-level variables and must therefore also be prone to the modifiable areal unit problem. It is also assumed that the relationships between the grouping variables and the aggregate variables are constant over space.

A similar approach has been taken by King (1997) who provides an interesting 'solution' to the ecological fallacy problem in a voting context. Here, data are available at different spatial scales (voting precincts and districts) and King's solution involves local discrete forms of regression and using constraints peculiar to voting data. The use of local forms of analysis in general might well provide a new way of thinking about the modifiable areal unit problem. Geographically Weighted Regression, as noted above, provides fuzzy zones around each regression point and presents a surface of parameters that might be more robust to the underlying scale of the units for which data are reported.

Yet another attempt to find a 'solution' to the modifiable areal problem involves the search for statistical methods, the results of which might be relatively robust to the definition of the spatial units for which data are recorded. This is the approach advocated by both Tobler (1989) and Fotheringham (1989). Tobler places the 'blame' for the modifiable areal unit problem not on the aggregation of spatial data but on the methods used to analyse the data. He advocates retaining only those methods that are robust in the face of alternative definitions of space. Fotheringham, while not being so prescriptive, investigates the use of fractal techniques as statistics that are relatively robust to scale changes. Interestingly, Anselin (1999: 71–2), in the same vein, advocates the construction of surfaces to combat scale sensitivity:

> Alternatively, one could argue that studies should be based on scale-invariant concepts or scale-invariant variables, such as densities and ***surfaces*** [our emphasis]. While this may have an intuitive attractiveness in the physical sciences, many processes in social science are discrete in nature, and modeling frameworks to deal with this characteristic are still in their infancy. The problem remains how to construct or estimate the relevant surfaces.

Following this line of reasoning, and King's use of local discrete models as a partial solution to the modifiable areal unit problem, we now briefly examine the sensitivity of GWR results to the level of aggregation of the underlying spatial data used to calibrate the models. We do not expect the surfaces of estimated localised parameters to be constant with respect to scale changes but we do wonder whether such surfaces are relatively more stable than the equivalent global model results. One reason for thinking this is that the parameter estimate surfaces are products of smoothing algorithms and spatial aggregation is itself a form of spatial smoothing. Another reason, at which Tobler (1989) hints, is that the modifiable areal unit problem may be in part a product of spatial non-stationarity in the relationships being measured. That is, if the relationships were constant over space, the zoning component of the modifiable areal problem should disappear and the scale problem may also be ameliorated.

6.3.2 An Experiment

Consider the London house price data described in Chapter 2 (Section 2.2.1). These data are available at the level of the individual property and hence this is the appropriate scale of analysis. However, suppose the data were only available at some spatially aggregated level – how different might the results of any analysis of these data be? We can shed some light on this issue by performing the aggregation ourselves and calibrating a hedonic price model similar to that formulated in Chapter 2 at each level of aggregation.[9] We can do this for both the global model and the local GWR model to examine how robust the latter is to scale variations.

The data set consists of 12 493 individual properties. These were aggregated to 6 595 enumeration districts and then to 745 wards, both sets of spatial units being defined by the UK Census of Population. The hedonic price model to be calibrated is based on the model given in equation (2.1) with a few variables omitted to increase the comparability of the results between spatial scales:

$$\begin{aligned} P_i = {} & \alpha_0 + \alpha_1 \text{FLRAREA} + \alpha_2 \text{BLDPWW1} + \alpha_3 \text{BLDPOSTW} \\ & + \alpha_4 \text{BLD60S} + \alpha_5 \text{BLD70S} + \alpha_6 \text{BLD80S} + \alpha_7 \text{TYPDETCH} \\ & + \alpha_8 \text{TYPTRRD} + \alpha_9 \text{TYPFLAT} + \alpha_{10} \text{GARAGE} \\ & + \alpha_{11} \text{BATH2} + \alpha_{12} \text{PROF} + \alpha_{13} \text{UNEMPLOY} + \alpha_{14} \log_e(\text{DISTCL}) + \varepsilon_i \end{aligned} \tag{6.6}$$

where

P_i is the price in pounds sterling at which a house sold.
FLRAREA is the area of the property in square metres.
BLDxxx is a set of dummy or indicator variables that depict the age of the property as follows:

BLDPWW1 is 1 if the property was built prior to 1914; 0 otherwise
BLDPOSTW is 1 if the property was built between 1940 and 1959; 0 otherwise
BLD60S is 1 if the property was built between 1960 and 1969; 0 otherwise
BLD70S is 1 if the property was built between 1970 and 1979; 0 otherwise
BLD80S is 1 if the property was built between 1980 and 1989; 0 otherwise

[9] Some variables had to be omitted from the original model formulated in Chapter 2. These were of two types: those variables for which there were very few data points, such as the presence of bungalows and houses without central heating, and those variables that did not have the same interpretation at the aggregate level such as the interaction terms between floor area and house type.

TYPxxx is a set of dummy variables that depict the type of house as follows:

> TYPDETCH is 1 if the property is detached (i.e. it is a stand-alone house); 0 otherwise
> TYPTRRD is 1 if the property is in a terrace of similar houses (commonly referred to as a 'row house' in the US); 0 otherwise
> TYPFLAT is 1 if the property is a flat (or 'apartment' in the US); 0 otherwise.

GARAGE is 1 if the house has a garage; 0 otherwise.
BATH2 is 1 if the house has 2 or more bathrooms; 0 otherwise.
PROF is the proportion of the workforce in professional or managerial occupations in the census ward in which the house is located. Each census ward has an approximate population of 5 250.
UNEMPLOY is the rate of male unemployment in the census ward in which the house is located.
DISTCL is the straight-line distance from the property to the centre of London (taken here to be Nelson's column in Trafalgar Square) measured in kms; $\log_e$ denotes a natural logarithm; and α denotes a parameter to be estimated.

To avoid multicollinearity problems, the following variables are excluded from the regression model:

BLDINTW which is 1 if the property was built between 1914 and 1939 and 0 otherwise
TYPSEMID is 1 if the property is semi-detached (i.e. it shares a common wall with one neighbour – often referred to as a 'duplex' in the USA); 0 otherwise.

A model very similar to this has already been calibrated with individual house price data and the results described in Chapter 2. In order to calibrate the model for a set of aggregated spatial units, each variable in equation (6.6) has to be summed over all the properties within each spatial unit and then divided by the number of such properties. Consequently, the aggregate equivalent of the individual house price is the mean house price within either an enumeration district or a ward; the aggregate equivalent of any continuous independent variable such as floor area is the mean of that variable; and the aggregate equivalent of any binary variable such as central heating is the proportion of houses in the spatial unit having that attribute. The results of calibrating a global hedonic price model with the three data sets are shown in Table 6.2.

These results exemplify the dangers of undertaking spatial analysis with data drawn from aggregate spatial units. If we were investigating the determinants of house prices we would reach some quite different conclusions if we used data for wards as opposed to enumeration districts. First, because of the reduced number of wards compared to enumeration districts, many of the determinants in our model at the ward level do not appear to be 'significant'. Second, where the variables are significant, we can draw different conclusions at the two spatial scales. For instance, at the ward level it would appear that there is a strong negative relationship

Table 6.2 Global regression parameter estimates

Variable	Individual data	Enumeration districts	Wards
Intercept	69 745 (32.0)	185 616 (23.4)	48 513 (5.1)
FLRAREA	654 (75.0)	717 (57.7)	784 (14.5)
BLDPWW1	−2 677 (−4.4)	−2 011 (−2.4)	−2 640 (−0.8)
BLDPOSTW	−2 953 (−3.2)	−3 685 (−2.9)	−3 642 (−0.7)
BLD60S	−5 140 (−4.9)	−5 649 (−3.7)	−9 974 (−1.5)
BLD70S	−1 987 (−1.7)	−4 482 (−2.6)	−17 171 (−2.2)
BLD80S	7 164 (7.7)	5 893 (4.2)	−10 376 (−2.0)
TYPDETCH	27 421 (21.8)	29 054 (15.9)	16 016 (1.7)
TYPTRRD	−10 070 (−16.0)	−11 868 (−12.9)	−13 358 (−2.9)
TYPFLAT	−8 027 (−10.3)	−5 900 (−5.2)	5 467 (1.0)
GARAGE	6 385 (11.2)	6 818 (7.9)	14 160 (3.1)
BATH2	23 386 (19.8)	26 018 (15.5)	90 575 (12.6)
PROF	73 (3.1)	116 (3.3)	336 (2.0)
UNEMPLOY	−169 (−5.6)	−202 (−4.5)	−782 (−3.7)
$\log_e$(DISTCL)	−17 668 (−29.0)	−17 699 (−22.9)	−14 793 (−8.0)
N	12 493	6 595	745
R^2	0.58	0.62	0.67

between average house price and the proportion of houses built during the 1980s across London. This relationship is entirely spurious as shown by the results at the enumeration district level and by the individual house price analysis. It is probably caused by higher levels of newer housing being built in some of the less desirable parts of London in the far west, the East End and the south-east. Relatively little new housing has been built in the more desirable and costly parts of central London and so a spurious negative relationship between house price and the proportion of houses built in the 1980s results from the aggregated data. A similar problem arises with the interpretation of the value of flatted properties which is significantly negative for both the individual level data and the enumeration district analysis but is positive (albeit not significantly so) for the ward-level analysis. Other problems arise in assessing the impact on house prices of variables such as property built during the 1960s, property built during the 1970s, the value of a garage, the value of having two or more bathrooms, and the undesirability of being in areas of high male unemployment.

Having described the problem of scale in the interpretation of global models calibrated with aggregated spatial data, we now examine if this problem is ameliorated in local analyses such as GWR. To do this, we calibrate the model in equation (6.6) by GWR with data drawn from the three spatial scales described above: individual houses; enumeration districts; and wards. In the case of the individual house data, the data are allocated to unit postcode centroids; in the case of enumeration districts and wards, the data are allocated to the centroids of these zones. In all three cases, distances are computed between the centroids in order to calculate the spatial weighting functions. As noted earlier, the data used in the three models are not identical – for instance, variables recorded as presence or absence at the individual level become proportions in the aggregate models – and this causes problems for the comparison of local parameter estimates across different scales. However, the problem is easily resolved by comparing t-surfaces where the local parameter estimates are divided by their local standard errors. This is another instance where t-surfaces are very useful in displaying local variations in relationships.

Examples of the local t-surfaces are provided in Figures 6.8–6.13. In each figure the top surface is that derived from the individual level data; the middle surface is derived from the enumeration district data; and the bottom surface is derived from the ward level (the most spatially aggregated) data. In each figure the scale of the shading is constant across the three surfaces to facilitate comparison. We are interested in how stable the surfaces are to data aggregation. One way to assess this is to consider the general conclusions that would be made about the spatial variations in the relationships depicted by each surface.

The general impression from this set of figures is one of considerable stability in the surfaces for each set of local parameter estimates. For instance, Figure 6.8 represents the spatial variations in the relationship between house prices and houses built between 1940 and 1959. The global t values were negative for all three levels of data but the global t value for the ward level data was only -0.7 suggesting a very weak relationship. The local t-surfaces indicate a great degree of similarity with significant negative relationships in a band between the boroughs of Richmond, Wandsworth, and Westminster. In all three cases, this band of negative relationships is bordered to the north by an area of positive relationships although these are weaker for the ward-level data than for the other types of data.

Figure 6.9 displays the surfaces for the relationship between house price and houses built during the 1980s. The global results for this relationship were highly dependent on scale with the parameter estimates derived from the individual level data and the enumeration districts being significantly positive and the parameter estimate for the ward-level data being significantly negative. The local surfaces show how the inconsistent global results arise. Essentially all three surfaces are

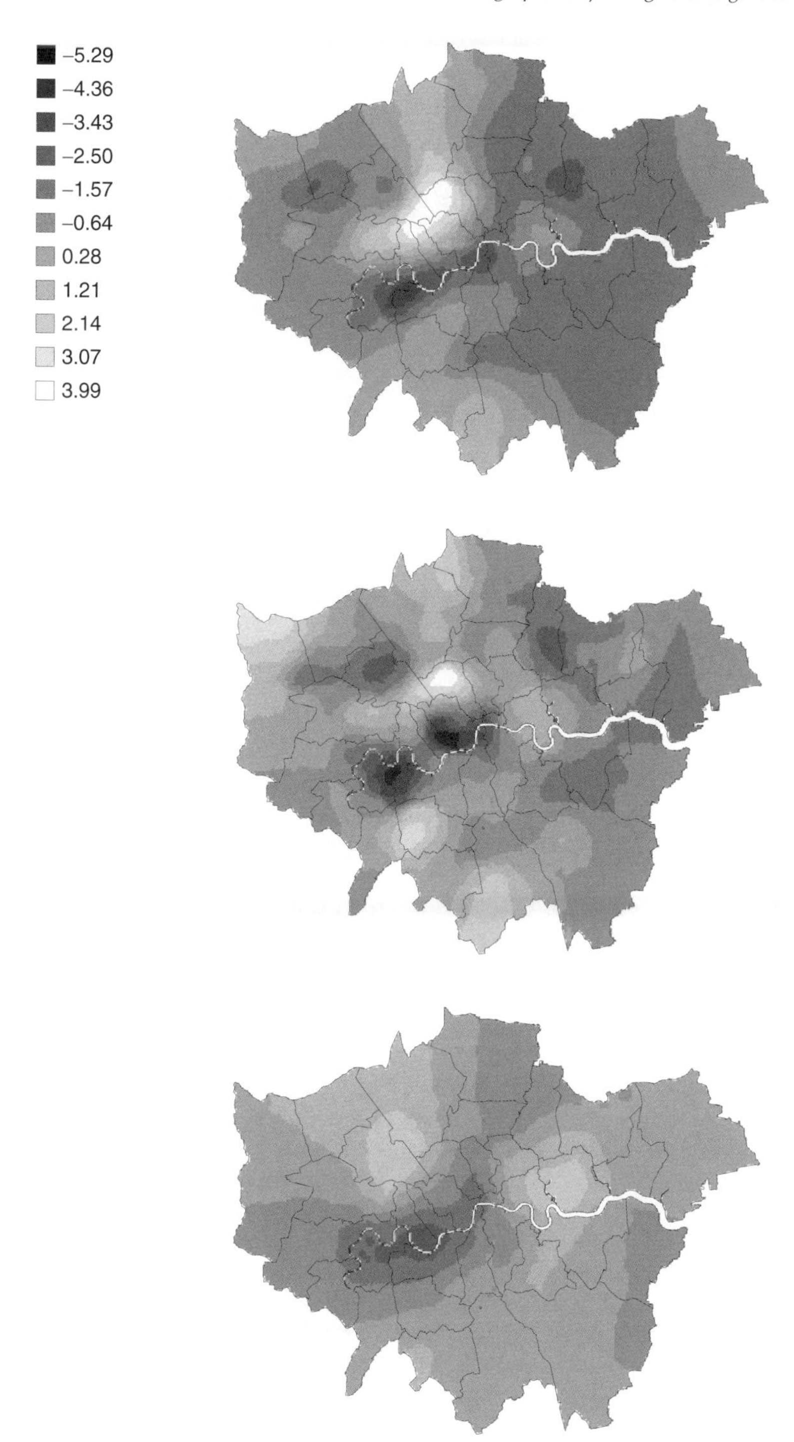

Figure 6.8 *Parameter: built post-war*

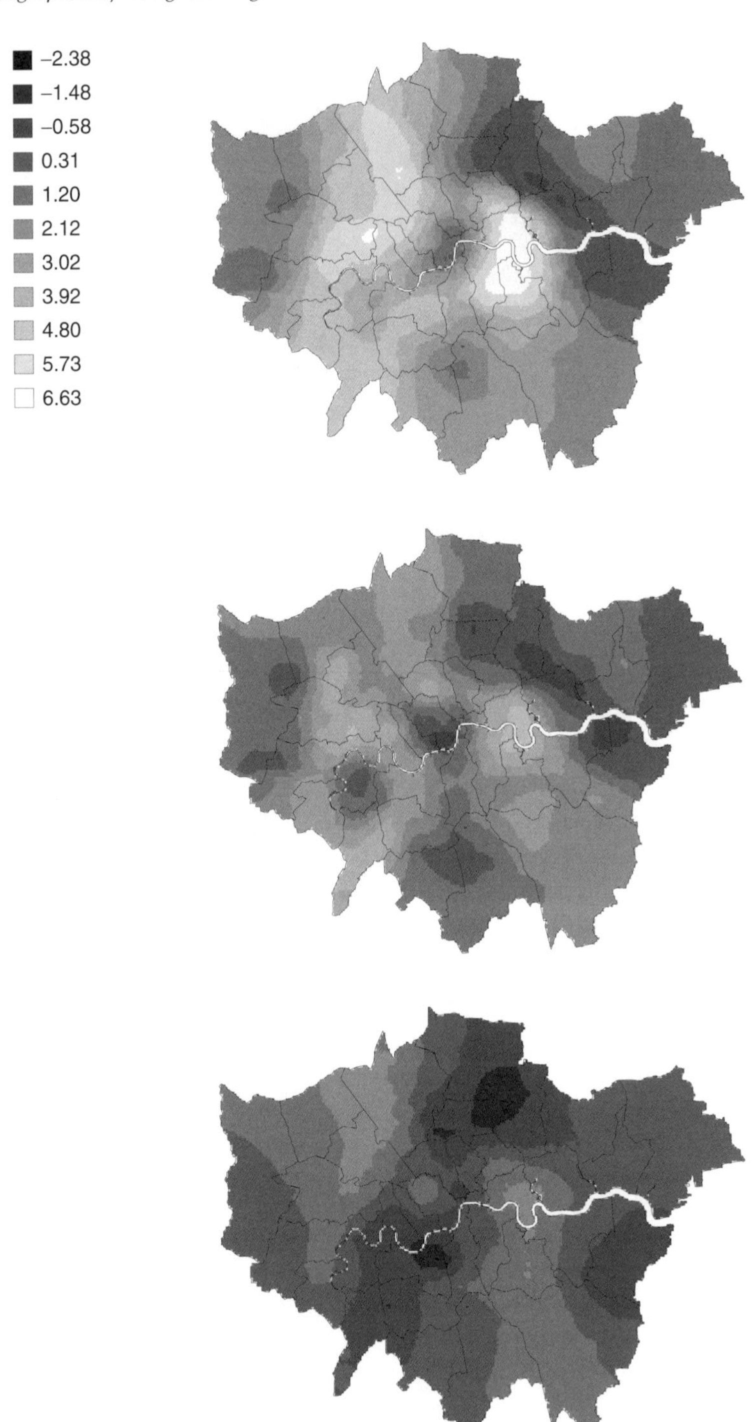

Figure 6.9 *Parameter: built in 1980s*

topologically similar but increasing data aggregation leads to the median level of the surfaces becoming increasingly negative.

Figures 6.10, 6.11 and 6.12 show the relationships between house prices and terraced housing, flats and the presence of a garage, respectively. All three sequences exhibit a great degree of stability. For the relationships between house prices and terraced and flatted properties there is a significant positive relationship centred on the boroughs of Kensington and Chelsea and Westminster with most of the rest of the maps exhibiting only weakly positive or negative relationships. For the relationship between house prices and the presence of a garage there is a distinctive band of significant negative relationships centred on the boroughs of Wandsworth and Lambeth with relationships being positive in most of the rest of the region.

The final sequence of surfaces in Figure 6.13 shows the local relationships between house prices and the proportion of people employed in professional occupations. For reasons not immediately obvious, this sequence exhibits less stability than the previous ones. At the level of individual house sales, the surface is fairly flat with generally weakly positive relationships. There is an interesting area of negative relative relationships in the far south-east of the region. At the enumeration district level, the most distinctive feature is perhaps the band of strongly negative relationships in the centre of London which does not feature on the individual surface. Finally, the ward-level surface shows yet another distinctive feature not on either of the other two maps which is a band of strongly positive relationships centred on the boroughs of Hammersmith, Wandsworth and Kensington and Chelsea. Just to the north-east of this is a band of negative local relationships.

Although these surfaces derived from data at three levels of spatial aggregation are by no means identical (it was never claimed they would be), there does appear to be some support for thinking that GWR surfaces might be more robust to scale issues than are the equivalent global results. However, this is an issue that demands a much greater amount of attention – the results here are only a beginning.

6.4 Summary

The scale and zoning problems in spatial analysis, commonly referred to jointly as the modifiable areal unit problem (MAUP), are exacerbated by: (i) trying to model continuous spatial processes with discrete zones; and (ii) assuming spatial stationarity in the relationships being examined and applying inappropriate global models. Both of these problems are removed in the application of GWR. However, this does not imply that GWR 'solves' the MAUP; it merely suggests that locally varying models may not be influenced by MAUP issues to the same extent as are more traditional global models. This suggestion needs to be explored further.

GWR does not imply a certain scale of analysis *a priori*. The zones over which a process is measured are determined endogenously. The likelihood of a data point belonging to a zone centred on a regression point is a function of the distance between the two points: the closer a data point is to the regression point, then the greater is the likelihood that the data point belongs to a region centred on the

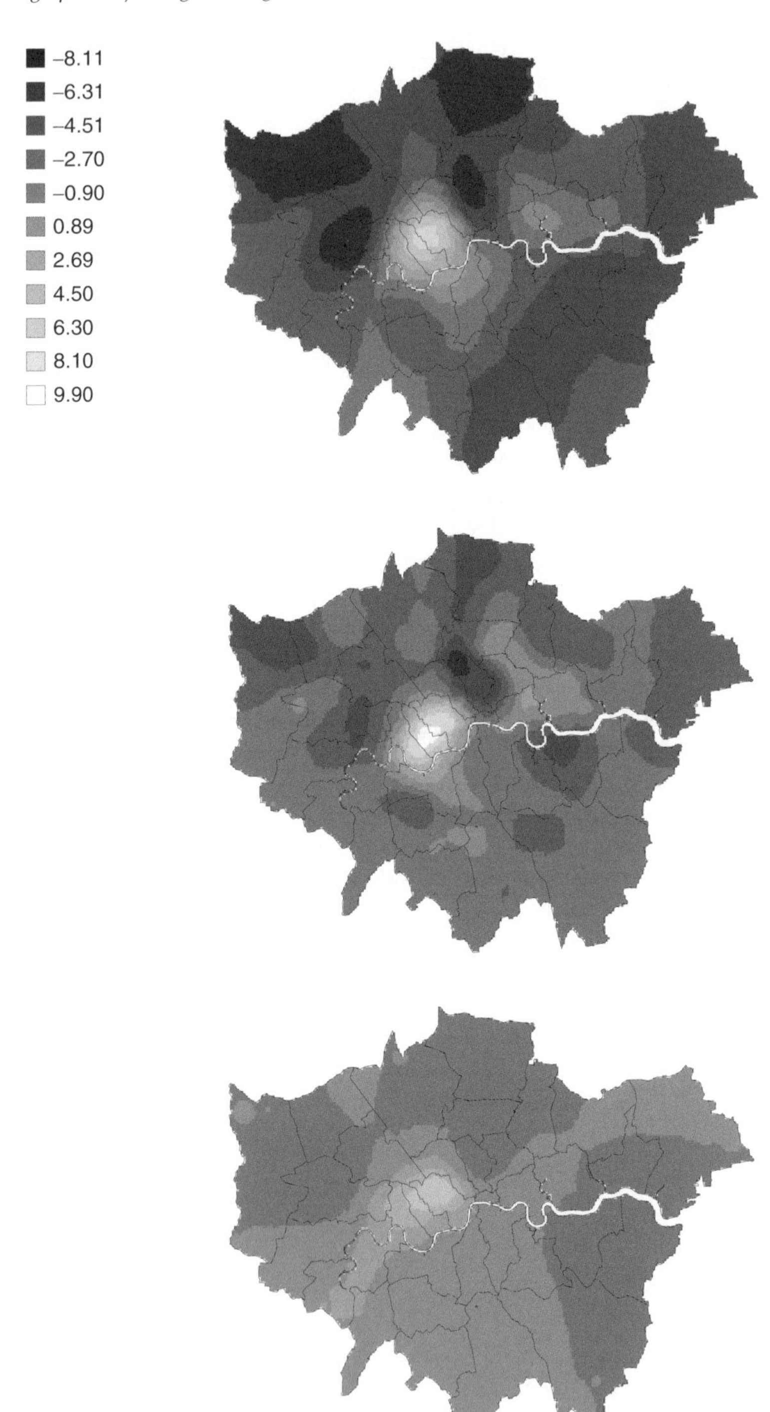

Figure 6.10 *Parameter: terraced*

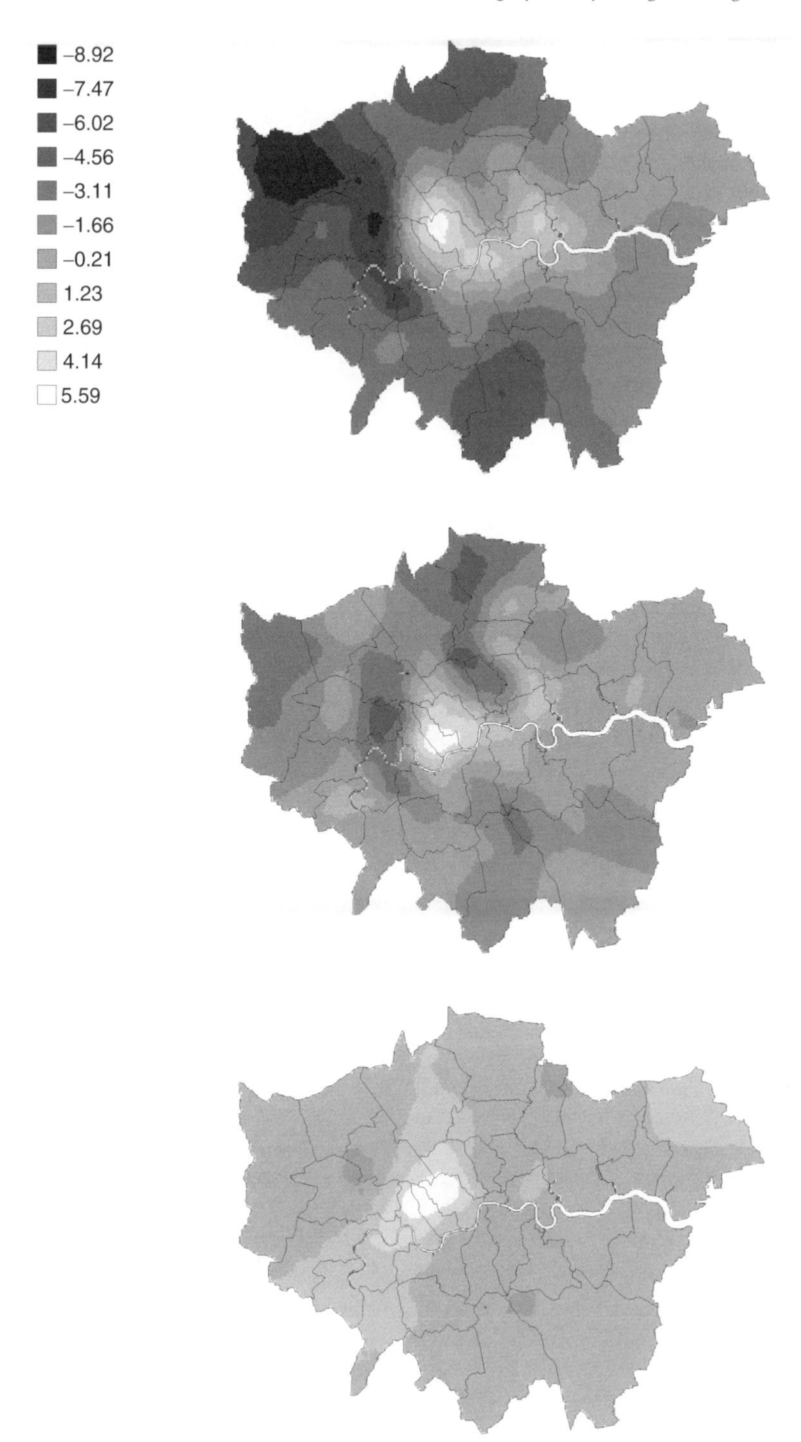

Figure 6.11 Parameter: flat

Figure 6.12 *Parameter: garage*

Figure 6.13 *Parameter: professionals*

regression point. This is much more compatible with continuous spatial processes than is the rather arbitrary definition of discrete zones in which spatial processes are often measured.

Of course, GWR still suffers from its own form of the modifiable areal unit problem in that the results are dependent on the bandwidth used, which is a measure of spatial scale. However, the particular bandwidth used in GWR is not arbitrary – the selection of an optimal bandwidth is part of the GWR calibration process (see Chapters 2 and 9). Alternatively, as demonstrated above, the GWR software described in Chapter 9 also allows the user to input a series of exogenously derived bandwidths to investigate the sensitivity of the GWR results to variations in spatial scale directly.

7
Geographically Weighted Local Statistics

7.1 Introduction

The subject matter of this chapter is perhaps more elementary than geographically weighted regression but it is still concerned with the idea of local statistics. Before any sophisticated statistical analysis takes place, it is usually a good idea to carry out some initial exploratory data analysis (EDA). As well as giving an overview of typical values and levels of variation for variables in the data set, EDA can help to identify outliers, detect general trends in the data, and identify potential problems with further modelling or more advanced statistical analysis. In addition to the graphical methods for EDA such as those cited in Fotheringham *et al.* (2000), summary statistics are also a useful tool. Typical summary statistics include the mean and standard deviation of continuous variables, frequency tables (or proportion tables) for discrete variables, and correlation coefficients between pairs of continuous variables.

As an example, we could consider a random sample of 347 cases from the Nationwide UK house price data set introduced in Chapter 2. The variable TYPDETCH denotes whether or not a given house is detached. This is a binary variable – it can only take the values 'Yes' and 'No' – and a useful summary statistic here is the proportion of detached houses in the data set. There are 66 yes responses and 281 no responses; hence the proportion of detached houses (that is, houses that are not joined to neighbouring buildings) in the data set as a whole is around 0.19. This provides some useful overview information – around one house in five is detached in England and Wales as a whole. However, much as with the earlier example of mean temperature in the United States (see Chapter 1), this information is rather general. Anyone who has travelled within the UK will be aware that it is a diverse place, and that the nature of its housing stock varies from locality to locality. For example, some more affluent areas may consist almost entirely of detached housing, but other equally affluent places, such as in central London, are dominated by luxury flats. As a (rather obvious) rule, there are more detached houses in sparsely populated areas. It would perhaps be more useful to divide the UK into a number of sub-regions (census districts for example), and to tabulate or map the proportion of detached houses in each of these. Although this approach would provide a more helpful summary than the single figure of 0.19 given earlier, it relies on the assumption that the choice of sub-regions reflects the spatial patterns in the housing stock. If a cluster of detached housing straddles two sub-regions without dominating either of them, then it may fail to show up in the above analysis. This is an example of a phenomenon first noted by Gehlke and Biehl (1934), termed the Modifiable Areal Unit Problem (MAUP) by Openshaw (1984), a topic addressed in Chapter 6.

An alternative approach is outlined in this chapter based on the notion of geographical weighting. The idea here is much the same as that used in GWR – a kernel is placed around a point (u_i, v_i) and a weighted proportion of all houses in this kernel that are detached, p_i, is calculated by:

$$p_i = \frac{\sum_j x_j w_{ij}}{\sum_j w_{ij}} \tag{7.1}$$

where w_{ij} is the weight assigned to house j for the calculation of the geographically weighted proportion at location i. The weight is obtained from a kernel function, such as the bi-square or Gaussian functions discussed in Chapter 2. The term x_j denotes a binary indicator variable taking the value 1 if house j is detached, 0 otherwise. Note that if all of the w_{ij}s were set to 1, p_i would simply be the number of detached houses divided by the total number of houses and would be a constant. More generally, if $0 \leq w_{ij} \leq 1$, then the weighting scheme can be thought of as turning some houses into fractional observations. For example, if $w_{ij} = 0.5$, we can think of house j as 'half' an observation for the calculation at location i. By allowing this window-based statistic to be computed at closely spaced regular grid points, a surface or pixel image of the spatial variation in the proportion of detached houses can be constructed. If the spacing of the grid is sufficiently close, then any cluster of detached houses should become apparent – at least one of the grid points should be reasonably close to the centre of the cluster. This should address the straddling problem discussed above.

Note that the above approach is a specific example of a much more general group of methods. Here we regard the proportion of detached houses as a summary statistic and we suggest exploring variation over geographical space with a geographically weighted summary statistic. However, we need not restrict this approach to computing proportions of binary variables. Most summary statistics are weightable, in which case it is possible to define a geographically weighted version of the statistic. Using these geographically weighted statistics it is possible, and

Table 7.1 A classification of localised statistics

Number of variables	Data type	Statistics
Univariate	Continuous	Mean, Standard deviation Skewness
	Binary	Proportion
Bivariate	Continuous	Correlation coefficient
	Binary	Odds ratio

often useful, to explore local trends in data. The aim of this chapter is to define various geographically weighted statistics and to give some indication of how they may be interpreted. A proposed classification of the statistics examined is given in Table 7.1; obviously a much larger set of statistics exists, but this gives a good idea of how geographical weighting can be applied to almost any statistic.

It is intended that this table will provide much of the structure for this chapter. After some more detailed consideration of the idea of geographically weighted summary statistics, each of the categories in Table 7.1 will be described in turn, together with some discussion of how they may be interpreted and some examples of their use. The chapter concludes by considering some further issues.

7.2 Basic Ideas

As suggested in Section 7.1, the key to the local statistics described here is geographical weighting. This provides a conceptual framework for deriving localised versions of a very general range of statistics. The computation of a statistic is localised to a point (u_i,v_i) by weighting each observation in the data set according to its proximity to this point. As with GWR, there is a parameter b, the bandwidth, which controls the rate at which this fall-off in weighting occurs as the distance to the point (u_i,v_i) increases. Any statistic weighted in this way becomes a continuous function of (u,v), and can thus be represented graphically by a surface or contour map.[1] The point (u_i,v_i) will be referred to as the *summary point* and may be regarded similarly to the *regression point* in GWR.

We may apply the same principle to all descriptive statistics which have a weighted version. For example, the sample mean may be replaced by the locally weighted mean:

$$\bar{x}_i = \frac{\sum_j x_j w_{ij}}{\sum_j w_{ij}} \tag{7.2}$$

Here, the x_js are observed values of some continuous variable of interest. Note that this equation is essentially the same as equation (7.1) – a sample proportion is

[1] Provided the weighting function is continuous.

just the sample mean of a binary variable. Equations such as (7.2) can also be simplified if the w_{ij}s are re-scaled to sum to one for each i, which they may be without loss of generality. In this case, we define the w_{ij} (new) to be w_{ij} (old) / $\Sigma_j w_{ij}$. We will assume that the w_{ij}s have been scaled to sum to unity throughout the rest of this chapter, unless otherwise stated. Thus, equation (7.2) may be re-written as

$$\bar{x}_i = \sum\nolimits_j x_j w_{ij} \tag{7.3}$$

and using these conventions, equation (7.1) may be written as

$$p_i = \sum\nolimits_j x_j w_{ij} \tag{7.4}$$

Of course, this is simply an interpolation formula (Ripley 1981). Given a set of points (u_j, v_j, x_j), the formula attempts to fit a surface of the form $x = f(u,v)$. However, the geographical weighting approach may be extended beyond this. For example, a geographically weighted standard deviation, s_i, may be defined as

$$s_i = \left[\sum\nolimits_j (x_j - \bar{x}_i)^2 w_{ij}\right]^{1/2} \tag{7.5}$$

Note the use of $\bar{x}_i$ in equation (7.5). Locally weighted variations around the localised mean are of interest here, not locally weighted variations around the unweighted global mean. This idea may be generalised to all moment-based descriptive statistics, that is, statistics based on expected values of powers of the observed variables. The general form for a geographically weighted moment-based statistic (in one or two variables) is

$$M_i(\alpha, n_1, n_2) = \left[\sum\nolimits_j w_{ij}(x_j - \bar{x}_i)^{n_1}(y_j - \bar{y}_i)^{n_2}\right]^{\alpha} \tag{7.6}$$

where x_j and w_{ij} are defined as before and y_j is a second continuous variable with local mean $\bar{y}_i$, and n_1, n_2 and α are constants that may be chosen to create a number of different statistics as shown in Table 7.2.

Not only do these definitions provide a number of useful statistics directly, but also further statistics may be derived by plugging these quantities in to some simple formulae. For example, a geographically weighted coefficient of variation may be obtained by dividing the geographically weighted mean by the geographically weighted standard deviation. Simple formulae also lead to the definitions of geographically weighted correlation coefficients and geographically weighted skewness as shown in Table 7.3. This covers all of the basic geographically weighted statistics, with the exception of the odds ratio, which will be dealt with separately in Section 7.6. Having defined these statistics, the following sections go on to consider the combinations of variable types presented in Table 7.1, giving practical examples based on the housing data and discussing how the various statistics may be interpreted.

Table 7.2 *Values of n_1, n_2 and α associated with moment-based statistics*

n_1	n_2	α	Statistic
1	0	1	Mean (also proportion)
2	0	1/2	Standard deviation
2	0	1	Variance
3	0	1/3	Skewness
1	1	1	Covariance

Table 7.3 *Moment-based statistics from equation (7.6)*

Statistic	Expression
Coefficient of variation	M(1,1,0) / M(½, 2, 0)
Correlation coefficient	M(1,1,1) / [M(½, 2, 0) . M(½ ,0, 2)]
Skewness	M(⅓, 3,0) / M(½, 2, 0)

7.3 A Single Continuous Variable

The first case we will deal with is that where we wish to investigate the spatial distribution of a single continuous variable, such as the price of a house. We may be interested in a number of statistics here. The most basic is the geographically weighted mean (GW mean). As stated earlier, the GW mean is essentially an interpolation operator, providing a moving window mean across a geographical study area. As Ord and Getis (2001) note, the G_i and G_i* statistics are also simply geographically weighted averages. As an example here, we compute the GW mean house price which is shown in Figure 7.1.[2]

Here, a bi-square kernel is used (that is, $w_{ij} = [1 - (d_{ij}/b)^2]^2$ for distances d_{ij} less than the bandwidth b, and $w_{ij} = 0$ if d_{ij} exceeds b) and the bandwidth is adaptive, chosen such that w_{ij} first becomes zero at the 35th nearest neighbour to the summary point. The use of an adaptive bandwidth is justified here, as the density of observations is not constant throughout the study area. There are distinct urban centres where there are notable clusters of houses. The rather arbitrary choice of the 35th nearest neighbour was used as this is roughly the 10% point of locations arranged in order of nearness to the summary point. From Figure 7.1, a number of features are apparent. Generally, prices are highest in the south-east, centred on London; prices are lowest down the east coast and in the south-west. The area of high house prices in the north-east is a sampling anomaly based on two houses which have unusually high prices.

However, although computing the GW mean has apparently led to these patterns, maps such as Figure 7.1 must be interpreted with some degree of caution. As outlined in Chapter 1, there are a number of reasons why localised means may vary across geographical space: one of these is the fact that house prices will always exhibit a degree of randomness. Even if there were no real change in the *mean*

[2] The shading intervals in Figure 7.1 and other similar figures are chosen to equalize approximately the areas shaded in each greytone.

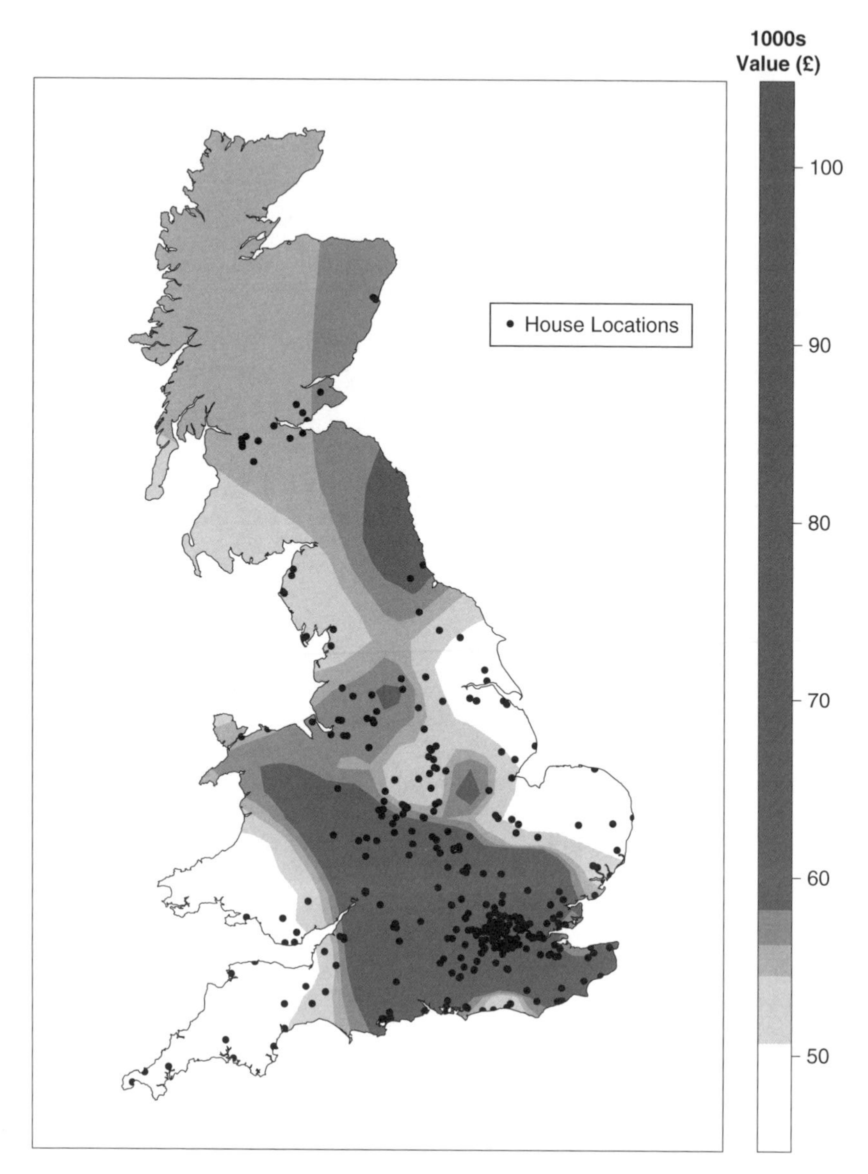

Figure 7.1 GW mean for UK house prices

house price across the whole of the UK, prices will fluctuate about this mean. In turn, this would suggest that localised means will also fluctuate around the mean value to some degree. Thus, care must be taken to check whether patterns in the GW mean of the data are due to genuine changes in local mean house price rather than the result of smoothing aspatial random variation in house prices.

In order to investigate this, we need to consider exactly how much variation would occur in GW means if in fact there were no change in the mean value across the study area. We may then compare the observed variation in GW means to this amount. To do this, we start by assuming that all of the x_js have an identical distribution, with a mean μ and a standard deviation σ. We will also assume that the x_js are independent. This is the situation that would arise if there were no spatial variation in house price distribution. Now we consider the distribution of the GW mean under this assumption. Using equation (7.3), and the convention that the w_{ij}s sum to one, we may obtain expressions for the mean and variance of the GW mean about the summary point i:

$$\mathrm{E}(\bar{x}_i) = \sum_j w_{ij}\mu = \mu \sum_j w_{ij} = \mu \tag{7.7}$$

$$\mathrm{Var}(\bar{x}_i) = \sum_j {w_{ij}}^2 \sigma^2 = \sigma^2 \sum_j {w_{ij}}^2 \tag{7.8}$$

With these relations we now know what values the GW mean and its standard deviation are likely to take if there is no genuine spatial variation in mean value. Under some circumstances we even have some idea of its distribution. If w_{ij} is non-zero for a sufficient number of data points (that is, if the bandwidth is sufficiently large), then the central limit theorem tells us that the distribution of the GW mean will be approximately Gaussian, regardless of the initial distribution of the data. We also know that the quantity

$$z_i = \frac{\bar{x}_i - \mu}{\sigma \sqrt{\sum_j {w_{ij}}^2}} \tag{7.9}$$

will have mean zero and standard deviation one. Essentially, this is a local z-score for the GW mean. Mapping this gives us some idea of where the GW mean varies more than one would expect under the assumption of no local variation in the mean. Traditionally, one would note areas where $|z_i| \geq 2$ as being potentially interesting. For the housing data with the kernel type and bandwidth used in Figure 7.1, the z_i surface is shown in Figure 7.2. Here, the values of μ and σ were estimated by taking the mean and standard deviation of the whole data set. We see that the statistic varies between approximately -2 and 4, indicating some unusually high variability in the local mean in certain places.

However, care should be taken with this interpretation of the results in Figure 7.2. Since we are evaluating z_i at several places, the probability that *one* of these several values will lie outside $[-2,2]$ is not 0.05, but somewhat higher. Think, for example, of throwing a die 100 times: the probability of getting a six *somewhere* in this sequence is far greater than one in six. One approach to resolving this problem of multiple hypothesis testing, as adopted in Chapter 6 and also in Getis and Ord (1992) and Ord and Getis (1995), is to apply Bonferroni corrections. The original significance level is adjusted by dividing by the number of multiple tests. This yields an approximate, although highly conservative, level which may be applied in each of the multiple tests (however, see the discussion in Chapter 10 for some cautionary notes about this approach). For the housing example, the geographically weighted computations were made on a 40×40 grid, so that a total of 1600 z_i values were

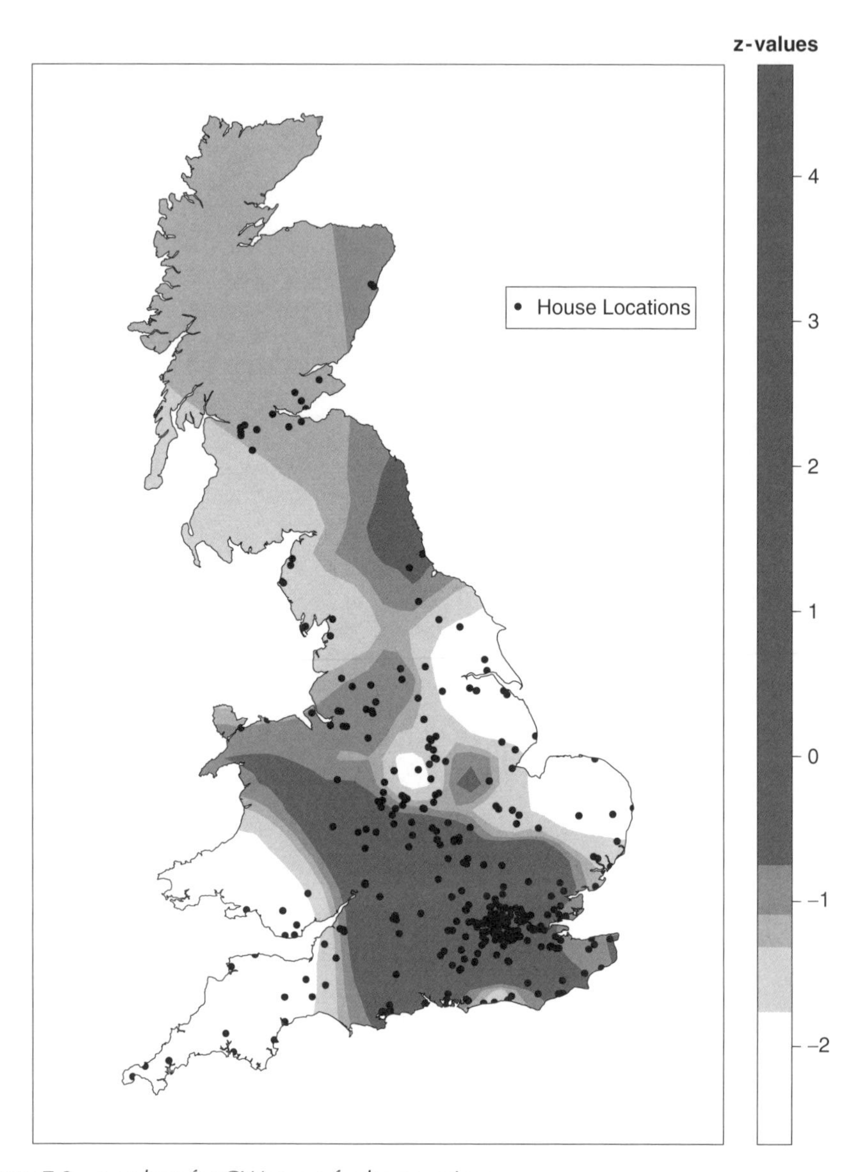

Figure 7.2 *z_i values for GW mean for house prices*

computed. The z-level of 2 corresponds roughly to the 0.025 significance level (recall, this is a two-tailed 5% test), so the correction here gives us an adjusted level of 0.025 / 1600 or 1.56×10^{-5}. The corresponding quantile on the normal distribution is 4.16 (or −4.16). Thus, according to this approach, results are interesting only when the z_i values exceed these limits.[3]

In fact, the limits derived from the Bonferroni correction are likely to be a little too conservative. First, although z_i values were computed across a rectangular grid, several points near the edges of the rectangle are outside the study area. Also, it is perhaps not necessary to carry out as many tests as there are points on the summary grid. Point pairs on the grid that are close together tend to be influenced by the same data points and so there is little utility in carrying out a test for both points. Computing values on a tightly spaced grid is a good strategy for producing mapped surfaces but that does not imply that this is the best approach for assessing the significance of the patterns. It can also be argued that the above approach, while useful in a formal statistical inference context, may not be so appropriate for an exploratory method. The aim here is to identify patterns that might be interesting and require further consideration: we are perhaps more concerned with making sure we do not miss any interesting patterns than with picking up some random patterns. This is very different from the conservative approach. Here we are more concerned with the risk of a false negative (that is, failing to spot a pattern that exists), whereas the conservative approach is more concerned with the risk of a false positive (spotting a 'pattern' that turns out to be an artifact of random variation).

Therefore, the approach recommended here is to use the local z-scores to assess the variability of the estimates, but not to apply the notion of p-values too stringently. This is in line with, for example, Bowman and Azzalini (1997) in their approach to kernel regression and kernel density estimation. Thus, we use basic 95% limits (plus and minus 1.96) to identify potentially interesting GW mean values and the variation in the GW mean in Figure 7.2 is sufficient to suggest that there are some interesting local variations.

A visualisation strategy which attempts to embrace this approach is now proposed. First, one draws a shaded surface for the GW mean, as in Figure 7.1. Next, one considers a different grid for the z_is. This is done using fewer summary points. For example, instead of the 40 × 40 grid used to create the GW mean surface, one could use a 10 × 10 grid. Finally, we check the z_i values against some threshold judged to be interesting (say, 1.96). If $|z_i| \geq 1.96$, then the point is plotted on the surface. Thus, the map now shows the GW mean surface, together with a series of data points that are considered interesting in that they lie outside the range in which they might reasonably expect to fall if there were no geographical variation in the mean value of house price. This conveys more realistic information than the z_i map, since actual house price values are plotted. Such a map is produced in Figure 7.3. Here we can see that a number of areas on the map are flagged as interesting, suggesting that there is a notable degree in variation in local house price mean.

[3] An alternative, although highly similar, correction procedure for multiple hypothesis tests is that derived by Sidak (1967) where the corrected alpha is $1 - (1 - \alpha)^{1/n}$ where n is the number of hypothesis tests conducted.

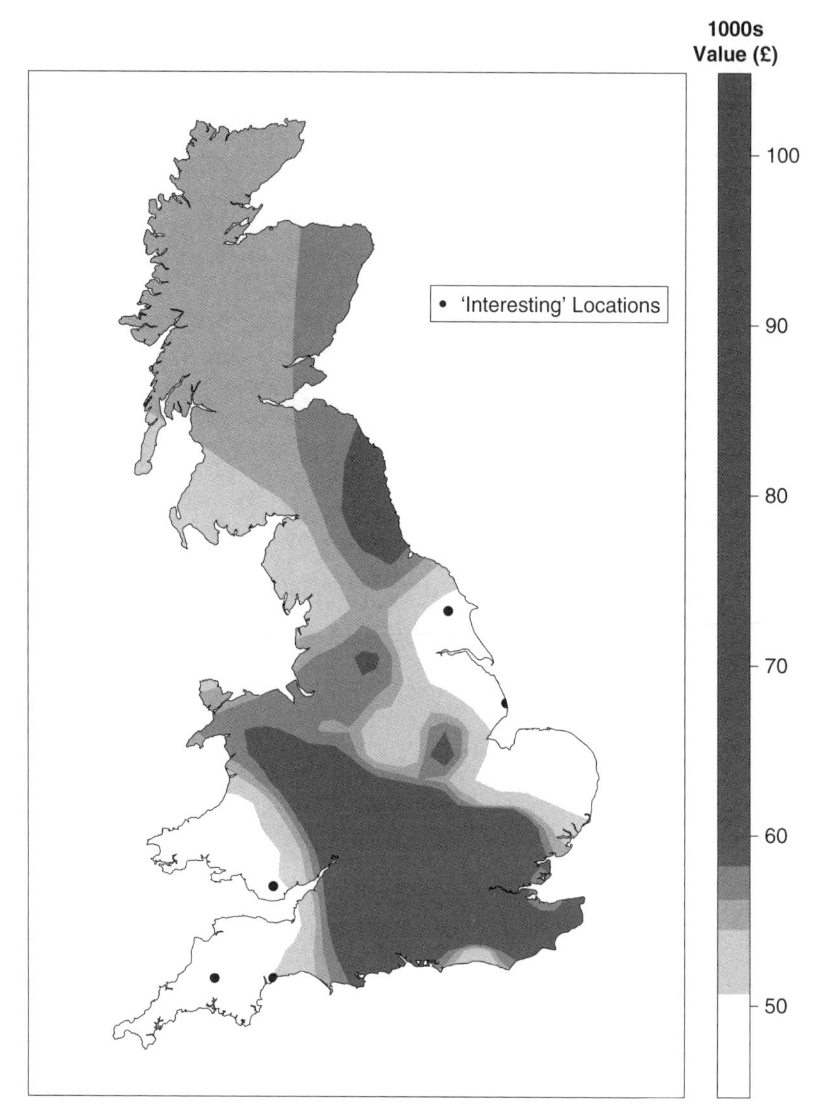

Figure 7.3 *GW mean surface for house prices, with significant local z-values*

Although we have only considered the local mean up to this point, Table 7.1 shows that there are several other statistics for continuous univariate data for which local values can be obtained by geographical weighting. For example, it can be informative to look at local degrees of spread in spatial variables. This can be done using a GW standard deviation. For a given area, the GW standard deviation gives some idea of the local variability of the statistic under investigation. In the house price example, this allows us to identify areas where there is a great deal of variation in house prices and areas where all houses cost roughly the same amount. This can be an interesting indicator as it shows the diversity of wealth in an area. It can also be an indicator of transition zones between low cost and expensive housing. When the summary point is located so that the kernel straddles the boundary between two zones of very different housing cost, a relatively high level of local standard deviation will be observed.

However, as with the GW mean, it is important to verify that the patterns observed are not simply artefacts of random variation in the data. Unfortunately, for statistics other than the mean, it is not so simple to derive an approximate distribution. One alternative approach is to consider methods based on randomisation. The randomisation hypothesis is that any pattern seen in $x_1, \ldots, x_n$ is a chance occurrence, so that any permutation of $x_1, \ldots, x_n$ amongst the data location points is equally likely. If we were to consider all $n!$ permutations, and compute the GW standard deviation at the summary point i for each of these, we could then compare our observed value (from the true assignment of the data to their locations) to this distribution and obtain some idea of how extreme it is. In practice this is impossible for any realistic sample size; for example, even a sample of size 300 would require more than 10^{614} evaluations of the local statistics at any summary point. However, we could use simulation and investigate the distribution of a large number (say, 10 000) of randomly chosen permutations. Then, to obtain for instance the 2.5% point, we find the value ranked 250th out of the 10 000 observations. Similarly, the 97.5% point will be the value ranked 9 750th. As with the mean, it is visually better to use a sparser grid to test for interesting values of local statistics than that used to produce a surface map. The GW standard deviation is shown in Figure 7.4. Several interesting locations are noted, suggesting there is some genuine variation in the variability in house prices across the UK. A number of bands of low variation may be observed, most notably in south Wales, the south-west of England and in East Anglia. A high degree of variation is shown in Wales and also in the north-east of England.

The GW standard deviation is also a useful diagnostic tool for GWR. In an ordinary regression, if one of the independent variables shows little or no variation, problems of model calibration may exist. In the case of zero variation the model cannot be calibrated without dropping the offending variable. If the variation is just very low, the model may be calibrated but the standard error for the regression coefficient will be unhelpfully large. In short, it is difficult to estimate how a response variable varies in relation to the predictor variable if the calibration data set does not have a reasonably wide range of values for the predictor variable. This problem will also affect GWR, but as GWR is a localised technique, problems can occur if an independent variable does not vary much *in some local region*. Thus a variable with a reasonable overall standard deviation may still be problematic if there are geographical pockets where its value is fairly constant. Mapping the GW standard deviation is a helpful diagnostic tool for identifying potential problems of this sort.

Other univariate continuous GW statistics which may be useful are the GW skewness and the GW coefficient of variation. The GW skewness referred to here

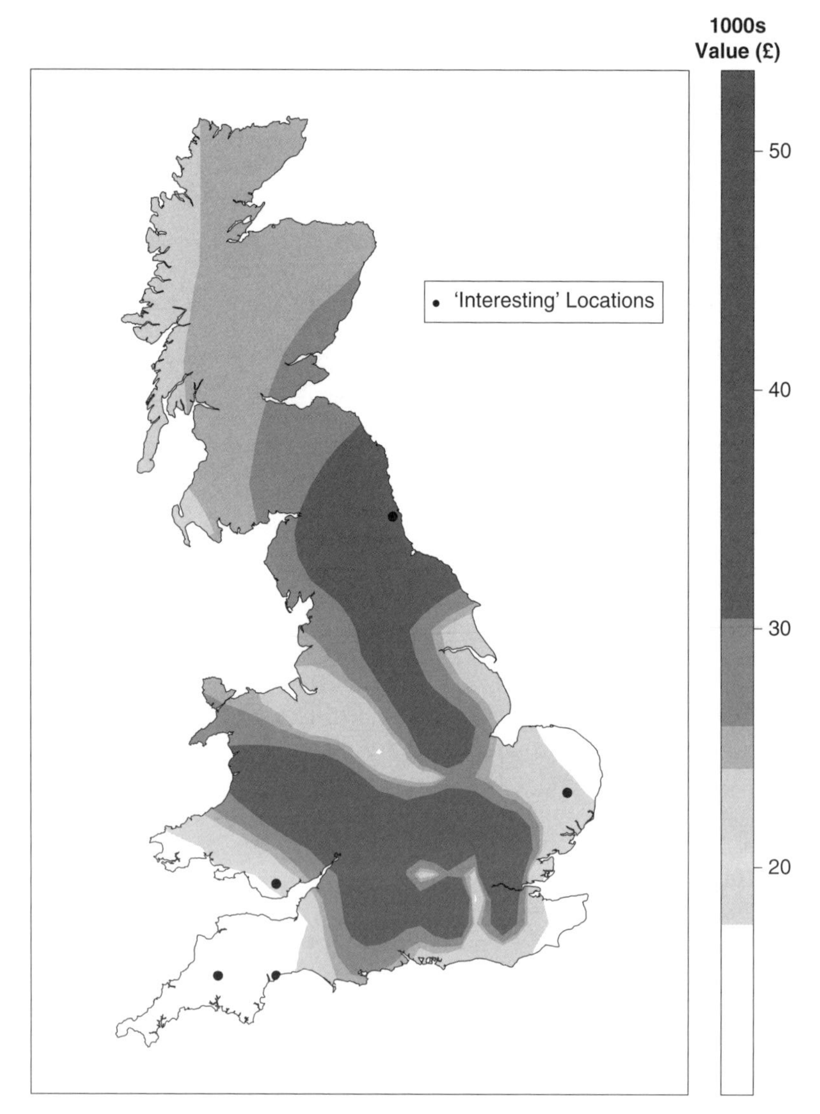

Figure 7.4 *GW standard deviation of house prices*

is the cube root of the geographically weighted third moment about the GW mean, divided by the GW standard deviation.[4] This gives us a dimensionless quantity (that is, its value does not change if the unit of measurement of the data alters, for example, from pounds sterling to dollars) which measures the symmetry of the

[4] This is one of several definitions of skewness, all of which can be geographically weighted.

local distribution. Positive values indicate a heavier tail to the right of the distribution, negative values a heavier tail to the left. In the house price example, a positive skewness measured at some summary point suggests that there is a minority of very highly priced houses in the area, but that this is not balanced by a similar cluster of low-cost houses. As with the GW standard deviation, there is no simple z-score formula to decide whether patterns are interesting, but the permutation approach may be used.

The GW skewness for the house price data is shown in Figure 7.5. In this case, there are no locations identified as interesting by the permutation test, suggesting that there is probably little variation in local skewness of house prices.

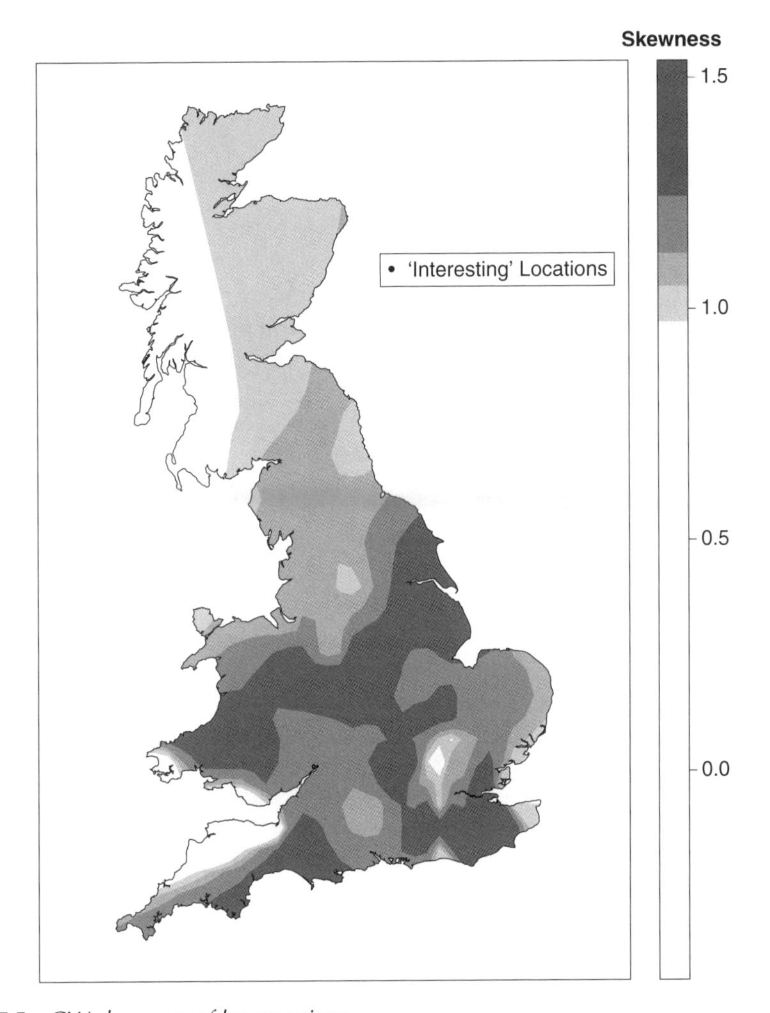

Figure 7.5 *GW skewness of house prices*

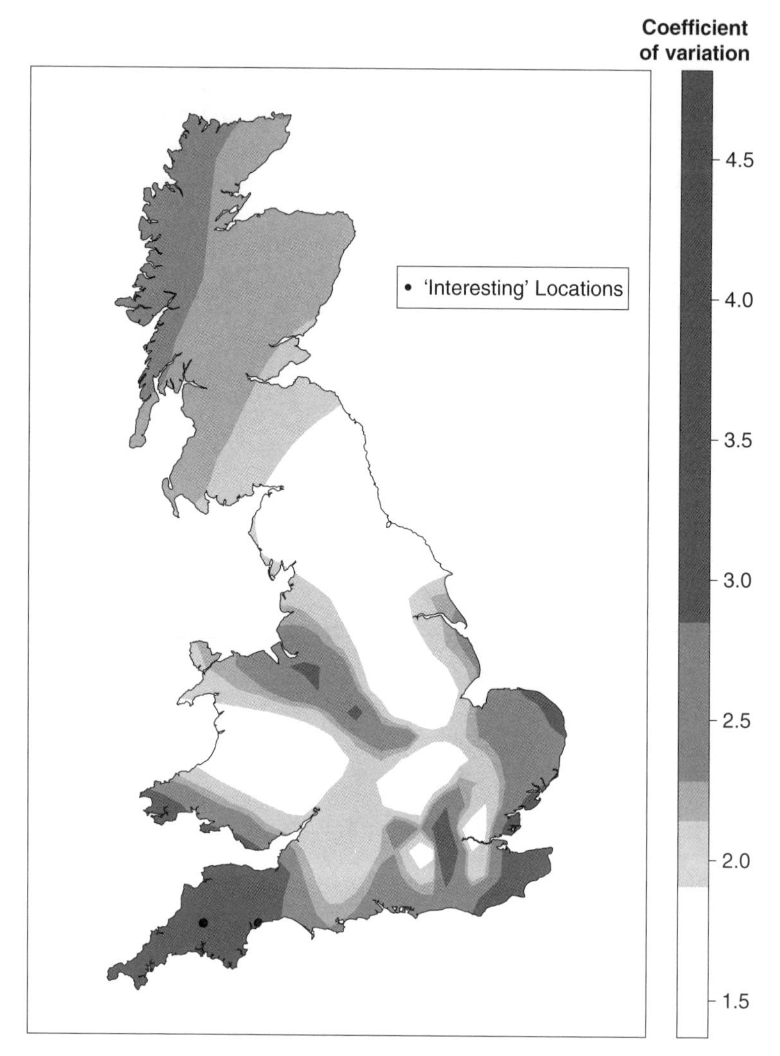

Figure 7.6 *GW coefficient of variation of house prices*

A final univariate statistic to be considered is the GW coefficient of variation (GW CV). This is the GW standard deviation divided by the GW mean. At a given summary point, this statistic indicates the local variability of the data but unlike the standard deviation, this quantity is dimensionless. Also, if the local mean value varies across the study area, this statistic considers the degree of variation in proportion to the changing mean. For example, with the house price data, a standard deviation of £20 000 when the mean price is £100 000 gives the same GW CV as a standard deviation of £10 000 when the mean price is £50 000.

In each case, the GW CV would be 0.2. A constant GW CV across the study area does not suggest that the *absolute* local variability of house prices is fixed across the study area. Rather, it implies that the degree of variability *as a proportion or percentage* is fixed. For the UK data, this is shown in Figure 7.6. Again, the south-west stands out here as having a high local variation in house prices, as do areas north and south of the Thames just west of London, parts of south Wales and the extreme south-east coast of England. Two other areas of relatively high local variation in house prices are just south of Manchester and a small area near Birmingham.

7.4 Two Continuous Variables

In this section we discuss GW summary statistics for two continuous variables. Perhaps the most important local bivariate summary statistic is the GW correlation coefficient, ρ_i, as defined in Table 7.3. A global bivariate correlation coefficient measures the degree of association between a pair of continuous variables such as house price and the year the house was built. A scatterplot of these two variables is shown in Figure 7.7 with points classified according to whether or not they are close to London (within 150 km). From this it can be seen that there is some evidence that the relationship differs for the two sets of points. Outside London there is arguably a stronger relation between the age of a house and its price than there is for houses in or close to London, with newer houses being more highly valued than older ones. However, this relationship is not particularly clear and there is a great deal of arbitrariness about the definition of the two geographical zones adopted. Also the image in Figure 7.7 is not in the form of a map. To investigate potential spatial variations in the relationship between the age of a house and its selling price, the GW correlation statistic is a useful tool.

As with the univariate statistics, it is possible to investigate interesting locations from a bivariate perspective. However, there is no simple closed-form expression for the random variability of localised correlation coefficients. Thus, assessment of 'interestingness' is best done using simulation-based methods. As with the univariate case, the random permutation approach is useful here. However, it should be noted that although the x and y variables should be randomly permuted amongst the locations, the data pairs (x_i, y_i) must be kept together. If these were to be scrambled, this would suggest that we were testing for global association between the variables. In this case, we are interested in testing the local variability in the GW correlation coefficient *given* the observed global value. The map of this is shown in Figure 7.8.

One extra feature of this map is the addition of a contour line for a correlation of zero. This delimits the areas where local correlation is negative. As can be seen, there is a notable u-shaped band of negative local correlation between house price and year built running across Wales and the English Midlands. Areas of positive correlation can be found in the south-west of England and also in the north. It is

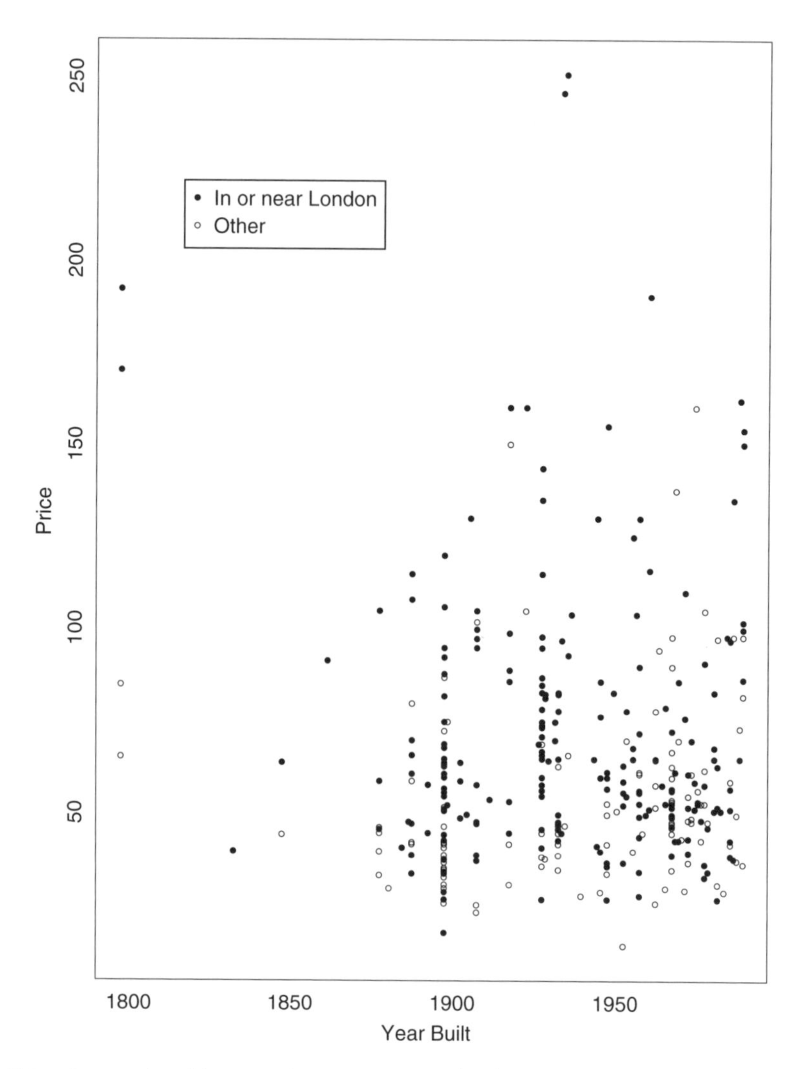

Figure 7.7 *Scatterplot of house prices against year built*

also worth noting that the GW correlation coefficient is a useful diagnostic tool for GWR. Here, mapping places where the local correlation coefficient is close to −1 or 1 for a given pair of independent variables identifies places where the local GWR calibration is likely to be problematic due to colinearity.

It is interesting to note that Wong (2001) has also recently devised a localised correlation measure which he calls a *location-specific cumulative distribution function*, although it depends on an arbitrary definition of a series of zones around the point at which the localised correlation is measured.

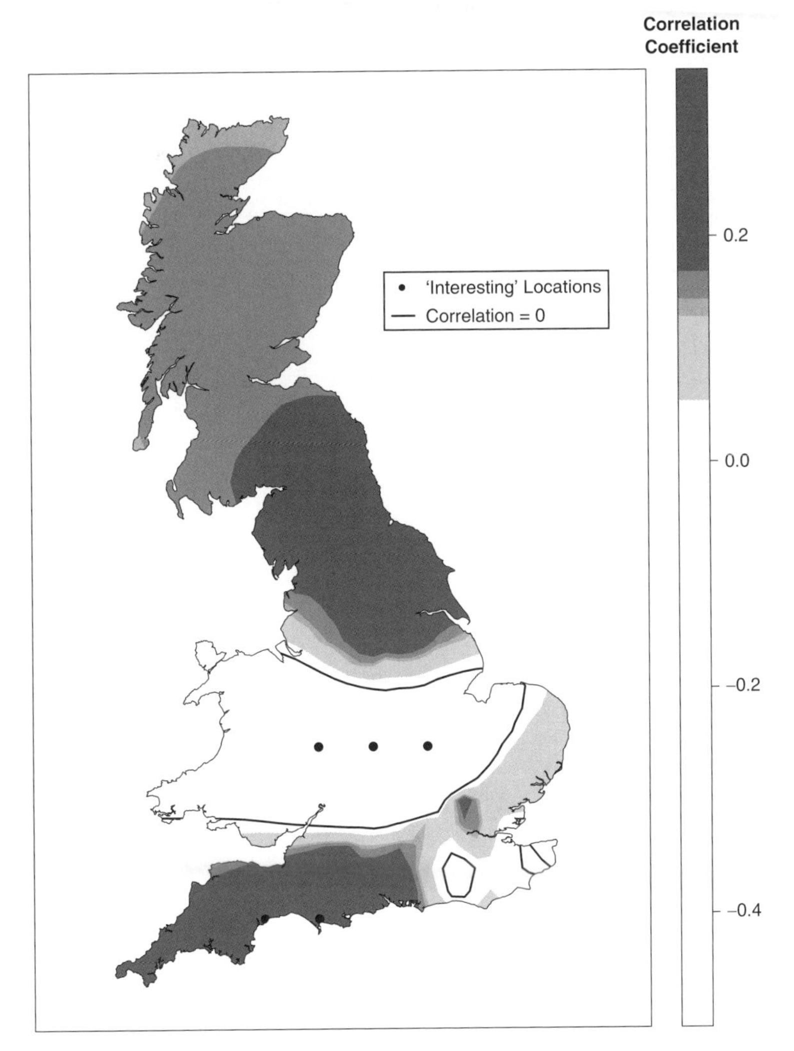

Figure 7.8 *GW correlation of house prices against year built*

7.5 A Single Binary Variable

In this section and the next, we consider binary variables. These are coded as either *true* or *false* and are more likely to be associated with categorical variables, for example, whether or not a house is detached, than with continuous variables, although they are also sometimes used for classifying the latter. An example of such a classification is an indicator as to whether a house was built after 1974. If the house was built after 1974, the variable is coded as *true*; otherwise it is coded as *false*. The main geographically weighted summary statistic for such variables is the GW proportion. This is simply a weighted estimate of the proportion (or percentage) of variables that are coded as *true* around some summary point. If we recode the variable as an indicator variable so that $x_i = 0$ for 'false' and 1 for 'true', then this is simply the geographically weighted mean of the recoded data, as defined in equation (7.3).

With this statistic, it is also possible to use a z-score approach to decide whether a local GW proportion is unusual under the assumption of no local changes in distribution. In this instance we assume independent identical Bernoulli distributions for all of the x_is. That is:

$$\begin{aligned} Pr(x_i = 1) &= \theta \\ Pr(x_i = 0) &= 1 - \theta \end{aligned} \tag{7.10}$$

Then

$$\mathrm{E}(x_i) = \theta \tag{7.11}$$

and

$$\mathrm{Var}(x_i) = \theta(1 - \theta) \tag{7.12}$$

so that if the GW proportion around the summary point i is denoted by θ_i, then

$$\mathrm{E}(\theta_i) = \theta \textstyle\sum_j w_{ij} = \theta \tag{7.13}$$

and

$$\mathrm{Var}(\theta_\mathrm{i}) = \theta(1 - \theta) \textstyle\sum_j {w_{ij}}^2 \tag{7.14}$$

Thus, we have expressions for the mean and variance of θ_i and so we may derive a z-statistic because it is well known that the binomial converges to the normal asymptotically. Comparing this statistic to ± 2 gives a fair, but arbitrary, guideline to assess whether variations in the local proportions are more extreme than might be attributed to chance. The parameter θ can be reasonably estimated as the global mean of the x_is.

For the variable TYPDETCH in the UK housing data sample, the GW proportion is mapped in Figure 7.9. Interestingly, there is an east–west divide, with the east having higher proportions of detached houses. However, against this trend there is also a spur with a higher proportion of detached houses running across an area just north of the London region.

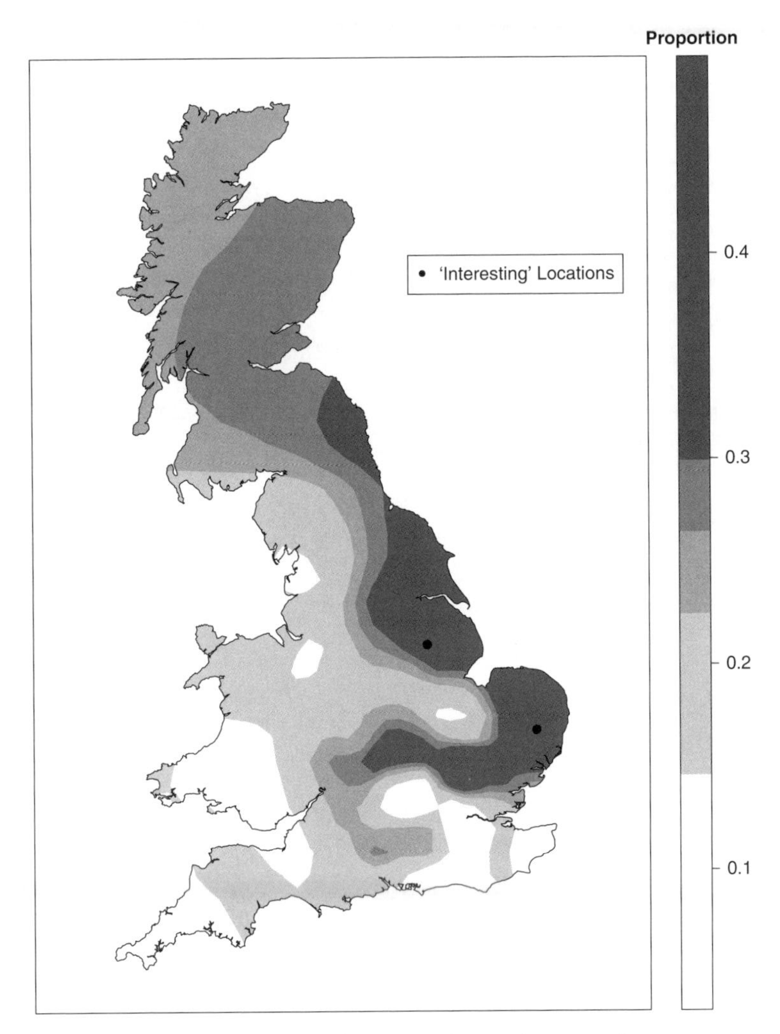

Figure 7.9 GW proportion of detached houses with significant z-values

7.6 A Pair of Binary Variables

Now we consider the bivariate case where there are two binary variables. In this instance, we can summarise the relationship between the variables using a 2×2 contingency table. For example, in addition to the TYPDETCH variable used in the last section, suppose we define a binary variable to indicate the relative newness of a house, which is 'true' if the house was built in or after 1975, and 'false' otherwise. In this case, the contingency table is as set out in Table 7.4.

From this table, it appears that for the UK as a whole, houses built in or after 1975 are more likely to be detached. This could be confirmed using a χ^2 test. Here, $\chi^2 = 15.2$ with 1 degree of freedom which is highly significant; the 95% point is

Table 7.4 *2 × 2 Contingency table for 'detached house' vs. 'built during or after 1975'*

	Detached	Not detached
Built 1975+	24	43
Built 1974–	42	238

3.54. The chi-squared test is therefore a single number summary that measures the association between the two binary variables. However, for 2 × 2 tables another, perhaps more intuitive, measure exists. This is the so-called odds ratio. According to the data, the odds of a house being detached if it were built during or after 1975 are 24:43. The same odds for a house built before 1975 are 42:238. We can express these odds as real numbers, by carrying out the implicit divisions: 24:43 = 24/43 = 0.558 and 42:238 = 42/238 = 0.176. We can now divide one set of odds by the other: 0.558/0.176 = 3.16. This tells us that the odds of a house being detached if it was built during or after 1975 are about three times what they are if the house were built before 1975. The quantity derived in this way is termed the odds ratio. Following Bishop *et al.* (1975), if a 2 × 2 contingency table is denoted as

$$\begin{array}{c|c} a & b \\ \hline c & d \end{array}$$

then the odds ratio, OR, is defined as

$$\mathrm{OR} = a/c \div b/d \text{ or } (ad)/(bc). \tag{7.15}$$

It is possible to carry out an approximate test of the null hypothesis of no association between the two variables using the odds ratio. Under this null hypothesis the mean of the natural logarithm of the odds ratio is zero, and the standard deviation is given by

$$s = \sqrt{a^{-1} + b^{-1} + c^{-1} + d^{-1}} \tag{7.16}$$

Dividing the log odds ratio by this quantity gives a z-score which is approximately Gaussian. Here, the log odds ratio is log(3.16) = 1.15, and from the above formula the standard deviation of this is $\sqrt{24^{-1} + 42^{-1} + 43^{-1} + 238^{-1}} = 0.3049$. Thus, the z-score is $1.15/0.3049 = 3.77$, very much higher than the 5% limit of 1.96.

The odds ratio is quite commonly used in medical statistics, but is perhaps less well known in other areas. For 2 × 2 tables it has the advantage that it is more intuitive than the χ^2 statistic. Statements such as 'The odds that a post-1975 house is detached are about three times as high as those for a pre-1975 house' are easily interpreted, whereas the interpretation of a χ^2 statistic of 15.2 is less immediately transferred to the real world. Another advantage is that the odds ratio gives an indication of the direction of the relationship between the variables – in this case whether a house being built during or after 1975 makes it more or less likely to be detached. The standard χ^2 only measures the level of association, not the direction. This problem is addressed using the signed-χ^2 measure (Visvalingham 1983), although this approach still suffers from the lack of directness of interpretation in comparison to the odds ratio. However, the χ^2 statistic does have the advantage that it may be applied to tables other than 2×2.

In this section we consider methods of geographically weighting these measures. The approach taken here will concentrate mainly on the GW odds ratio rather than the χ^2. This is for a number of reasons. First, Rogerson (2001) has already considered the idea of localised χ^2 statistics in depth. Second, we feel that the odds ratio is a more appropriate statistic in the special case of 2×2 tables: the aim of this chapter is more to consider localised summary statistics rather than inferential statistics and in terms of a reasonably interpretable single quantity, the odds ratio provides the better solution.

To define a GW odds ratio, we first consider the idea of a localised 2×2 contingency table. For each binary variable pair (x_j,y_j) we can define four indicator variables, say A_j, B_j, C_j, and D_j, one for each of the four possible states that (x_j,y_j) can take, as set out in Table 7.5. Then it may be checked that $a = \sum_j A_j, b = \sum_j B_j, c = \sum_j C_j$, and $d = \sum_j D_j$. If we have a summary point i, then it is possible to redefine a, b, c and d in terms of weighted sums of A_j, B_j, C_j and D_j around i. Thus, localised versions of the contingency table elements are obtained by $a_i = \sum_j A_j w_{ij}$, $b_i = \sum_j B_j w_{ij}$, $c_i = \sum_j C_j w_{ij}$ and $d_i = \sum_j D_j w_{ij}$ where w_{ij} represents the weight of data point j for the calculation of the statistics at summary point i. These may then be combined to produce a local odds ratio around the summary point i:

$$\mathrm{OR}_i = a_i d_i / b_i c_i \tag{7.17}$$

We can then map this local odds ratio. Here, we map the log odds ratio rather than the odds ratio itself. The reason for this is that although the odds ratio gives some indication of direction of the effect, for odds ratios exceeding 1 there is an infinite range of measure, whereas for those less than 1 measurement is confined to the range between 0 and 1. However, taking logs resolves this problem, as the latter range is mapped onto the infinite transformed range $(-\infty,0)$, and the former range onto the transformed range $(0,\infty)$. This provides directional symmetry; indeed this is partly the reason why the global log odds ratio statistic described above may be reasonably approximated using the Gaussian distribution.

Table 7.5 *Definition of A_j, B_j, C_j and D_j in terms of x_j and y_j*

x_j	y_j	A_j	B_j	C_j	D_j
0	0	1	0	0	0
1	0	0	1	0	0
0	1	0	0	1	0
1	1	0	0	0	1

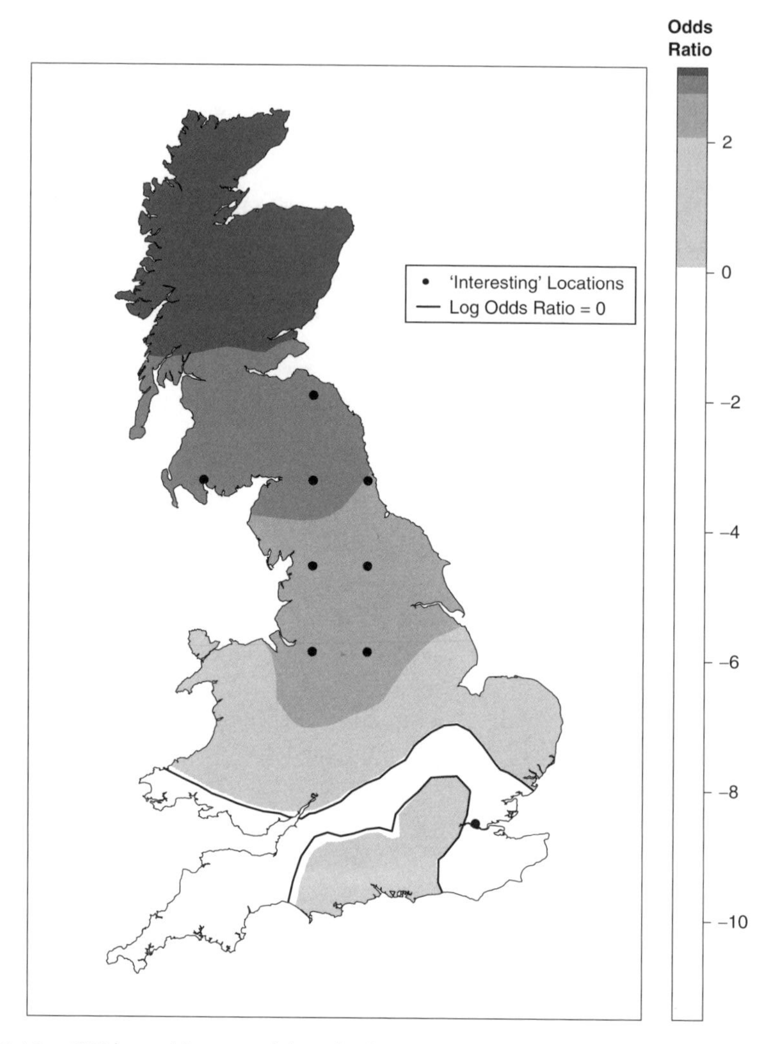

Figure 7.10 *GW log odds ratio of detached against built 1975 or later*

For the data set used throughout this chapter, the GW log odds ratio is shown in Figure 7.10. In general, the odds ratio gets larger the further north one goes. This implies that newer houses in the north of the country are more likely to be detached than are newer houses in the south. Presumably this is because the greater population density and higher house prices in the south encourage multiple dwelling units to be built. As with the other local statistics, it is possible to tag interesting areas. One way of doing this is to use the random permutation approach, as was employed to identify the interesting areas in Figure 7.10. Another approach is to use the z-statistic based on equation (7.16) with the geographically weighted contingency table. To some extent this latter approach is questionable since the original equation is based on the assumption that a, b, c and d have Poisson distributions, whereas the geographically weighted versions will not even be integers. However, the method can provide an informal indication of 'interestingness', in which case the weights should not be re-scaled to sum to one (as is conventional in the other methods).

7.7 Towards More Robust Geographically Weighted Statistics

The previous sections have provided GW descriptive statistics for a number of situations. However, one important issue that has yet to be dealt with is the problem of outliers. Essentially, outliers are individual observations in a data set that are very different from most of the other observations in the same data set. In the univariate case, this implies unusually high or unusually low values – for example, very expensive or very cheap houses.[5] The main problem with outliers for geographically weighted summary statistics is that they may have a distorting effect. The sample of data around a summary point may be relatively small, and a single very high or very low observation could have a distorting effect on a statistic such as the GW mean or the GW standard deviation. In the bivariate case, it is also true that unusually high or low observations may have a distorting effect. However, there is another kind of outlier that should be considered in this case – it is possible that there may be an (x_j,y_j) pair such that the x_j and y_j values are unremarkable when considered in isolation, but whose combination of values is unusual. For example, in the scatterplot in Figure 7.7 there are a number of points which stand clear of the main cloud of observations, but are not on the extreme ends of either the x- or y-scale on the plot. Finally, the issue is complicated by the fact that GW statistics are not only distorted by global outliers but values which are only unusual *in the neighbourhood* of the summary point i can also have distorting effects.

All of these observations serve to underline the importance of considering outliers when computing local statistics. In the global case, a number of possible strategies for handling outliers exist. These are often referred to as robust methods. For univariate statistics, such as means or standard deviations, trimming may be used. This simply involves automatically identifying and removing (or downweighting) extreme data values. For example, for a given data set, the approximate 97.5th and 2.5th percentiles could be computed, and all observations lying outside this range of values excluded from the computation of the mean; this is sometimes referred to as a trimmed mean. For a GW trimmed mean, this calculation could be carried out locally. That is, for a given summary point i, this operation could be carried out on all of the data within some search radius of i. The idea of downweighting as a method of handling outliers in GWR is described in Chapter 3. An alternative is to consider order-based summary statistics instead of moment-based

[5] Note that this implies that the notion of outliers in the binary univariate case is fairly meaningless – observations may only take the value 0 or 1. The only possible way a binary variable might be considered an outlier is if virtually all observations had the same value and only a handful had the other value.

ones. For example, the median is often used to provide a more robust measure of central tendency than the mean. Similarly, the interquartile range is a more robust measure of spread than the standard deviation. These may therefore make suitable candidates for alternative GW summary statistics. However, before doing this, the notion of weighting needs to be considered for these statistics. Note that both the median and the interquartile range are functions of the general p-quantile for various values of p. That is, the median is the quantile for which $p = 0.5$ and the interquartile range is the difference of the quantiles for $p = 0.25$ and $p = 0.75$. Thus, deriving localised p-quantiles is the key to obtaining the weighted statistics. We need the discrete distribution expression for the p-quantile, which is the minimum solution for q of the equation

$$\Pr(x < q) = p \tag{7.18}$$

Suppose that our sample $x_1 \ldots x_n$ is a mass-point distribution (MPD) that is, the value x_i occurs with probability w_i (recall that the w_is are scaled to sum to one so that weights and probabilities are synonymous here). If this MPD is derived using a local kernel, such as that above, we term it a local mass-point distribution (LMPD). We can then write equation (7.18) as

$$\sum_{x_i < q} w_i = p \tag{7.19}$$

This expression may be best understood if we label the x_is in ascending order and label the corresponding w_is accordingly. Regard the left-hand side of equation (7.18) as a function of q. This takes the value of the sum of the set $\{w_1, w_2, \ldots, w_j\}$, where j is the index of the largest x_i not exceeding q. This function jumps by an amount w_i each time q exceeds a value x_i. Thus, the left-hand side only takes one of the n values w_1, $w_1 + w_2$, $\ldots w_1 + w_2 + \ldots + w_n$ (note that the last of these is equal to one). Unless one of these values happens to equal p, equation (7.19) has no solution. For some j we will have $w_1 + \ldots + w_j < p$ and $w_1 + \ldots + w_{j+1} > p$. A problem therefore arises: when this happens, it appears that the LMPD has no p-quantile. We overcome this difficulty by extending the definition of a p-quantile in this situation by interpolation. When $q = x_j$ we have $\Pr(x \leq q) = w_1 + \ldots + w_j = w_j^*$, say, and when $q = x_{j+1}$ we have $\Pr(x \leq q) = w_1 + \ldots + w_{j+1} = w_{j+1}^*$. If we were to assume that $\Pr(x \leq q)$ were a linear function between $q = x_j$ and $q = x_{j+1}$, rather than the discontinuous jump it actually is, then the solution to equation (7.19) would be

$$q = x_j + (x_{j+1} - x_j)\frac{p - w_j^*}{w^*_{j+1} - w_j^*} \tag{7.20}$$

which is just the standard linear interpolation formula. Finally, noting that $w^*_{j+1} - w_j^* = w_{j+1}$, the result may be simplified to

$$q = x_j + w^{-1}{}_{j+1}(x_{j+1} - x_j)(p - w^*{}_j) \tag{7.21}$$

If $p = 0.5$, and all w_is are equal to $1/n$, then we obtain the standard expression for the sample median. Once again, we obtain an expression which can be thought of as the geographically weighted generalisation of a global summary statistic.

Thus, via equation (7.21) we obtain localised versions of the median and the interquartile range. These complement the moment-based descriptors derived earlier. Often, both types of statistic are of use: typically, the quantile-based estimates are more robust to outlying values (as stated earlier), but the moment-based estimates tend to be smoother. Note that this list is not exhaustive. For example, one could go on to consider skewness or kurtosis (based on the fourth moment). However, we feel that these statistics provide a useful grounding for an exploratory analysis of local distribution shape.

Examples of the GW median and GW interquartile range are given in Figures 7.11 and 7.12, for the UK house price data. One important difference between the GW mean and the GW median is that the former has a higher value region in the north-east of England. However, the local statistics in this area are based on relatively sparse data, and it could be the case that one unusually expensive house has distorted the GW mean surface. The GW median is more robust to this kind of influence.

7.8 Summary

In this chapter, we have shown how a number of techniques may be geographically weighted to explore local patterns in data. The emphasis has been on exploration; although we have introduced some ideas of identifying unusual local pockets of data and relationships, these are not intended to be regarded as formal significance tests. Indeed, the aim here is to be more risk averse. In the exploratory stages of data analysis we consider that it is more costly to obtain false negative results, where a genuine pattern is not detected, than false positive results, where a pattern which could have occurred at random is labelled interesting.

As a side issue, it should be noted that it is possible to obtain significant local results when the global result is not significant. It is also possible to obtain no locally significant results when the global result is significant. Although the latter seems counter-intuitive, it is an example of Simpson's paradox (Simpson, 1951) described in Chapter 1 whereby local results can run counter to the global one.

Although we have selected a particular subset of descriptive statistics to demonstrate the concept of geographical weighting, the methodology adopted could be thought of as a more general framework for creating GW descriptors. As argued in

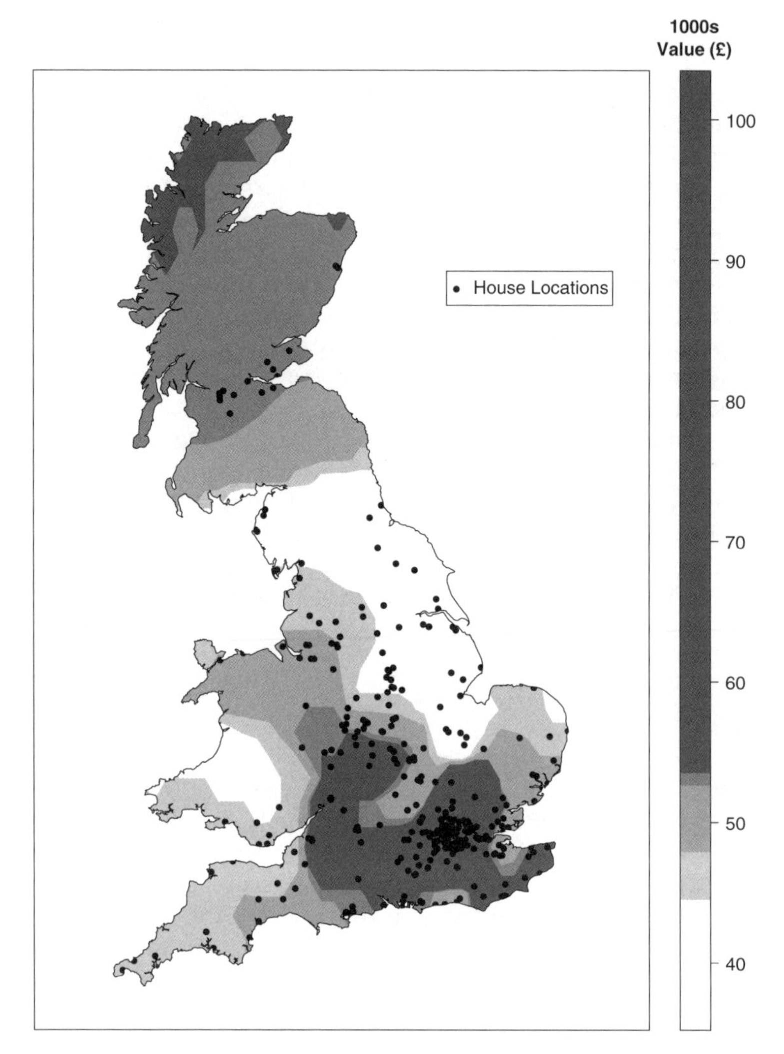

Figure 7.11 *GW medians of house prices (£ 000)*

Brunsdon *et al.* (1996), if a technique is weightable, then it is geographically weightable. Applying the technique in some cases may lead to an existing technique, albeit with a slightly new interpretation. For example, a geographically weighted

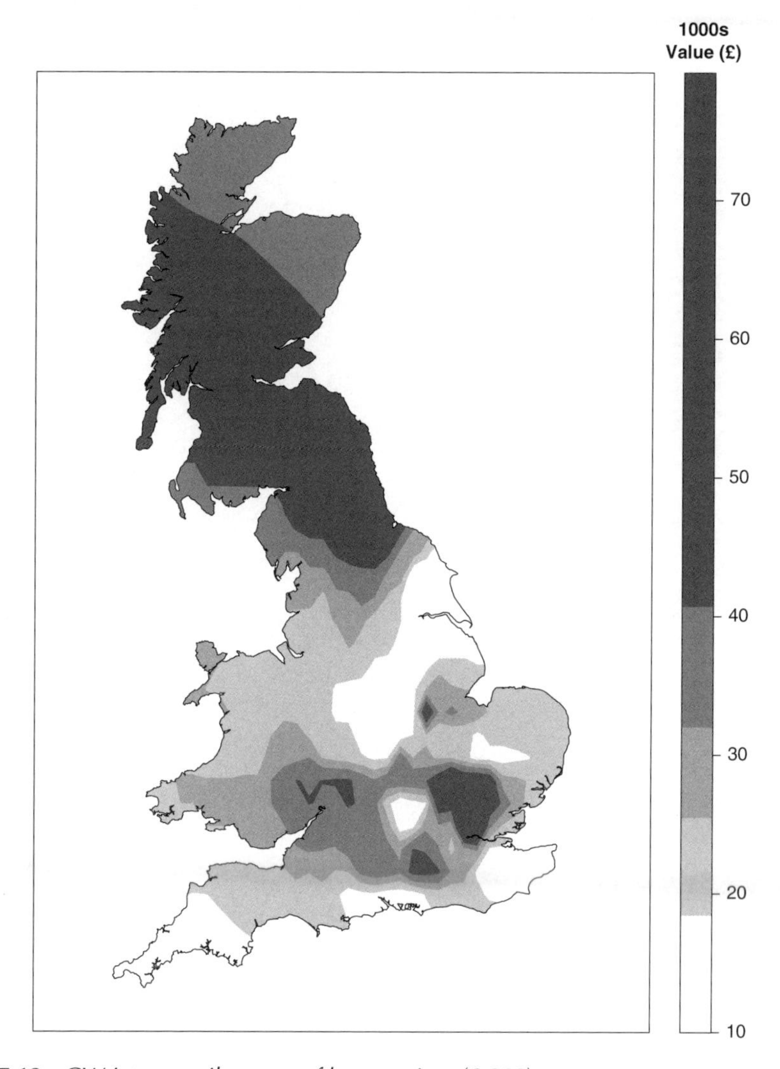

Figure 7.12 *GW interquartile range of house prices (£ 000)*

mean is a basic moving window filter. However, in other cases, for example, the geographically weighted odds ratio, a much more innovative method of data exploration may be discovered.

8

Extensions of Geographical Weighting

8.1 Introduction

In this chapter, we look beyond the basic model employed in GWR to three extensions where the concept of geographical weighting can also be applied.

1. First, we will consider regression models where the error term is non-Gaussian. This is an important extension because in many spatial models, the dependent term is not a continuous variable capable of taking both negative and positive values. For instance, many data take the form of a count of individuals or items. In such a situation, the basic Gaussian model is clearly inappropriate – not only does it suggest that the dependent variable can take negative values, it also suggests that non-integer values are possible. It is more helpful in this case to provide a count-based distribution model, such as those based on the Poisson or Binomial distributions, but the concept of geographical weighting is equally applicable to such models. We demonstrate this by applying geographical weighting to generalised linear models.
2. We also look to more extreme departures from GWR. The extension to either Poisson or Binomial models still involves a regression approach, where we attempt to predict one variable on the basis of others. However, we can also extend the geographical weighting concept in the general direction of local variance structure information. Local principal components provide a means of interpreting local covariance matrices in multivariate data sets and in a sense build on the idea of local correlation coefficients proposed in Chapter 7.
3. The final extension discussed is that of local probability density estimation. This is an alternative approach to local univariate descriptive statistics: rather than attempting to describe the local distribution of some variable by local

indices of distribution shape, such as mean, median, standard deviation or skewness, we attempt to reconstruct the local density function itself.

8.2 Geographically Weighted Generalised Linear Models

The basic linear regression model assumes a Gaussian distribution for the dependent (y) variable, which has a number of properties: for example, it is symmetrical about some mean value and it admits values anywhere in the interval $(-\infty, \infty)$. These properties also hold for the basic GWR model (see Chapter 4). However, there are situations where these properties are not good models of reality. For example, the distribution of y may not be symmetrical, or y may only be defined for positive values (the idea of a negative house price seems unlikely, for example). Indeed, the y variable may only be defined for integers (such as the number of assaults occurring in an area over a given time interval), or be a logical variable (for example, whether or not a house has a garage).

In the non-geographical case, this issue was addressed in the early 1970s by *Generalized Linear Models* (Nelder and Wedderburn 1972). These extend the basic regression model, where y has a Gaussian distribution, to the case where the distribution of y is a member of the *exponential family* of distributions

$$f(y|\theta, \phi) = \exp\left\{\frac{y(\theta) - b(\theta)}{a(\theta)} + c(y, \phi)\right\} \tag{8.1}$$

where ϕ is a 'scale' parameter, θ is a 'shape' parameter and $a(\ )$, $b(\ )$, and $c(\ )$ are arbitrary functions constrained only in that their choice in any given case must result in a well-defined distribution function $f(\ )$. Note that appropriate choices of these functions can yield, among others, the Gaussian, Binomial, Poisson and Gamma distributions. This defines the distribution of the y variable, but we also need to define its relationship with the independent variables $(x_1, x_2, \ldots, x_m)$. The expected value of y is specified in terms of a linear function of the x variables:

$$g(\mathrm{E}(y_i)) = a_0 + a_1 x_{i1} + a_2 x_{i2} + \ldots + a_m x_{im} \tag{8.2}$$

Here, the additional i subscript denotes the case number, ranging from 1 to n. The function $g(\ )$ is specified in advance and is referred to as the link function. In the basic regression model it is simply the identity function: $g(x) = x$. $\mathrm{E}(y_i)$ will depend on ϕ and θ_i. ϕ has no i subscript since it is the same for all observations. In most cases, emphasis is placed on estimating $\mathrm{E}(y_i)$ via the linear coefficients $(a_0, a_1, \ldots, a_m)$, rather than θ_i. The scale parameter ϕ is generally regarded as a nuisance parameter which must be estimated in order to make inferences about the parameters of interest but is typically not of interest in itself. In some cases, such as the Poisson and Binomial, $a(\ )$ is a constant function and $c(\ )$ depends only on y so there is no scale parameter. This reflects the fact that in both of these cases, the variance of y is a function of $\mathrm{E}(y)$.

To calibrate the model, one must estimate $(a_0 \ldots a_m)$. This is achieved using the Fisher scoring procedure, a form of Iteratively Reweighted Least Squares (IRLS). As its name suggests, this is an iterative procedure, requiring an initial guess at the regression coefficients – say, $(a_0^{(0)}, \ldots, a_m^{(0)})$. Typically, this can be obtained by an ordinary least squares regression of $g(y)$ on the x-variables. Then, the IRLS procedure is as below:

1. Use $(a_0^{(0)}, \ldots, a_m^{(0)})$ to obtain initial estimates of the expected values of the y_is: $E_0(y_i) = g^{-1}(v_o)$ where $g^{-1}(\ )$ is the inverse function of $g(\ )$, and $v_0 = a_0^{(0)} + a_1^{(0)} x_{i1} + a_2^{(0)} x_{i2} + \ldots a_m^{(0)} x_{im}$.
2. Construct the *adjusted dependent variable* z_i: $z_i = v_0 + (y_i - E_0(y_i))g'(E_0(y_i))$ where $g'(\ )$ is the first derivative of $g(\ )$.
3. Define a set of weights $w_i = 1/(g'(E_0(y_i)^2 V_{i0})$, where V_{i0} is the modelled variance of y_i given by the current set of coefficient estimates.
4. Regress z on the xs with weights w_i. Obtain a new set of parameter estimates $(a_0^{(1)}, \ldots, a_m^{(1)})$.
5. Return to step 1, substituting the latest coefficient estimates. Continue to loop in this way until the estimates converge.

The above is a very general algorithm. Some more specific cases are listed in Table 8.1. With this information, it is possible to calibrate a wide range of models. However, until now nothing has been said about *geographically weighted* models. A geographically weighted generalised linear model (GWGLM) can be specified by allowing the linear coefficients in equation (8.2) to be arbitrary functions of points (u, v) in geographical space:

$$g(E(y_i)) = a_0(u, v) + a_1(u, v)x_{i1} + a_2(u, v)x_{i2} + \ldots + a_m(u, v)x_{im} \tag{8.3}$$

For a specific point (u_0, v_0) we may apply the local likelihood arguments of Chapter 4 to obtain estimates of $(a_0(u_0, v_0), \ldots, a_m(u_0, v_0)$.) For the Fisher scoring procedure, this is achieved by multiplying the w_is in the iterative algorithm by the geographical weights, as defined by a typical geographical kernel function such as the Gaussian or bi-square kernel. Allowing (u_0, v_0) to scan across the study area, coefficient surfaces may be constructed as with basic GWR. There are a number of parallels between this approach and the calibration of Generalized Additive Models (Hastie and Tibshirani, 1990).

Table 8.1 *z_i and w_i values for some common generalised linear models*

y_i distribution	Link (V_i)	z_i	w_i
Gaussian	$E(y_i)$	y_i	1
Binomial (θ_i, k_i)	$\log(\theta_i/(1-\theta_i))$	$v_i + (y_i - k_i\theta_i)/(k_i\theta_i(1-\theta_i))$	$(k_i\theta_i(1-\theta_i))$
Poisson	$\log(E(y_i))$	$v_i + (y_i - E(y_i))/E(y_i)$	$E(y_i)$
Gamma	$\log(E(y_i))$	$v_i + (y_i - E(y_i))/E(y_i)$	1

Thus, we now have a method for calibrating GWGLMs, based on iteratively reweighted least squares. Note that many of the ideas applied to Gaussian GWR may also be applied here. For example, we can use mixed models, where some coefficients are global and others vary geographically, by applying the mixed model calibration procedure (taking into account the Fisher score weighting) to the z_is. However, it should be noted that on occasion the computational overhead for fitting a GWGLM can be quite large. First, for each iteration of the model we require $E(y_i)$ to be estimated, so that each observed data point must also be a regression point. This is also the case when, for example, we wish to compute the Akaike Information Criterion (see Chapters 2 and 4), but in this case several iterations may be necessary. The intention here is not to discourage the use of such models, but to point out that, at the time of writing, lengthy computation times may occur if fitting models to large data sets (say 10 000 or more cases). In the following sections, two examples of GWGLM will be given, one using a Poisson model and one using a Binomial (logistic) model. In both examples, data on house prices from the Nationwide Building Society sample will be used.

8.2.1 A Poisson GWGLM

In this section, we consider the number of bathrooms in a house to be the dependent variable. Obviously, this is typically a low-valued integer, most frequently 1 in the UK, although 2 or even 3 bathrooms do occur a notable number of times. There are cases of *no* bathrooms (when houses are in a poor state of repair) although this is unusual. Here we model this variable as a Poisson random variate and regress this against the price of the house as a predictor variable. This is a very simplistic model, and one where the direction of the dependency can be questioned, but it helps us to describe the application of GW Poisson models. We can also explore the relationship between the amount paid for a house and the number of bathrooms one might expect the house to have, and how this relationship varies across the UK. It therefore provides a simple exploration of the variation in value for money potential house owners experience across the UK. The model here is

$$\log(E(B_i)) = a_0(u_i, v_i) + a_1(u_i, v_i)P_i \tag{8.4}$$

where B_i is the number of bathrooms for house i, P_i is the price of house i, and $a_0(\ )$ and $a_1(\ )$ are geographically varying regression coefficients as defined in equation (8.3). As stated earlier, B_i is an integer with a Poisson distribution. The model is calibrated using the geographically weighted IRLS algorithm described above, using the 347 observations in the random sample of house sale completions across Great Britain taken from the Nationwide Building Society as described in Chapter 7. The calibration converged after 4 iterations. The results are shown in Figures 8.1a–d. The coefficient function $a_1(u, v)$ relating to price is perhaps of most interest here. This coefficient becomes larger towards the north and also towards the south-west and south Wales. In these areas, a marginal increase in house price is more likely to result in a greater number of bathrooms than in other areas of the

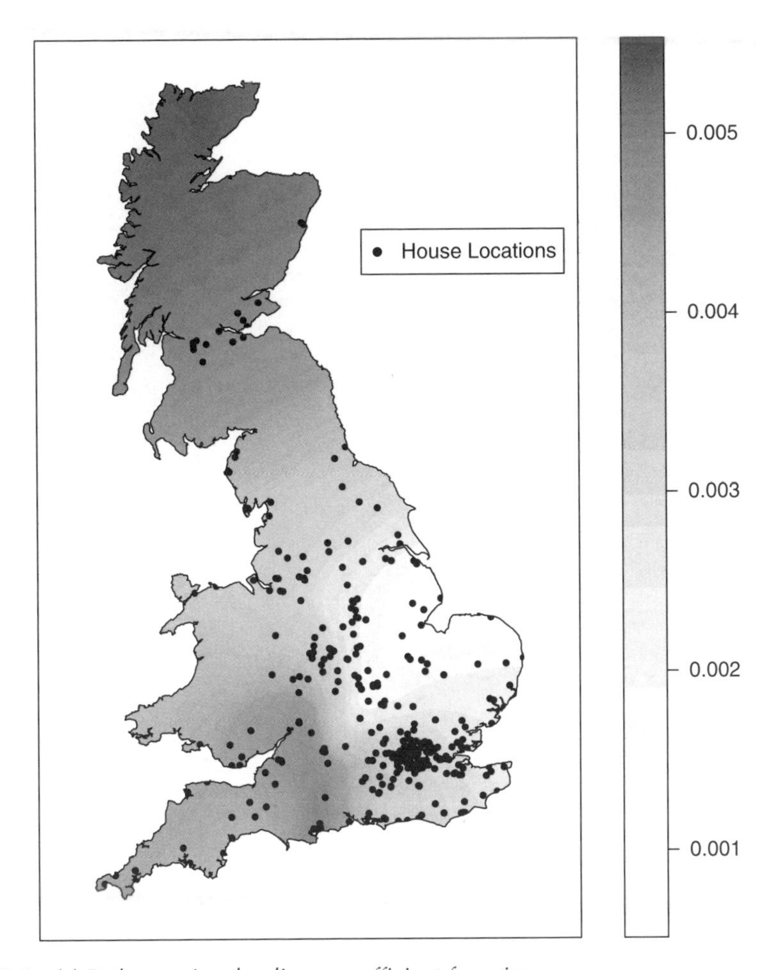

Figure 8.1 *(a) Baths ~ price: log-linear coefficient for price*

country. The map of $a_0(u,v)$ shows the general trend in number of bathrooms regardless of price. This has a notably low value to the south-west of London, stretching upward to the Bristol area.

Perhaps the best way to understand these results is to 'plug in' a specific house price into equation (8.4) and then use the model to predict the expected number of bathrooms a house will have in different parts of the country. For example, if we choose £150 000 as our price, then we have

$$\hat{\mathrm{E}}(B) = \exp(\hat{a}_0(u,v) + 150\,000\hat{a}_1(u,v)) \tag{8.5}$$

Note the subscript i is missing from the u and v here; the model is not restricted to the locations of the observations (u_i,v_i). The hats (e.g., $\hat{a}_0$) denote that the expressions underneath are estimates based on the data. This 'plugging in' of a house

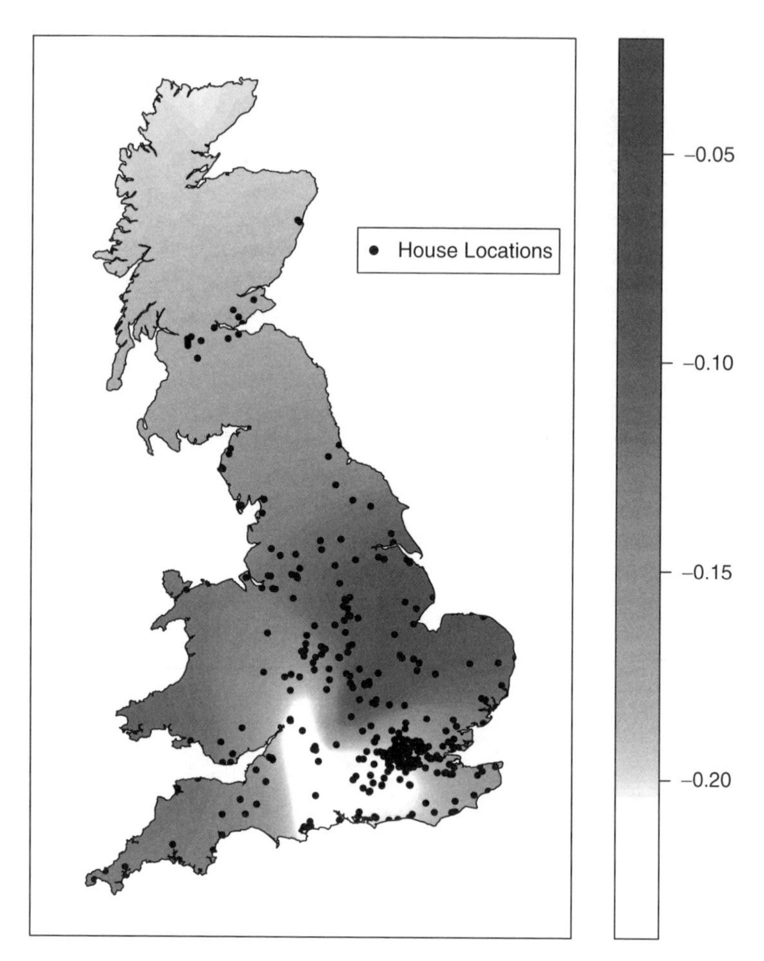

Figure 8.1 *(b) Baths ~ price: log-linear coefficient for intercept term*

price value generates an 'expected number of bathrooms' surface over the UK for houses costing £150 000. A very simple, yet effective, way to display this information is to draw a contour line on a map of the UK, representing the boundary between the region where one bathroom (or zero) is the most probable value of B, and the region where two or more bathrooms are more likely. This contour line should be drawn when $P(B=0)+P(B=1)=\frac{1}{2}$. Recalling that B has a Poisson distribution, this occurs when $\exp(-E(B))+E(B)\exp(-E(B))=\frac{1}{2}$. This may be solved numerically, giving $E(B)=1.678$. Thus, on the 'expected number of bathrooms' surface, the contour line(s) should be drawn at this value. For the Nationwide sample, such a map is shown in Figure 8.2. As can be seen, places where one might expect to find a house with two or more bathrooms for £150 000 are Scotland, South Wales and the South West of England.

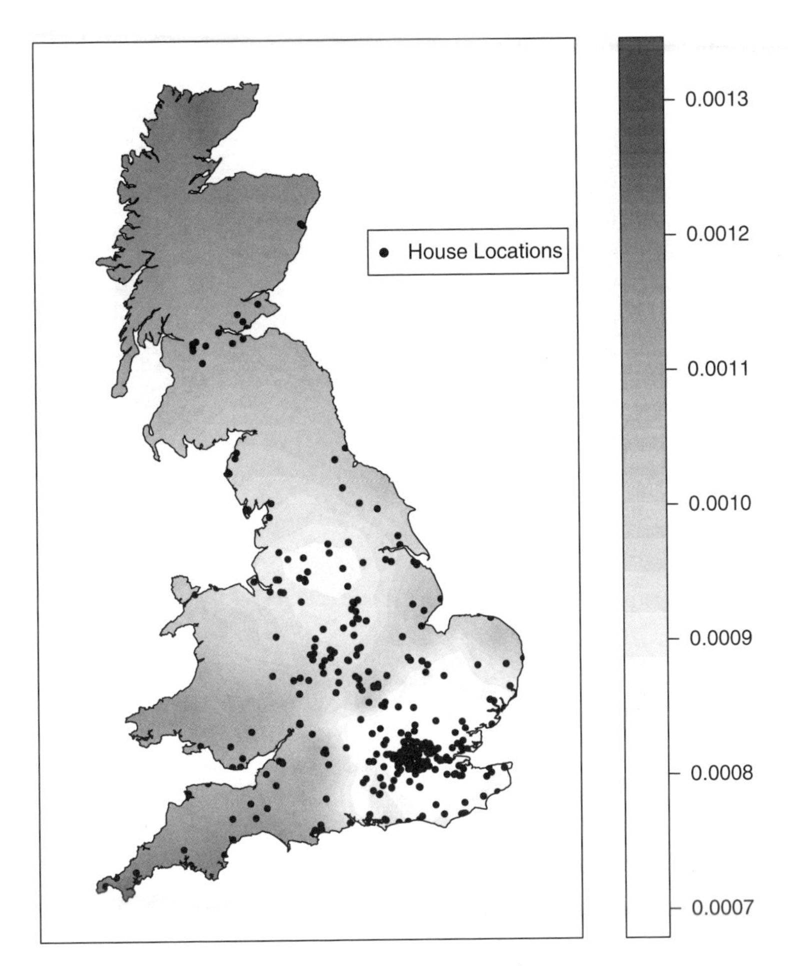

Figure 8.1 *(c) Baths ~ price: SE of log-linear coefficient for price term*

8.2.2 A Binomial GWGLM

This section could equally be entitled 'A Logistic GWGLM' because we wish to model a logistic function of the binary variable TYPDETCH (see Chapter 2) as a linear function of whether the house was built after 1974. Both variables are binary and the model is

$$\log\left(\frac{\mathrm{E}(D_i)}{1-\mathrm{E}(D_i)}\right) = a_0(u_i, v_i) + a_1(u_i, v_i)M_i \tag{8.6}$$

where D_i is an indicator variable set to 1 if house i is detached, and 0 otherwise, M_i is an indicator variable set to 1 if house i was built after 1974, and 0 otherwise. Note that $\mathrm{E}(D_i)$ in this case is simply $\mathrm{P}(D_i = 1)$ so that we are modelling the logistic function of the probability that a house is detached. However, the binomial

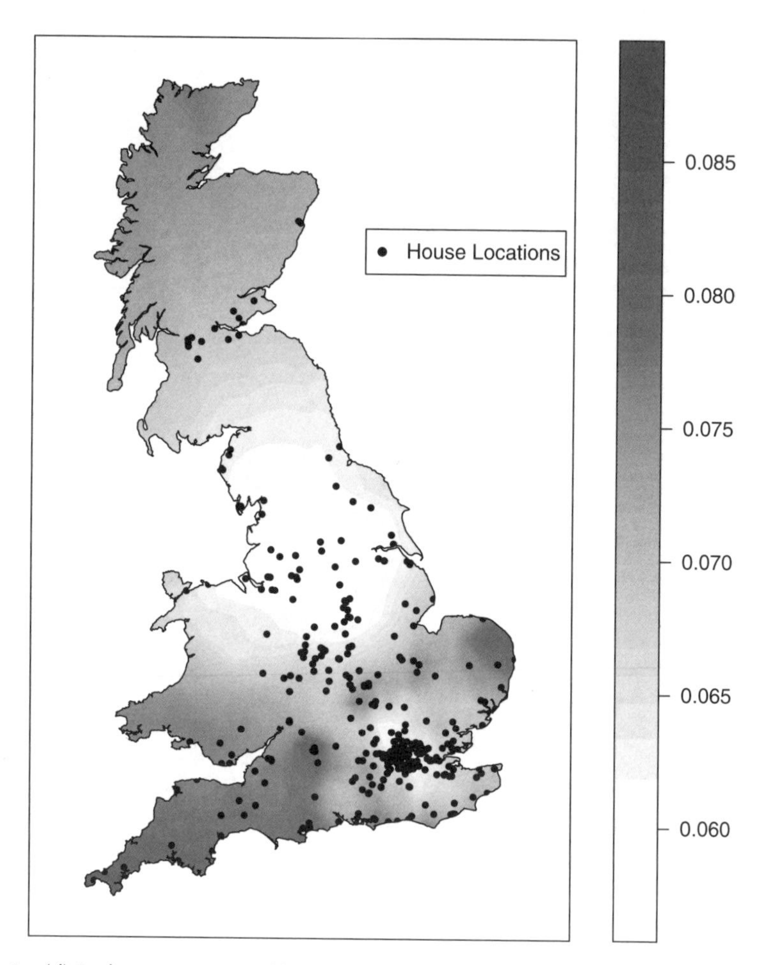

Figure 8.1 *(d) Baths ~ price: SE of log-linear coefficient for intercept term*

tag is also justified because we are simply modelling binomial variates where the number of trials is always 1 and we can use the IRLS algorithm for binomial models given in Table 8.1 with k_i set to 1 for all i.

Here $a_0(u,v)$ can be interpreted as the logarithm of the odds that the house is detached if it was not built after 1974, where the odds of an event E are defined as $\Pr(E)/(1 - \Pr(E))$. The quantity $a_0(u,v) + a_1(u,v)$ can now be interpreted as the logarithm of the odds that a house is detached if it *was* built after 1974. Subtracting the former expression from the latter gives the logarithm of the ratio between these two odds, and is referred to as the log odds ratio (see also Chapter 7). In terms of the coefficients, the log odds ratio is simply $a_1(u,v)$. The odds ratio itself is therefore $\exp(a_1(u,v))$. This quantity has a fairly intuitive interpretation. For example, an odds ratio of 2 suggests that the odds of a house being detached if it were built after 1974 are twice that if it were built in 1974 or before. Although we have not done so here, it is possible to add several more explanatory variables in such

Figure 8.2 Number of bathrooms likely for £150 000

a model. Since we are using a geographically weighted variable here, this ratio is allowed to vary geographically. Calibrating equation (8.6) using the Nationwide Building Society data required seven iterations, giving the results shown in Figures 8.3a and 8.3b.

The distribution of the local log odds ratio is shown in Figure 8.3a while Figure 8.3b shows the standard error of the local logistic coefficient. The contours on Figure 8.3a show where the log odds ratio is equal to zero. This occurs when the 'after 1974' variable has no relationship with whether a house is detached. The two contour lines on the map show a band across the southern UK, including the south-east and south-west but excluding London. Within this band, the log odds ratio is below zero, implying that a house being built after 1974 is associated with it *not* being detached. For the rest of the country, the opposite effect is observed, where being built after 1974 makes a house more likely to be detached. This gives

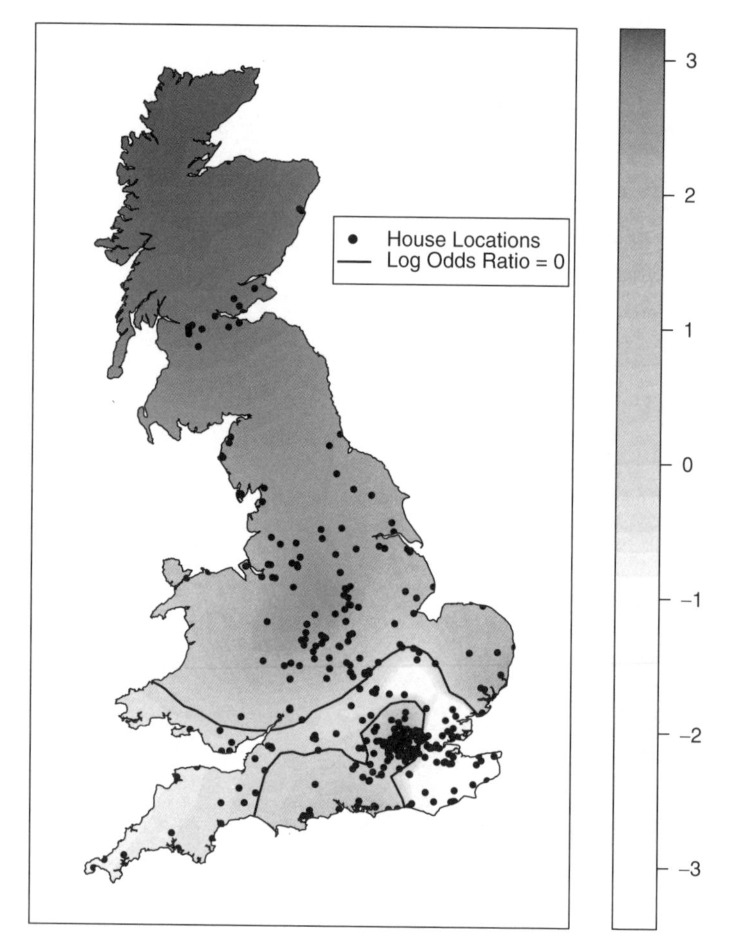

Figure 8.3 (a) Detached ~ after_1974: logistic coefficient for after_1974 term

some indication of house building trends throughout the UK in the past quarter century. Figure 8.3b displays the local standard error of this estimate which is relatively high in the south-east and relatively low in northern England.

8.3 Geographically Weighted Principal Components

8.3.1 Local Multivariate Models

In this section, a departure is made from the standard regression model. The standard model, even in the generalised form of equation (8.3), is essentially univariate, in the sense that although several predictor variables exist, there is just one random variate, the response variable y. Although there are obviously several xs taken into consideration, equation (8.3) specifies the conditional distribution of y

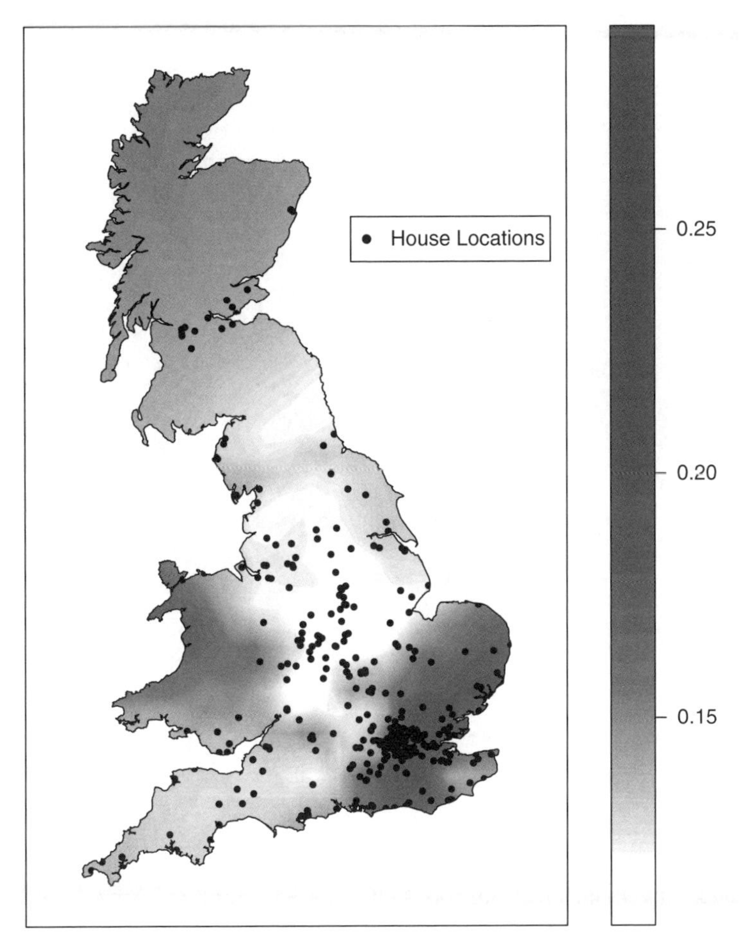

Figure 8.3 *(b) Detached ~ after_1974: SE of logistic coefficient for after_1974*

given all of the xs (and the location (u,v)) and it is the conditional distribution of this one variable we attempt to model. Here we move to a multivariate approach – we no longer single out one variable to be the y variable. Instead, we consider the joint probability distribution of all of the variables under examination, conditional on location (u,v).

We denote all of the observed x-variables at location i by the vector $\boldsymbol{x}_i$. If we assumed a multivariate Gaussian model for these variables, we would write

$$\boldsymbol{x}_i \sim N(\boldsymbol{\mu}, \boldsymbol{\Sigma}) \tag{8.7}$$

That is, $\boldsymbol{x}_i$ has a multivariate Gaussian distribution with mean vector $\boldsymbol{\mu}$ and variance–covariance matrix $\boldsymbol{\Sigma}$. However, to consider local effects, the model may be extended so that both of these parameters vary over space. In this case, we put

$$\boldsymbol{x}_i|(u,v) \sim N(\boldsymbol{\mu}(u,v), \boldsymbol{\Sigma}(u,v)) \tag{8.8}$$

Note that $\boldsymbol{\mu}$ and $\boldsymbol{\Sigma}$ are now functions of u and v and on the left-hand side of equation (8.8) $\boldsymbol{x}_i$ is now denoted as being conditional on u and v. Regarding $\boldsymbol{\mu}$ and $\boldsymbol{\Sigma}$ as functions may be unfamiliar notation: however this simply implies that each element of $\boldsymbol{\mu}(u,v)$ and $\boldsymbol{\Sigma}(u,v)$ is a function of u and v. That is

$$\boldsymbol{\mu}(u,v) = \begin{pmatrix} \mu_1(u,v) \\ \mu_2(u,v) \\ \vdots \end{pmatrix} \tag{8.9}$$

and

$$\boldsymbol{\Sigma}(u,v) = \begin{pmatrix} \sigma_{11}^2(u,v) & \sigma_{12}(u,v) & \cdots \\ \sigma_{21}(u,v) & \sigma_{22}^2(u,v) & \cdots \\ \vdots & \vdots & \ddots \end{pmatrix} \tag{8.10}$$

Also, $\boldsymbol{\Sigma}$ must be a symmetric, positive definite matrix, so that $\sigma_{ij}(u,v) = \sigma_{ji}(u,v)$.

8.3.2 Calibrating Local Multivariate Models

Using the local likelihood approach set out in Chapter 4, calibrating $\boldsymbol{\mu}(u,v)$ and $\boldsymbol{\Sigma}(u,v)$ is achieved by computing locally weighted means around a centre (u,v) and then computing locally weighted variances and covariances around these means. This is much the same procedure as that used in Chapter 7 for computing locally weighted univariate and bivariate summary statistics. Indeed, it is possible to compute the locally weighted correlation matrix from $\boldsymbol{\Sigma}(u,v)$. If the locally weighted correlation matrix is denoted by $\boldsymbol{\rho}(u,v)$ then we have

$$\boldsymbol{\rho}(u,v) = \begin{pmatrix} 1 & \frac{\sigma_{12}(u,v)}{\sqrt{\sigma_{11}(u,v)\sigma_{22}(u,v)}} & \frac{\sigma_{13}(u,v)}{\sqrt{\sigma_{11}(u,v)\sigma_{33}(u,v)}} & \cdots \\ \frac{\sigma_{21}(u,v)}{\sqrt{\sigma_{11}(u,v)\sigma_{22}(u,v)}} & 1 & \frac{\sigma_{23}(u,v)}{\sqrt{\sigma_{22}(u,v)\sigma_{33}(u,v)}} & \cdots \\ \frac{\sigma_{31}(u,v)}{\sqrt{\sigma_{11}(u,v)\sigma_{33}(u,v)}} & \frac{\sigma_{32}(u,v)}{\sqrt{\sigma_{22}(u,v)\sigma_{33}(u,v)}} & 1 & \cdot \\ \vdots & \vdots & \vdots & \cdot \end{pmatrix} \tag{8.11}$$

Note that although these estimates for $\boldsymbol{\mu}(u,v), \boldsymbol{\Sigma}(u,v)$ and the related quantity $\boldsymbol{\rho}(u,v)$ are derived from the local likelihood calibrations for the Gaussian model, they may also be regarded as estimates of the first and second moments of any distribution. At this level we will not consider estimating the standard errors of these estimates.

8.3.3 Interpreting $\boldsymbol{\Sigma}$ and $\boldsymbol{\rho}$

The elements of $\boldsymbol{\mu}(u,v)$ may be interpreted as smoothed trends in the variables and the off-diagonal elements of $\boldsymbol{\Sigma}(u,v)$ and $\boldsymbol{\rho}(u,v)$ may be interpreted as second-order trends, showing the local variability of the association between all pairs of variables in the data set. Typically the elements of $\boldsymbol{\mu}(u,v)$ may be mapped as surfaces, giving a total of m maps (if there are m variables) which may be interpreted with relative ease. However, when considering $\boldsymbol{\Sigma}(u,v)$ or $\boldsymbol{\rho}(u,v)$ there is some danger of information overload if every element of the matrix is mapped as a surface. Even allowing for the fact that the matrix is symmetrical, there are $m(m-1)/2$ distinct $\sigma_{ij}(u,v)$-surfaces to be mapped. In a relatively simple situation where there are 6 variables, 15 maps would be needed to represent local second order interactions between the variables. It is no mean task to unpack the information in such a set of maps. It is for this reason that we move on to consider the use of principal components.

Typically, principal components analysis may be used to interpret large variance–covariance matrices; for a thorough discussion refer to Mardia *et al.* (1979). In brief, the first principal component, p_1, of a multivariate distribution is a linear combination of the variables: $p_1 = a_1x_1 + a_2x_2 + \ldots + a_mx_m$. The coefficients (or loadings) are chosen to maximise the variance of p_1 subject to the constraint $a_1{}^2 + a_2{}^2 + \ldots + a_m{}^2 = 1$. The constraint is needed since otherwise the variance would increase without bound as the values of $a_1, a_2, \ldots a_m$ increase. This may be written more concisely in vector form:

$$\begin{aligned} p_1 &= \boldsymbol{a}^{\mathrm{T}}\boldsymbol{x} \text{ where} \\ \boldsymbol{a} &= \operatorname{argmax}_a(\operatorname{Var}(\boldsymbol{a}^{\mathrm{T}}\boldsymbol{x})) \text{ subject to } \boldsymbol{a}^{\mathrm{T}}\boldsymbol{a} = 1 \end{aligned} \tag{8.12}$$

The second principal component, p_2, is defined in the same way as p_1, except a further constraint is added that $\operatorname{Cov}(p_1, p_2) = 0$. Similarly, the third component, p_3, is defined in the same way as p_2 except the extra constraints this time are that $\operatorname{Cov}(p_1, p_3) = 0$ and $\operatorname{Cov}(p_2, p_3) = 0$. This definition continues until the mth principal component, p_m, is obtained which is subject to the constraints $\operatorname{Cov}(p_1, p_m) = 0$, $\operatorname{Cov}(p_2, p_m) = 0$, and so on up to $\operatorname{Cov}(p_{m-1}, p_m) = 0$. Thus, the result of computing all principal components provides a linear transformation from $(x_1, x_2, \ldots, x_m)$ to $(p_1, p_2, \ldots, p_m)$ where $(p_1, p_2, \ldots, p_m)$ have the following properties:

- $(p_1, p_2, \ldots, p_m)$ are a set of uncorrelated variables;
- p_1 is a composite linear index of variables identifying the greatest degree of variation in the data;
- p_j is a composite linear index of variables identifying the greatest degree of variation in the data, with the further property that it is uncorrelated to $p_1, \ldots p_{j-1}$.

Thus, the principal components are a set of indices which help to identify the structure of the variability in a data set. The first principal component, p_1, represents the linear combination of variables highlighting the strongest degree of variability, and so on. Also important is the corresponding set of coefficients $a_1 \ldots a_m$. By noting which of the coefficients in each set have the largest absolute values, we can find which variables contribute most to each principal component. For example, by identifying the variables with the highest absolute weighting in the first principal component, we effectively find the variables that account for the greatest degree of variability in the data set. We may also plot the principal components on a case-by-case basis, identifying the high and low scorers for these key variables.

Computing the principal components is relatively simple, once the variance–covariance matrix is known. If the variance–covariance matrix is $\boldsymbol{\Sigma}$, then find its eigenvalues, $(\lambda_1, \lambda_2, \ldots, \lambda_m)$ say, and the corresponding eigenvectors $(z_1, z_2, \ldots, z_m)$. If the indices for the eigenvalues and eigenvectors are arranged in descending order of the λ's, and the eigenvectors are re-scaled so that $z_j'z_j = 1$, the re-scaled z_i then becomes the loading coefficient vector $\boldsymbol{a}$ for the jth principal component.

In a standard principal components application, only the global $\boldsymbol{\Sigma}$ would be considered. However, the technique may prove useful when attempting to consider the *local* variance–covariance matrices of equation (8.8). Here, we can compute the principal components based around a point (u,v) by substituting $\boldsymbol{\Sigma}(u,v)$ in the above computational method. Thus, as $\boldsymbol{\Sigma}(u,v)$ varies over space, it is possible to examine how the component loadings vary. Looking at how these loadings alter gives an idea of which particular variables show the greatest degree of variability in different localities. In particular, it is possible to see whether the 'most influential variable' in the first principal component varies over space.

8.3.4 An Example

Here, we consider the southern USA data on levels of educational attainment and five explanatory variables introduced in Chapter 4. These data consist of 1990 US census data tabulated at county level – a total of 1425 counties in all. Six variables were derived from the census and the definitions of these are reproduced from Chapter 4 in Table 8.2.

Table 8.2 *Variables derived from 1990 US Census used in this example.*

Variable name	Description
PCTRURAL	Percentage of the county population classed as 'rural'
PCTBACH	Percentage of employable adults with a Bachelor's degree
PCTELD	Percentage of county population classed as elderly
PCTFB	Percentage of county population foreign born
POVPCT	Percentage of county below the poverty level
PCTBLACK	Percentage of county who are black

The first local principal component at point (u,v) was computed for a series of local grid points covering the southern United States. Here, we initially consider the variable having the highest loading across the entire grid, as shown by Figure 8.4. Also identified is the relative difference between the largest and the second largest loading. When the largest loading exceeds the second largest by 5% or more, the grid square behind the point in question is shaded in grey.

A number of patterns may be identified. In much of Florida, the variable with the highest weighting is POVPCT, suggesting that the key variation in characteristics of counties is one linked to poverty. However, around Maryland, PCTBACH appears to dominate. Towards the west, it is proportions of elderly people which seem to vary most between counties. However, the observation relating to Florida should be tempered by the observation that the highest variable loading does not exceed the second highest by more than 5%. Bearing this in mind, together with the observation that principal components are generally a mixture of characteristics, the variables with the second largest loading are shown in Figure 8.5. From this map it becomes clear that the second most influential variable in much of Florida is PCTRURAL. Indeed this variable, although hardly appearing at all in the map of the most influential loading, appears over a wide geographical expanse as the second highest.

This analysis could be continued for lower ranks of loading and also for principal components other than the largest. However, although quite a large amount of information has not yet been considered, the use of principal components has enabled the analysis to focus on geographical changes in the key features of variation. Thus, an impression of the general trends in multi-way variation in the data set has been identified using only a small number of maps.

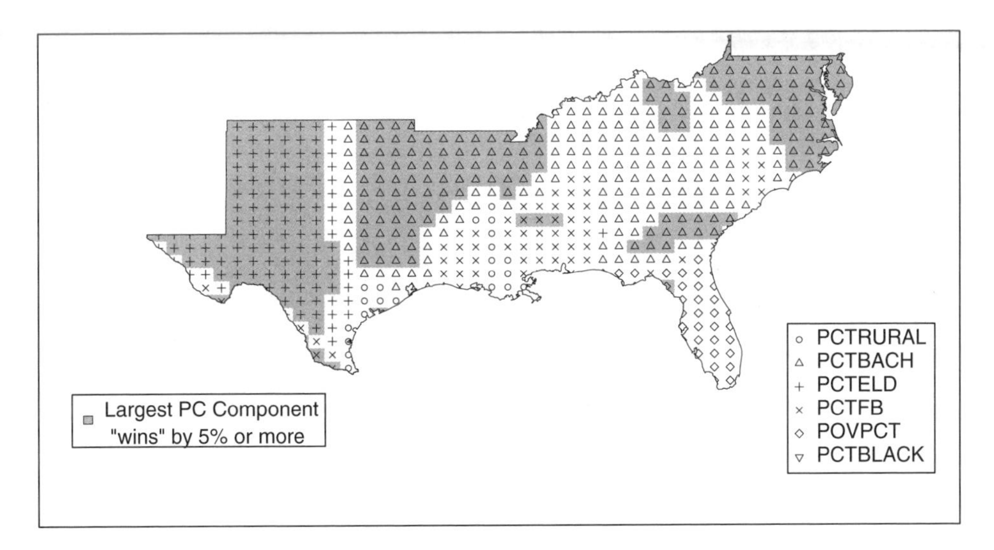

Figure 8.4 *Variables having the largest loading in a local principal components analysis*

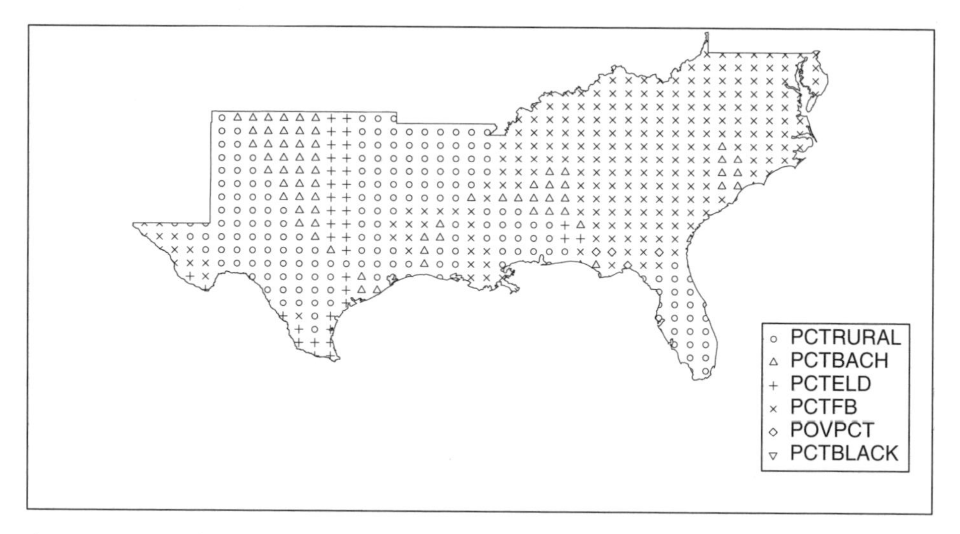

Figure 8.5 Variables having the second highest loading in a local principal components analysis

8.4 Geographically Weighted Density Estimation

8.4.1 Kernel Density Estimation

We now consider an extension of the geographical weighting approach that is related to the idea of geographically weighted summary statistics introduced in Chapter 7. Here, rather than using summary statistics as indicators of local distributional space, we attempt to reconstruct the distribution itself. This is done by extending the definition of kernel density estimation (Silverman 1986; Brunsdon 1995). In its usual form, kernel density estimation is an aspatial method of estimating the probability density function of some variable x from a data set of observed values $\{x_1, x_2, \ldots, x_n\}$. If the true distribution is $f(\,)$ then the kernel density estimate of $f(\,)$, $\hat{f}(\,)$ say, is given by

$$\hat{f}(x) = \frac{1}{nh}\sum_{j=1}^{n} k\left(\frac{x - x_j}{h}\right) \tag{8.13}$$

where $k(\,)$ is some probability density function (often a Gaussian, but others could be used – typically symmetrical and centred on zero), and h is a constant termed the bandwidth. The characteristics of the estimate depend on the values of h and $k(\,)$. When h is too small, $\hat{f}(\,)$ becomes very 'spiky' and when h is too large, $\hat{f}(\,)$ is oversmooth; in either case the genuine features of $f(\,)$ can become obscured. Typically, the choice of $k(\,)$ is less critical, provided the constraints stated above are met. For brevity we do not include a more advanced discussion of kernel density estimates, but instead recommend the reader to Chapters 4 and 5 of Fotheringham *et al.* (2000), or Chapter 1 of Bowman and Azzalini (1997), where among other issues, methods for choosing h are discussed.

8.4.2 Geographically Weighted Kernels

To this point, the discussion of kernel density estimation has been entirely aspatial and the kernel density estimates simply produce what is essentially a continuous version of a histogram that describes the frequency distribution of a particular attribute. To see how the idea might be extended into a geographically weighted framework, note from equation (8.13) that $\hat{f}$ () is the arithmetic mean of the terms $k/h[(x - x_j)/h]$ and that the simple arithmetic mean could be replaced by a geographically weighted mean. In particular, if each observed value of x, x_j, is associated with a location (u_j,v_j), then for a summary location i we could create a set of weights $w_{ij} = g(d_{ij})$ where $d_{ij}^2 = (u_i - u_j)^2 + (v_i - v_j)^2$. Clearly, d_{ij} is the distance between the summary point i and the location at which x_j is measured. The function $g()$ takes values between 0 and 1 and decreases as d_{ij} increases. One example that has been used in previous chapters is $g() = \exp(-d_{ij}^2/b^2)$ where b is a geographical bandwidth (distinct from the kernel density bandwidth).

This whole operation may be viewed in terms of conditional distributions. For each observation j we have the trio of observations (x_j, u_j, v_j) and we may consider the trivariate joint distribution $f(x,u,v)$. Indeed, using a three-dimensional version of (8.13) we may obtain an estimate of this joint distribution. However, here the distribution of interest is not the joint trivariate distribution but the conditional distribution of x *given* (u,v) and we regard the local kernel outlined above as an estimate of this conditional distribution. Thus, replacing (u,v) by i for convenience, we may write

$$\hat{f}(x|i) = \left(h\sum_{j=1}^{n} w_{ij}\right)^{-1} \sum_{j=1}^{n} w_{ij} k\left(\frac{x - x_j}{h}\right) \tag{8.14}$$

This will create a geographically weighted kernel density estimate. In essence, data points are weighted both according to their proximity in attribute space (the regular aspatial kernel) and according to their proximity in geographical space (the spatial kernel). The estimated geographically weighted kernel densities obtained from (8.14) could be used in a number of ways. One simple approach is to take two distinct locations and to compare the density estimates at these different locations graphically. A more sophisticated approach might be created by the use of 'brushing' (see, for example, Brunsdon 1998). Here, as the mouse pointer is moved across a map of the study area in a window on a PC screen, the local kernel density estimate is updated in another window. This would require a fairly powerful computer because the local kernel density estimate would need to be updated in real time. However, some degree of pre-processing and some computational short-cuts would allow a technique of this sort to be implemented.

8.4.3 An Example Using House Prices

The methods described above were applied to the 347-item house price data set used in Chapter 7. Specifically, local models of house price distribution in two locations, one centred on London in the south-east and the other on Newcastle in the north-east, were obtained. For these two areas, local kernels were computed, as shown in Figure 8.6. Here, the kernel bandwidth was selected using the method suggested by Bowman and Azzalini (1997). The chosen geographical bandwidth was deliberately wide, around 100 km, to allow a comparison of price distributions at the regional level. These highlight the difference in distribution of house prices in the two areas and provide evidence of a north/south split in house prices. Although both distributions are clearly skewed, the one for Newcastle peaks at a lower price than that for London.

It is also possible to plot the quantity $\log(\hat{f}_1(x)/\hat{f}_2(x))$ where $f_1()$ and $f_2()$ refer to the local kernels for London and Newcastle respectively. This is a useful plot, as it provides a single curve identifying the price ranges for which either London or Newcastle has a greater concentration. There are a number of reasons for selecting the use of the logged density ratio as a comparison tool here. First, a ratio is useful as it takes into account the relative frequencies of houses in different price ranges. However, working with an untransformed ratio has the problem that all of the regions of the curve where the density is higher in Newcastle would be in the range 0–1, whereas when the density is higher for London, the range would be $1 - \infty$. This asymmetry can be addressed by taking logs so that in this case positive values of the index indicate values where the density of prices is greater in London and negative values indicate that the density of prices is greater in Newcastle. Here logs are taken to the base e although there is no reason not to use any other base if preferred. Also, if logs are taken to the base e, then the log density ratio is essentially the quantity that is averaged in order to find the Kullback–Leibler information distance between the two distributions.

The log density ratio plot for this example is shown by the darker line in Figure 8.7. Over most prices, the density of housing centred on London is greater than for Newcastle. Exceptions occur at the very low end of the price range and also around a value of £150 000.

A final use of the log density ratio is as a semi-formal test for the differences between the two curves. The results suggest that the distribution of house prices around Newcastle is bimodal with a greater density of houses than London at both

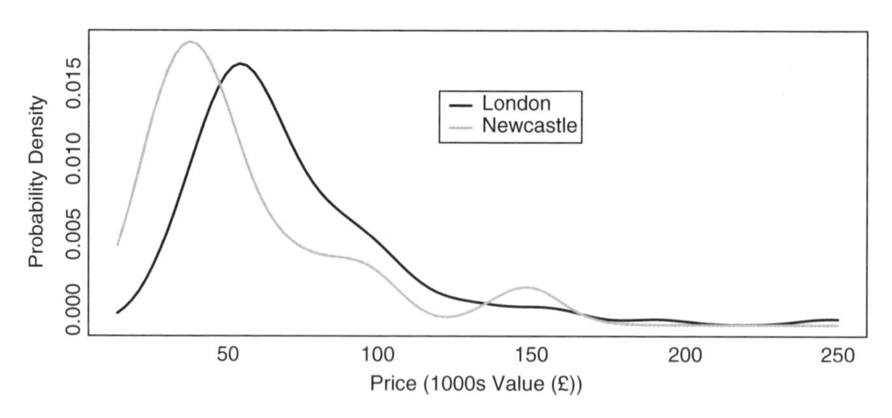

Figure 8.6 *Local probability density curves for London and Newcastle*

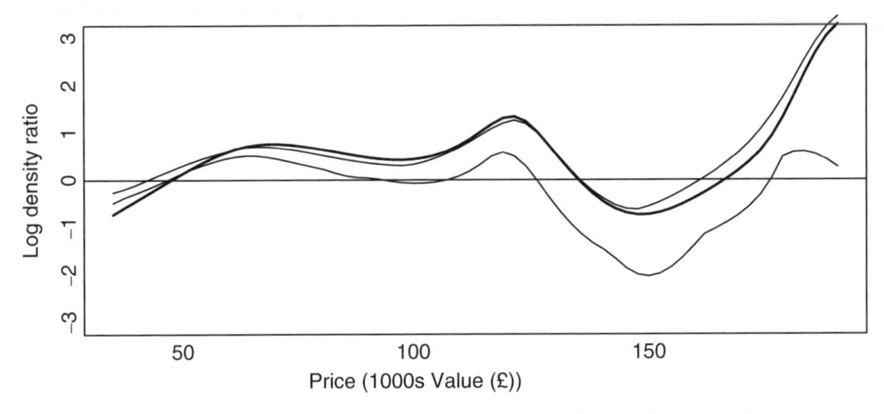

Figure 8.7 *Log density ratio of house prices between London and Newcastle*

the lower and upper ends of the price range. Those familiar with the UK housing market may find this surprising and it is likely that this finding is a spurious result, due mainly to the relatively small sample of house prices used in the analysis, particularly around Newcastle. One way of testing this is to consider the null distribution of the log density ratio in this (and other) house price ranges under the hypothesis that there is no difference in the house price distributions for the two areas. This would be impossible analytically, as we are working on the assumption that the two distributions are unknown arbitrary functions. However, we may use a permutation test to solve the problem.

By randomly permuting the house prices amongst the locations a large number of times (in this case 1000), and finding the 5th and 95th percentiles, we identify upper and lower limits between which we would expect the log density ratios to remain under the null hypothesis of no difference in the distributions. If we calculate these limits for a range of prices, we obtain upper and lower envelope curves between which one would usually expect the actual data curve to lie. Noting whether this curve strays outside the limits gives some indication as to whether an effect is genuine, or just an artefact of the data sample. It should be noted, however, that this is not a simple significance test – a better idea is to interpret the envelope in a similar way to that used in *k*-function comparison (Diggle 1984). The lighter curves in Figure 8.8 depict this envelope, and it can be seen that although the actual curve strays outside the envelope below around £50 000, and also between £60 000 and £100 000, it is confined within it in the £140 000 to £160 000 range – where the 'Newcastle bump' is located. This suggests that although the other observed differences between the two distributions seem likely to be real, the upper-tail bump in the Newcastle distribution appears to be an artefact of the small sample size.

8.5 Summary

In this chapter, three distinct geographically weighted methods have been described. Although they appear to have little in common, they do share one common factor: in each case, some kind of distributional model exists. For the generalised linear model approaches, we model the distribution of a dependent variable, y, given at a set of explanatory variables and a location in geographical space. For the local principal components, we model a multivariate Gaussian distribution, given a location in geographical space. Finally, for the local probability density estimates, we model the density function of a single variable, given a location in geographical space. The common phrase in each of the above one-line descriptions is 'given a location in geographical space'. All of the above rely on some kind of estimate of a conditional probability distribution, where the conditional variables include spatial location. From a statistical viewpoint, this is a helpful way of looking at local modelling. In its most general form we can think of some spatial process as a joint probability distribution of some attribute vectors $\boldsymbol{x}$ and $\boldsymbol{y}$ and a location vector $\boldsymbol{u}$ – say $f(\boldsymbol{x},\boldsymbol{y},\boldsymbol{u})$. Then, to model $\boldsymbol{y}$ we are concerned with deriving the conditional distribution $f(\boldsymbol{y}|\boldsymbol{x},\boldsymbol{u})$. In some cases, $\boldsymbol{x}$ may be the empty set (as in the principal components and density estimation examples above), but in all cases, $\boldsymbol{u}$ becomes a conditioning variable. This demonstrates one important characteristic of these techniques, also considered in Chapter 4: the formal modelling approach used in geographically weighted methods is concerned with a single model in which the likelihood of events depends on location, rather than a series of local models.

9

Software for Geographically Weighted Regression

9.1 Introduction

In this chapter we consider some computational aspects of geographically weighted regression. In Sections 9.2 to 9.8 we outline some general issues concerned with the implementation of GWR on a computer. Section 9.9 contains what is essentially a users' manual for GWR 2.0, specialised software for GWR, which is available from the authors.[1,2] This section also contains a worked example of GWR 2.0. Finally, in Section 9.10 we consider some of the ways of visualising the output from GWR.

Our aim in developing GWR 2.0 was to produce a user-friendly program that would allow users to specify and run a GWR model on their own data and have the results presented in a suitably flexible manner. The software produces the same results whether it is run under Windows on a PC or under UNIX on a workstation. The mode of running, however, varies slightly by platform: Windows users have the benefit of a GWR editor written in Visual Basic to create the model specification and run the program, whereas UNIX users have to write a control file. Both versions of GWR 2.0 will produce output files that can be imported into either a spreadsheet or a mapping package such as ArcInfo, ArcView or MapInfo.

[1] The software is made available to users on condition that it is not used for commercial purposes, and also that suitable acknowledgement will be made. A small charge is made to cover the costs of packing and postage. The software is shipped on CD-ROM. Details are available at *http://www.ncl.ac.uk/geography/GWR* or by email to Stewart.Fotheringham@ncl.ac.uk

[2] Users who already have an earlier version of the GWR code, GWR 1.0, should note that significant changes have been made to the software and they are strongly advised to upgrade to the new version.

9.2 Some Terminology

The data used to calibrate a GWR model may represent observations relating to individual points or to areas. If they relate to areas, some representative location of that area should be selected: commonly, this is a centroid, either the geometric centroid of the coordinates of the area boundary, or one weighted by another variable such as population. Regardless of whether the data refer to points or areas, we refer to these observations as being the *data points*.

The goal of GWR is to provide predictions and parameter estimates at a set of locations. These locations we refer to as *regression points*. Often the data points and regression points will be the same locations. If this is the case, then a range of diagnostic tools are available to help you evaluate your model. However, one interesting feature of GWR is that it is possible to obtain a predicted y and local parameter estimates at regression points which are not data points. When this is the case, some diagnostic tools are unavailable because y_i is then unknown. Within GWR 2.0 the user specifies whether or not the regression points are the data points and if they are not, the user can input a set of coordinates for the regression points.

9.3 The Data File

This section describes the layout of the file containing the data on which to run GWR 2.0. The data can be prepared in a text editor, a word-processor, a spreadsheet, or another program. The organisation of the file is described below. As an example, the data file shown in Figure 9.1 contains the first few lines of a data set taken from the 1990 US census. The data refer to counties in Georgia and the following variables are included in the data set:

ID	An identification number for each county which will be used later in ArcView to link the data back to the Georgia shapefile for mapping the results.
Latitude	The latitude of the county centroid
Longitud	The longitude of the county centroid
TotPop90	Population of the county in 1990
PctRural	Percentage of the county population defined as rural
PctBach	Percentage of the county population with a bachelor's degree
PctEld	Percentage of the county population aged 65 or over
PctFB	Percentage of the county population born outside the US
PctPov	Percentage of the county population living below the poverty line
PctBlack	Percentage of the county population who are black

The first line of the data file consists of the set of variable names; these *must* be separated by commas and should be no more than eight characters in length. Two of the variables must describe the location of each data point in some Cartesian coordinate system. In the data set described in Figure 9.1 these two variables are 'Latitude'

```
ID, Latitude, Longitud, TotPop90, PctRural, PctBach, PctEld, PctFB, PctPov, PctBlack
13001, 31.753389, -82.285580, 15744,75.6,8.2,11.43,0.635,19.9,20.76
13003, 31.294857, -82.874736, 6213,100.0,6.4,11.77,1.577,26.0,26.86
13005, 31.556775, -82.451152, 9566,61.7,6.6,11.11,0.272,24.1,15.42
13007, 31.330837, -84.454013, 3615,100.0,9.4,13.17,0.111,24.8,51.67
13009, 33.071932, -83.250851, 39530,42.7,13.3,8.64,1.432,17.5,42.39
13011, 34.352696, -83.500539, 10308,100.0,6.4,11.37,0.340,15.1,3.49
13013, 33.993471, -83.711811, 29721,64.6,9.2,10.63,0.922,14.7,11.44
13015, 34.238402, -84.839182, 55911,75.2,9.0,9.66,0.816,10.7,9.21
13017, 31.759395, -83.219755, 16245,47.0,7.6,12.81,0.332,22.0,31.33
13019, 31.274242, -83.231790, 14153,66.2,7.5,11.98,1.194,19.3,11.62
...   ...   ...   ...
```

Figure 9.1 *Example of a GWR 2.0 data file: census data for Georgia counties*

and 'Longitud'. Each succeeding line in the data file contains the data for each observation and there should be the same number of values on each line as there are variables defined in the first line. The data values may be separated either by commas or spaces. No alphabetic data should appear in the data lines; if such data are present, the program is likely to come to a premature end. There should not be any blank lines in the file, and the last line of data must be on the last line of the file.

An important component of GWR modelling is the computation of distances between pairs of points. In GWR 2.0 all the distance measurements are made assuming that the coordinates being used are in a Cartesian system with 1 unit of distance in the x-direction being equivalent in length to 1 unit of distance in the y-direction. This means that there will be no directional bias in the weighting schemes used in the models. If using latitude and longitude, it should be noted that although 1 degree along a meridian always represents the same distance north–south along the earth's surface, 1 degree along a parallel represents a different distance depending on the latitude of the observation. In high latitudes the difference can be quite marked and the weighting functions will then tend towards an elliptical shape. Naturally it is the user's responsibility to obtain an appropriate distance metric.

Note that it is possible to have more variables in the data file than will actually be used for any one GWR model. The GWR software has facilities to permit the selection of a subset of variables which allows users to experiment with more than one model formulation with the same set of data.

How does the user create the data file? Much depends on circumstance. The data may be stored on a computer already, although perhaps not in the form needed here. On a PC it is convenient to use a spreadsheet program such as Excel and save the worksheet as a comma-separated-variable (*.csv*) file. UNIX users may have to do some manipulation with *awk* and *vi* to get the desired result.

9.4 What Do I Need to Specify?

Consider the following generic GWR model:

$$y_i = \beta_0(u_i, v_i) + \sum\nolimits_k \beta_k(u_i, v_i)x_{ik} + \varepsilon_i \tag{9.1}$$

where the dependent variable y is regressed on a set of independent variables, each denoted by x_k, and the parameters are allowed to vary over space. To calibrate this model, we need to specify:

- the dependent variable;
- the independent variable(s);
- the variables describing location.

As in a standard weighted least squares procedure, in certain circumstances that might produce heterogeneous error variances, we may also want to include a variable that allows each observation to have a different, non-geographical, weight in the model (so that the error variance is reasonably constant).

Finally, we will need to specify the geographical weighting scheme – this requires the specification of a kernel shape and a bandwidth. Two weighting schemes are implemented in GWR 2.0: one uses a constant bandwidth around every regression point and the other allows the bandwidth to vary spatially. The user specifies which weighting scheme is to be used in the calibration of the model as described below. Further discussion of kernels and weighting can be found in Chapter 2.

9.5 Kernels

The parameter estimates at any regression point are dependent not only on the data supplied by the user but also on the kernel chosen and the bandwidth for that kernel. Two types of kernel can be selected in GWR 2.0. The first assumes that the bandwidth at each regression point is a constant across the study area. We refer to this as a fixed kernel. In this case we use a Gaussian function where the weight of the jth data point at the ith regression point is calculated by:

$$w_{ij} = \exp[-(d_{ij}/b)^2] \tag{9.2}$$

where d_{ij} is the Euclidean distance between the regression point i and the data point j, and b is the bandwidth. At the regression point the weight of a data point is unity and the weights decrease as distance from the regression point increases. However, the weights are non-zero for all data points, no matter how far they are from the regression point.

The second available kernel permits use of a variable bandwidth. Where the regression points are widely spaced, the bandwidth is greater than where the regression points are more closely spaced. The particular implementation of a variable kernel in GWR 2.0 uses a bi-square function where the weight of the jth data point at regression point i is given by:

$$\begin{aligned} w_{ij} &= [1 - (d_{ij}/b)^2]^2 && \text{when } d_{ij} \leq b \\ w_{ij} &= 0 && \text{when } d_{ij} > b \end{aligned} \tag{9.3}$$

At the regression point i, the weight of the data point is unity and falls to zero when the distance between i and j equals the bandwidth. When the distance is greater than the bandwidth, the weight of the data point is zero. The bandwidth is selected so that there are the same number of data points with non-zero weights at each regression point.

9.6 Choosing a Bandwidth

The choice of bandwidth has a large impact on the results obtained from GWR. It is possible to think of the bandwidth as a *smoothing* parameter, with larger bandwidths causing greater smoothing. An oversmoothed model will produce parameters that are similar in value across the study area and an undersmoothed model will produce parameters with so much local variation that it is difficult to determine whether there are any patterns at all. The 'best' bandwidth is that which provides a happy medium between these two extremes. GWR 2.0 allows the user to choose one of three methods of bandwidth selection:

1. providing a user-supplied bandwidth;
2. selecting the bandwidth that minimises a cross-validation function;
3. selecting the bandwidth that minimises the Akaike Information Criterion (AIC)

We now consider these three options in turn.

9.6.1 User-Supplied Bandwidth

Perhaps the easiest way of using GWR 2.0 is to specify the size of bandwidth directly. For the fixed kernel option this will be a distance in the same units as the location variables. For example, an application in Great Britain might have coordinates specified in metres relative to the origin of the National Grid of the Ordnance Survey; the bandwidth would then be specified in metres. For the variable-width kernel option, the bandwidth is expressed as the number of data points to be included in the kernel.

There may be good reasons for supplying a bandwidth directly. The analyst might have already obtained a suitable value for the bandwidth from a previous run of GWR 2.0 with a particular data set or model. Alternatively, the analyst may have a good theoretical reason for supplying a particular bandwidth. Finally, in cases where the data and regression points are not identical, this is the only way of supplying a bandwidth to the model.[3]

[3] Note that this is not as restrictive as it might seem. There might be a situation for which estimates at non-data points are desired but estimates at data points can be obtained. For example, the data might have been collected at irregularly spaced locations and the user requires estimates at the mesh points of a regular grid for mapping. One approach here is to find the 'best' bandwidth using the data points as the regression points and using the methods outlined in either 9.6.2 or 9.6.3. This value can then be used as an exogenously supplied bandwidth as described in 9.6.1 for GWR at the set of user-supplied regression points.

9.6.2 Estimation by Cross-validation

In cases where the bandwidth is unknown or there is no prior justification for supplying a particular bandwidth, the analyst may let the software choose an appropriate bandwidth. Two methods are available in the software to accomplish this. Cross-validation is a technique in which the optimal bandwidth is that which minimises the following score:

$$\mathrm{CV} = \sum_{i=1}^{n} (y_i - \hat{y}_{i \neq i})^2 \tag{9.4}$$

where n is the number of data points and the prediction for the ith data point is obtained with the weight for that observation set to zero (so that it is omitted from the computation). Clearly, the calculation of the CV statistic is only possible when the regression point locations are the same as the data point locations. The minimisation is carried out using a Golden Section search technique (see below) and the 'best' bandwidth is reported on the output listing. The user has the option of obtaining a listing of the search process.

9.6.3 Estimation by Minimising the AIC

Adjusting the bandwidth changes the number of degrees of freedom in the model. If the analyst chooses to use cross-validation as the bandwidth selection criterion, then the score for any given bandwidth refers to a slightly different model. An alternative strategy is to find the model which minimises the Akaike Information Criterion (AIC) described in Chapters 2 and 4. The AIC takes into account the different number of degrees of freedom in different models so that their relative performances can be compared more accurately. A model with a lower AIC than another is held to be a 'better' model. The AIC used in GWR is computed as:

$$\mathrm{AIC}_c = 2n \log_e(\hat{\sigma}) + n \log_e(2\pi) + n\left\{\frac{n + \mathrm{tr}(\boldsymbol{S})}{n - 2 - \mathrm{tr}(\boldsymbol{S})}\right\} \tag{9.5}$$

where $\mathrm{tr}(\boldsymbol{S})$ is the trace of the hat matrix (see Chapters 2 and 4) and n is the number of observations. Again the function minimisation procedure uses the Golden Section method and the intermediate stages of this process may be listed in the GWR 2.0 output.

9.6.4 The Golden Section Search

The process of finding a bandwidth which minimises the cross-validation score or AIC is known to numerical analysts as function minimisation. Typically the process involves evaluating the function with three different values, say, a, b, and c, where $a < b < c$. The values of the function at these points are $f(a)$, $f(b)$, and $f(c)$ and, collectively, these are known as a 'triplet'. The function is then evaluated again at a new value d which can be chosen so that it is either between a and b or between b and c to give a new value $f(d)$. We must discard either a or c to form a new triplet. The following rules are employed:

$$\text{if } f(b) < f(d)\text{: the new triplet is } a < b < d$$

$$\text{if } f(b) > f(d)\text{: the new triplet is } b < d < c$$

The process continues until two successive values of $f(d)$ are almost the same. The value of b is chosen such that its value is $(1-0.618)$ times the distance between a and c. A comprehensive treatment of function minimisation using the Golden Section is given in Press *et al.* (1989: 309–18). In GWR 2.0, the a, b, c and d values correspond to bandwidths and $f(a)$, $f(b)$, $f(c)$ and $f(d)$ correspond to the cross-validation scores or AICs for these bandwidths. The software allows the analyst to list these values so that they may be graphed in some suitable package such as Excel. The main problem with function minimisation is being sure that the method chosen has found the global minimum and has not fallen into a local minimum. Experience with GWR suggests that the cross-validation and AIC functions are reasonably well behaved in this regard.

9.7 Significance Tests

In calibrating a GWR model, it is useful to consider whether the local model offers an improvement over the global model. A simple procedure is to determine whether any of the local parameter estimates are significantly non-stationary. Two tests for this are available in GWR 2.0: a Monte Carlo significance test and a test attributed to Leung *et al.* (2000a). Both are rather computationally intensive. Details on both tests can be found in Chapter 4; here we provide simple descriptions.

As the theoretical distributions underlying the local parameter estimates are unknown, one approach to examining the validity of any inferences drawn from the local results is to use a Monte Carlo significance test. In this test, the observed value of a test statistic is compared with $n-1$ simulated ones. The results are sorted, and the rank of the observed test statistic is determined. The p-value for the test is obtained by subtracting the ratio rank/n from unity. More details can be found in Hope (1968). In the current version of the GWR software, the number of local model calibrations is set to 100. For each variable, the observed variance of the local parameter estimates is computed and stored. Following this, 99 sets of variances are obtained for each variable based on different randomisations of the observed data. The p-value is then computed for the local parameters associated with each variable, as described above.

The Monte Carlo significance testing procedure employs a pseudo-random number generator to reallocate the observations across the spatial units. Many operating systems offer a random number generator which can be called from an application program. However, the statistical properties of many of these are open to doubt; the UNIX online documentation, for example, suggests that the only advantage of its generator `rand` is that is it widely available. The generation of random numbers in GWR 2.0 is based on the Wichmann–Hill method (Wichmann and Hill 1982; 1984). This uses the fractional part of the sum of three multiplicative congruential generators as the source of its random number stream and will produce rectangularly distributed pseudo-random numbers between 0 and 1. The generator will produce 6.95×10^{12} pseudo-random numbers before repeating. The generator requires separate seeds for each of its three constituent generators. Wichmann and Hill (1982) state that the seeds may be in the range 1–30 000 'preferably chosen at random'. In GWR 2.0 the first two are set at 13 579 and 24 680 while the third is computed from the current time in seconds since midnight modulo 30 000.[4]

The other significance test available in GWR 2.0 is that based on Leung *et al.* (2000a) with some slight modifications. The following steps are used:

1. Compute $\hat{\theta} = \text{RSS}_{\text{GWR}}/(n - \text{tr}(\boldsymbol{S}))$ where RSS_{GWR} is the GWR residual sum of squares, n is the number of sample points and $\text{tr}(\boldsymbol{S})$ is the trace of the hat matrix.
2. For each variable:
   (a) Compute the variance of the GWR coefficients for the variable, V_{GWR}.
   (b) Form the matrix $\boldsymbol{B}$ such that $B\mathbf{y} = \mathbf{c}$, where $\mathbf{y}$ is the vector of observed y values and $\mathbf{c}$ is the vector of coefficients for the variable in question.
3. Compute $f = (V_{\text{GWR}}/\text{tr}(\boldsymbol{B}))/\hat{\theta}$ which is tested as an f statistic with $\text{tr}(\boldsymbol{B})$ and $n - \text{tr}(\boldsymbol{S})$ degrees of freedom.

9.8 Casewise Diagnostics for GWR

Seven casewise statistics can be reported as an option in GWR 2.0. The first three of these are self-explanatory and are simply the observed value, the predicted value from the GWR, and the residual. The remaining four are now described.

9.8.1 Standardised Residuals

One product of local or global regression analysis is a vector of predicted values, $\hat{\mathbf{y}}$. From this we can create a vector of residuals, $\mathbf{e}$, and the unbiased estimate of the residual variance is given by

$$\sigma^2 = \hat{e}^{\mathrm{T}}\hat{e}/(n - \text{tr}(\boldsymbol{S})) \tag{9.6}$$

[4] It should be noted that no current machine-driven random number generator produces purely random numbers *ad infinitum* but the Wichmann–Hill generator is better than most. McLeod (1985), for example, draws attention to a problem with the Wichmann–Hill generator on a PRIME 400 computer with single-precision arithmetic. Our implementation uses double precision (64 bit) arithmetic so this problem is unlikely to arise.

where n is the number of observations and tr($\boldsymbol{S}$) is the trace of the hat matrix. In a global OLS regression the denominator is n-p where p is the number of parameters in the model. In GWR the effective number of parameters is given by the trace of the hat matrix (i.e. the sum of its leading diagonal).

It would be helpful to have some notion of whether a residual is 'large' in some sense compared with the others. The problem with the residuals is that each has a different standard error, so we cannot easily compare them. In this case we can use standardised residuals, sometimes known as *internally* Studentised residuals:

$$r_i = \hat{e}_i/(\hat{\sigma}\sqrt{1 - s_{ii}}) \tag{9.7}$$

where s_{ii} is the ith element of the leading diagonal of the hat matrix (see Chapters 2 and 4). Some statistical programs will compute Studentised residuals, which are sometimes known as *externally* Studentised residuals. In this case the residual variance is computed repeatedly without the ith observation before the residual is Studentised. The GWR software currently provides a calculation of the standardised residuals. Residuals that are larger than -3 or $+3$ are considered to be unusual and the analyst is advised to examine such observations.

9.8.2 Local r-square

In global regression modelling, a leading diagnostic statistic is that of r-square, or coefficient of determination, which measures the proportion of variance in the observed data explained by the model. Local variations of the r-square statistic can be computed to give a 'feel' for how well a local model can replicate the data recorded in the vicinity of the regression point. The formula for the local r-square statistic, r_i^2, is

$$r_i^2 = (\mathrm{TSS}^w - \mathrm{RSS}^w)/\mathrm{TSS}^w \tag{9.8}$$

where TSS^w is the geographically weighted total sum of squares, defined as

$$\mathrm{TSS}^w = \textstyle\sum_j w_{ij}(y_j - \bar{y})^2 \tag{9.9}$$

and RSS^w is the geographically weighted residual sum of squares, defined as

$$\mathrm{RSS}^w = \textstyle\sum_j w_{ij}(y_j - \hat{y}_j)^2 \tag{9.10}$$

where w_{ij} describes the weight of data point j at regression point i. For further examples of this type of geographically weighted descriptive statistic, see Chapter 7.

Although the local r-square statistic is reported in the GWR output, it needs to be interpreted carefully. Essentially, the statistic measures how well the model calibrated at regression point i can replicate the data in the vicinity of point i. Bearing in mind that we have a model that is potentially non-stationary, we might not expect the model calibrated in one location to replicate the data at other locations particularly well unless the processes being modelled are relatively stable. Consequently, local r-square statistics reflect a mixture of two issues: how well the model replicates the data and how stationary are the processes being modelled. While the former probably dominates the determination of the local r-square value, we cannot be sure of the influence of the latter. Hence, the local r-square cannot be interpreted with quite the same confidence as the global measure.

9.8.3 Influence Statistics

The final two statistics printed as casewise diagnostics are both measures of the influence an observation has on the model calibration. Chatfield (1995: 286) recommends that it is 'always wise to find out why and how an observation is influential'. The first of these is the leading diagonal of the hat matrix, s_{ii} (see Chapter 4). It is reported in GWR 2.0 as 'Influence' but it is also known as a leverage value. For any observation, if the value of this statistic is large, the influence of that observation on the model is likely to have been large. As with all these casewise statistics, it is possible to map the distribution of them to examine possible spatial patterns in the statistics which might lead to further understanding of the processes being modelled.

An associated measure is Cook's Distance, D, which also indicates the influence of an observation. It is calculated as

$$D_i = r_i^2 s_{ii}/p(1 - s_{ii}) \tag{9.11}$$

where r_i is the ith internally Studentised residual and p is the number of parameters. Influential observations will have values that exceed 1; they may be unusual in terms of the dependent variable (high residual, low leverage) or in terms of the independent variables (low residual, high leverage).

9.9 A Worked Example

In this section, we present a worked example of the GWR 2.0 software using the Georgia data set described in Section 9.3. In effect, this section acts as a users' manual for the software. Because the output from GWR is largely location-specific, in a later section we describe various ways in which the output can be visualised.

9.9.1 Running GWR 2.0 on a PC

We assume that the reader has installed GWR 2.0 and wishes to undertake the calibration of a GWR model.[5] This section shows how to set up and run a GWR model using the Visual Basic GWR Editor.

[5] Installation notes are provided with the software.

Name	Size	Type	Modified
data		File Folder	12/06/01 10:43
listing		File Folder	12/06/01 10:43
model		File Folder	12/06/01 10:43
results		File Folder	12/06/01 10:43
gwr.exe	121KB	Application	08/01/01 18:38
Gwr2.dep	4KB	DEP File	10/03/01 23:45
GWR2.exe	356KB	Application	10/03/01 23:35
run_GWR.bat	1KB	MS-DOS Batch ...	20/07/01 14:50
St5unst.log	5KB	Log File	12/06/01 10:43

Figure 9.2 *Organisation of GWR folders and application files*

The organisation of the GWR folders and application files within the GWR software folder is shown in Figure 9.2. The name of the GWR Model Editor is **GWR*n*.exe** where n is the current version of the program. A link can be created to the file **GWR*n*.exe** from the desktop or in a toolbar which will create a GWR icon. The program assumes that the folders contain the following material for each problem

data:	the data files
model:	the model control files
listing:	the listings for printing
results:	the result files for loading into other software.

These folders are created by the set-up file that comes with the GWR software. Once the GWR software is available on the PC, a specific data file with a **.dat** extension should be placed in the data directory. Then the appropriate icon is selected to run the program.

The main GWR program window is shown in Figure 9.3; it has three items in the menu bar, 'File', 'GWRModel' and 'Help'. The program assumes that the user will wish to proceed with one of three options, and provides a 'Wizard' for guidance through the process.

The first step is to create a new model to use with the Georgia data. If there is an existing model control file, then this can be run or the model editor can be invoked to change the variables or some other control parameters. At this point the user has the option of clicking on 'Go' to proceed with the new model, 'Cancel' to close the Wizard, or 'Help' to obtain some assistance on what to do next. In this example, we have checked 'Create a new model' and clicked on 'Go'. The prompt shown in Figure 9.4 will appear.

Before a new model can be created, a data file must be selected from the data folder (see Section 9.3 for details of the data file structure). The model editor will extract the names of the variables from the first line of the data file that is selected. Click on OK to proceed.

The form shown in Figure 9.5 will now appear – it is a standard Windows type 'File Open' form. There are only two data files in the data folder, one for the

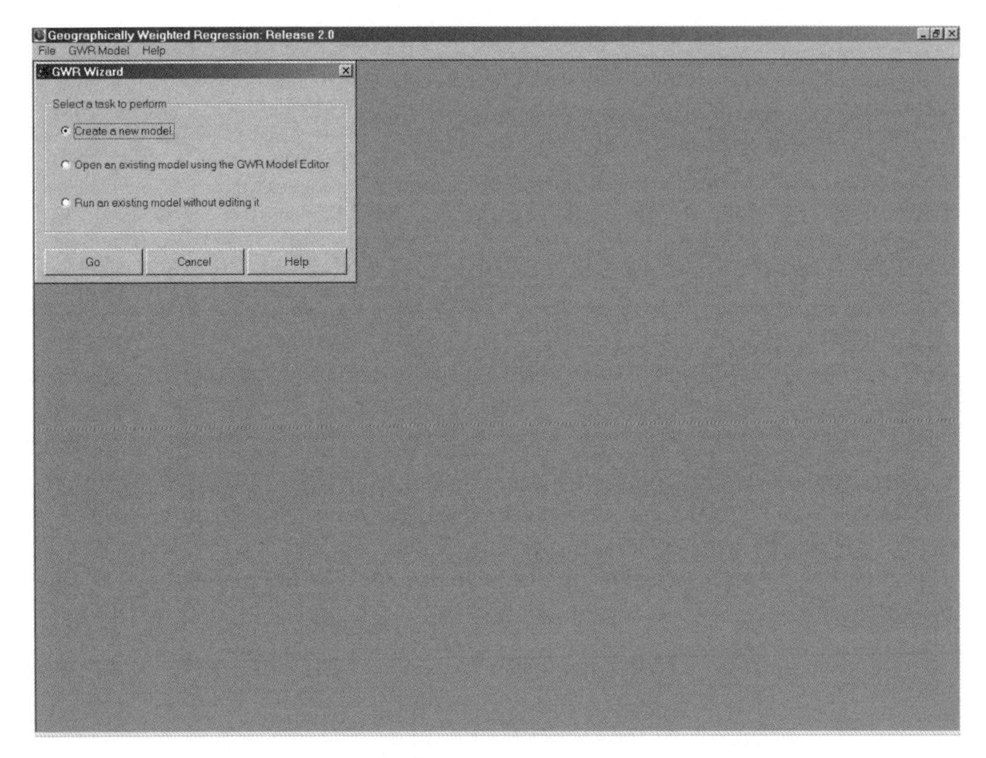

Figure 9.3 Main GWR program window

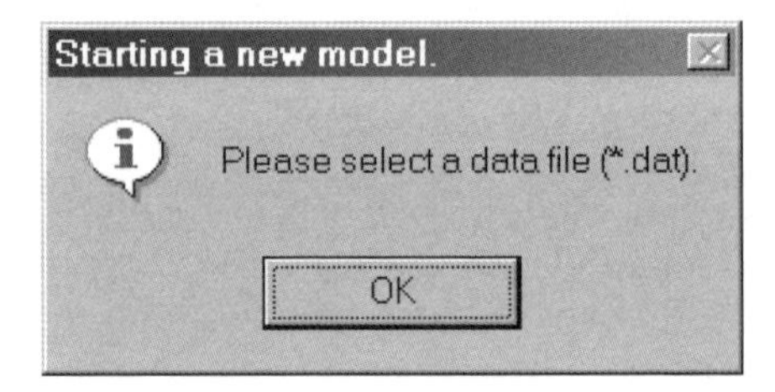

Figure 9.4 Prompt for a data file

example we are using and one which is supplied with the software. Click on the relevant data file name to highlight it and click on 'Open' to proceed.

GWR estimates may be produced at locations other than those at which data are sampled. Locations where observations are recorded are referred to as data points and the locations at which the estimates are produced as regression points. In most instances, the regression points and the data points will be the same. However, there is an option in GWR 2.0 to produce estimates of local parameters at locations other than those at which data are recorded, for example at the mesh points of a regular grid. The prompt shown in Figure 9.6 appears to allow the user to make this decision. In this instance, we click on 'Yes'. Clicking 'No' brings up

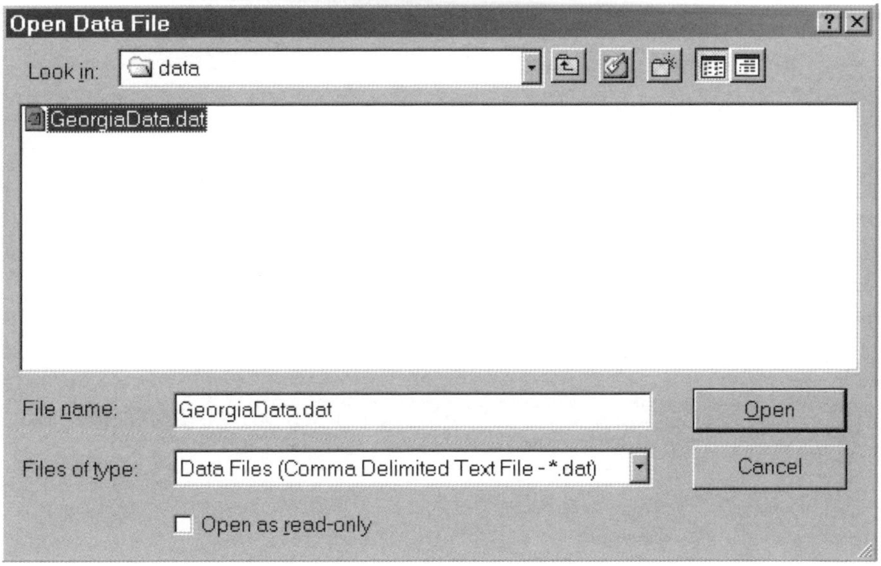

Figure 9.5 *Data files*

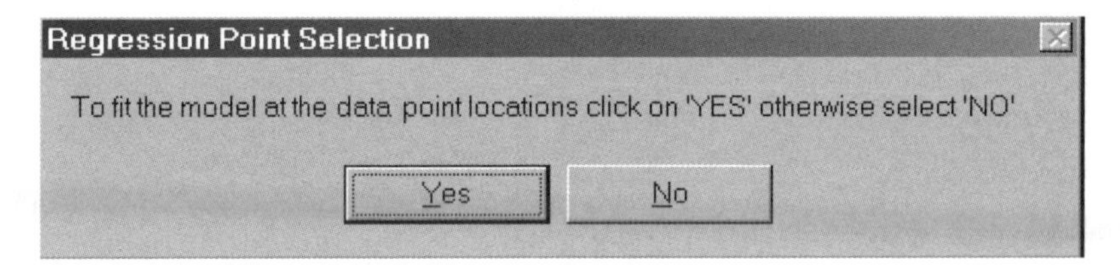

Figure 9.6 *Prompt to fit the model at data points*

another form to allow the user to select a separate file of regression point locations. Note that using this second option means the automatic bandwidth selection and a range of diagnostic statistics will not be available.

Next, the name of the file into which the results will be written must be specified. This file can be in one of several formats (comma-separated variables, ArcInfo uncompressed export, and MapInfo Interchange). This file will eventually appear in the Results folder, specified using the form in Figure 9.7. The user also needs to specify the appropriate filetype: **.e00** for an ArcInfo export file, **.csv** for a comma-separated variable file, and **.mif** for a MapInfo Interchange File. If no output is required, do not make any changes here.

The Model Editor Window appears next and is shown in Figure 9.8. It allows a GWR model to be created, saved and run. The **Title** box allows the user to input a title which will then appear in the output listing. The list of **Variables** is read automatically from the comma-separated list on the first line of the data file that

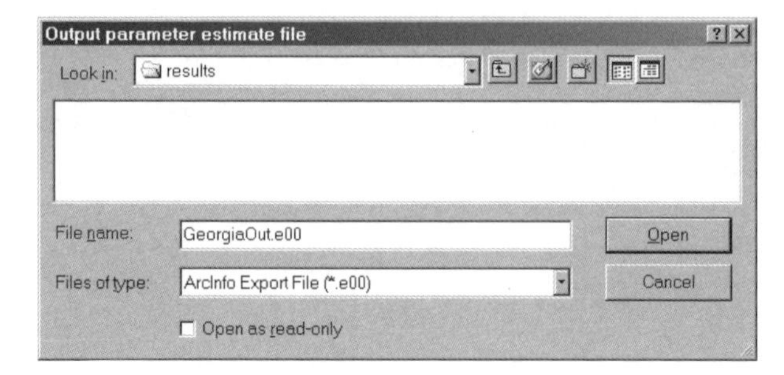

Figure 9.7 *Output parameter estimate file*

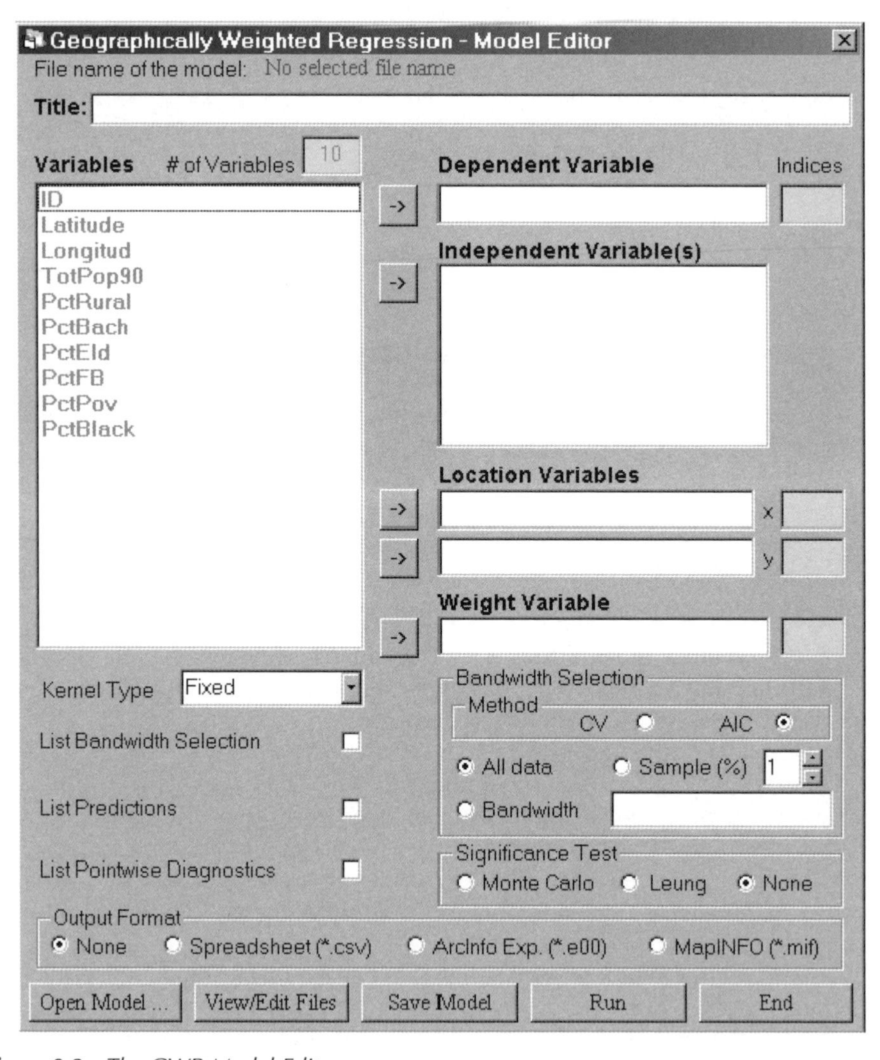

Figure 9.8 *The GWR Model Editor*

has been specified. From this, a **Dependent Variable** and one or more **Independent Variable(s)** are selected by highlighting a variable and moving it with the appropriate arrow key. Next, two variables representing the coordinates of the data points, the **Location Variables**, need to be assigned, and an optional **Weight Variable** can be selected. Note that this weight variable is *not* a geographical weight but simply allows data points to be weighted by some attribute reflecting different levels of uncertainty about the measurements taken across the data points. In most cases, this will be left empty.

Once the variables have been selected, which essentially defines the model, the **Kernel Type** is chosen for the GWR. The choices, as described above, are either 'Fixed' (Gaussian) or 'Adaptive' (bi-square). The kernel bandwidth is determined by either cross-validation (**CV**) or AIC (**AIC**) minimisation. Alternatively, an *a priori* value for the bandwidth can be entered by clicking on the **Bandwidth** option and entering the bandwidth in the window. With a very large data set, bandwidth selection can be made using a sample of data points in order to save time. This is achieved by clicking on **Sample (%)** and entering the desired percentage of the data used for the bandwidth selection procedure. The default is that the procedure will use **All data**.

Other options on the model editor include the type of output required and the type of significance test to be employed on the local parameter estimates. Apart from the default output listings (described later), the user has the option of outputting **List Bandwidth Selection, List Predictions** and **List Pointwise Diagnostics**. Examples of these are shown below. The significance testing options are: **Monte Carlo, Leung**, or **None** (see above). Finally, the format of the output file needs to be specified which needs to be compatible with the previous selection of an output filetype (see above).

A completed example of the GWR Editor is shown in Figure 9.9. The dependent variable is the proportion of the county population with education to degree level. Suppose we are interested to see how this is related to total population within each county, the percentage rural, the percentage elderly, the percentage foreign born, the percentage below the poverty line and the percentage black. We would also like to see if there are any geographical variations in the relationships between educational attainment and these variables.

The sample point location variables are 'Longitud' (x) and 'Latitude' (y). There is no aspatial weight variable. We have chosen an adaptive kernel and the bandwidth will be chosen by AIC minimisation using all the data. A Monte Carlo significance testing procedure is selected for the local parameter estimates. Printing of a range of diagnostics has been requested and the output will be written to an ArcInfo export file. Some of the output will, by default, also be written to the screen.

Before the model can be run, it must be saved. Clicking on **Save Model** will open the standard window shown in Figure 9.10 which depicts the contents of the model folder where the model control files are stored. Type the name of the file in the **File name** box or click on an existing filename and then click on **Save**. Once the model has been saved, it can be run. Simply click on the **Run** button in the Model Editor window and this brings up the form shown in Figure 9.11.

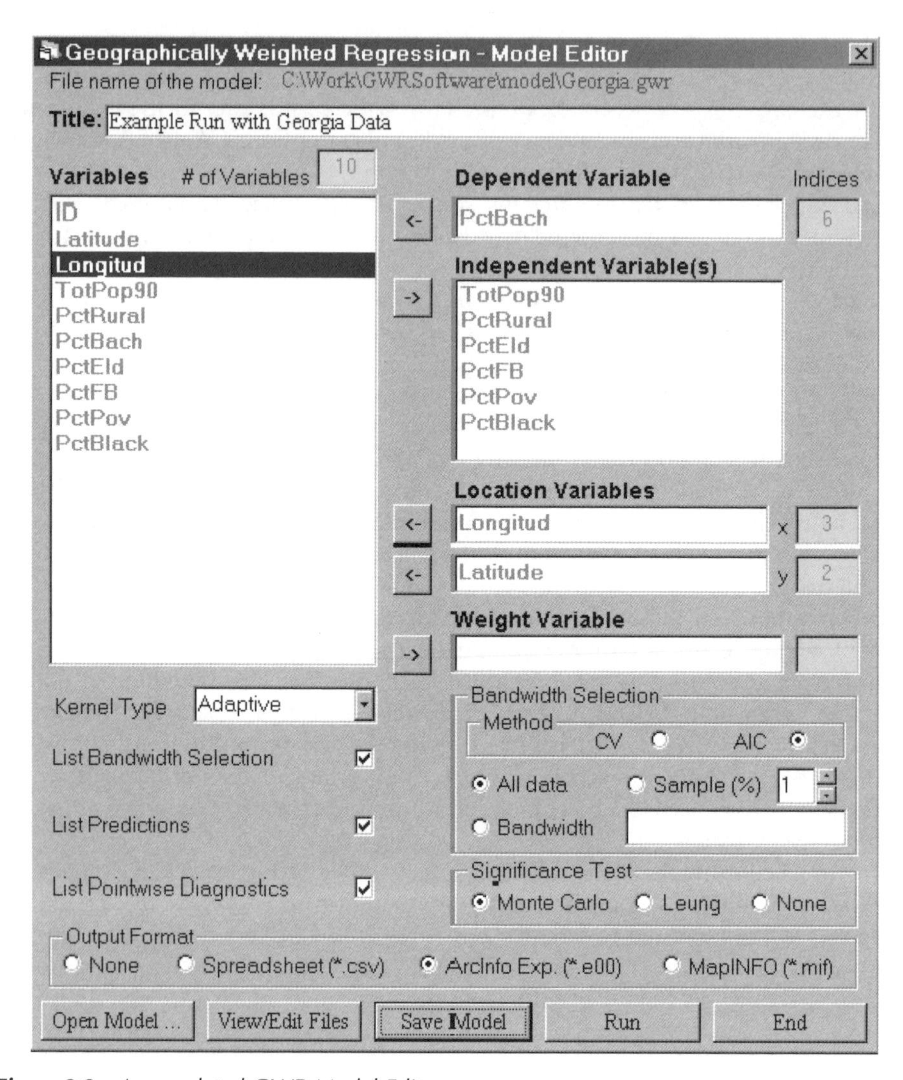

Figure 9.9 *A completed GWR Model Editor*

A name must be specified for the **Model Listing File (.txt)**. This file will be placed in the listing folder. To specify a filename click on the [...] button to the right of the filename box. Once this is done, click on the **Run** button. The model control file is now passed to the GWR program and the program is invoked and run in a DOS window as in Figure 9.12.[6]

[6] You may need to make a small alteration in your Windows set-up so that the DOS box closes on program termination.

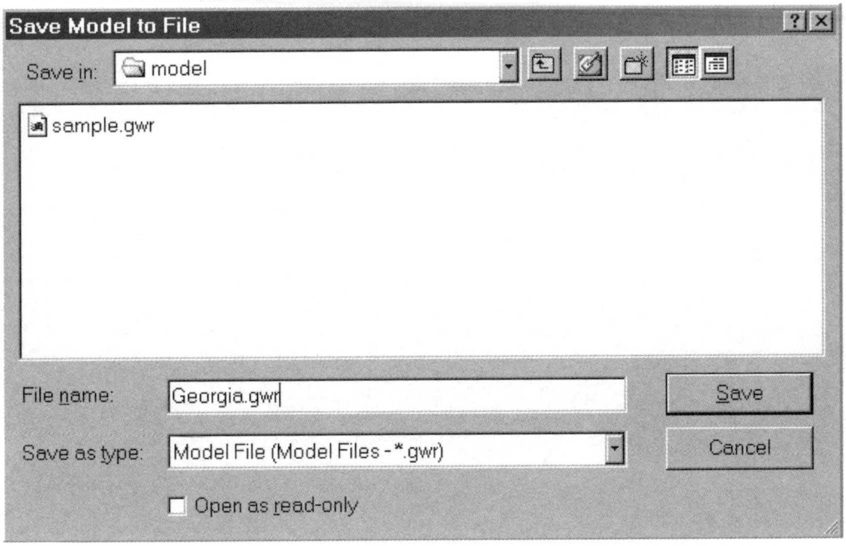

Figure 9.10 *Saving the model*

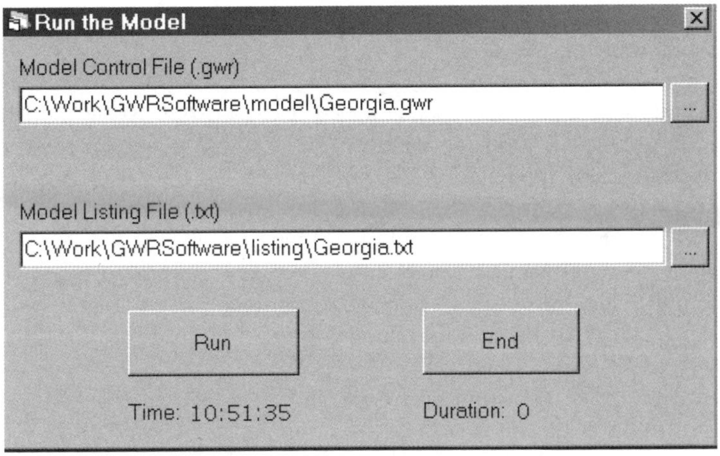

Figure 9.11 *Running the model*

With small data sets and simple models, the program runs very quickly. For instance, calibrating a bivariate GWR model using the 159 counties of Georgia on a Pentium III PC took less time than it has taken to type this sentence. However, the time requirements increase rapidly as both model complexity and the number of data points increase. One solution to very slow calibration times is to use the option in the Model Editor which allows the user to supply a percentage of the data points on which to base the calibration procedure.

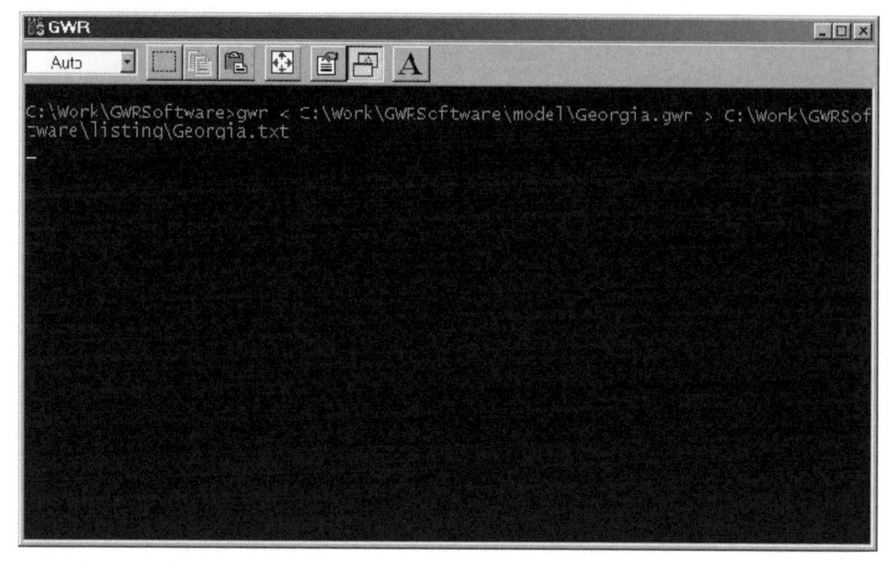

Figure 9.12 The DOS window

9.9.2 The Outputs

Once the program has run, the user is asked if the output listing is to be viewed. This listing appears in a separate window; an example of which for the Georgia educational attainment model is shown in Figure 9.13. The user can scroll down the file to view other sections. The listing file is a text file with the filetype **.txt** so that it can also be opened in MS Word for viewing or printing.

Following a description of the model that has been calibrated (see Figure 9.13), the first section of the output from GWR 2.0 contains the parameter estimates and their standard errors from a global model fitted to the data. This is shown in Figure 9.14.

There are two parts to output from the global model. In the first panel, some useful diagnostic information is printed which includes the residual sum of squares ($e^T e$), the number of parameters in the global model, the standard error of the estimate (σ), the Akaike Information Criterion and the coefficient of determination (the global r-square). In the second panel the matrix contains one line of information for each variable in the model. The columns are: (a) the name of the variable whose parameter is being estimated; (b) the estimate of the parameter; (c) the standard error of the parameter estimate; and (d) the t statistic for the hypothesis $\alpha = 0$. These global results suggest that educational attainment is positively related to total population and percentage foreign born and is negatively related to percentage rural and percentage below the poverty line. Educational attainment does not appear to be related to the remaining two variables, percentage elderly and percentage black. The model replicates the data reasonably well (65% of the

```
Model Listing
**************************************************
*      Geographically Weighted Regression        *
*               Release 2.0                      *
*           Dated: 11-ii-2002                    *
*                                                *
*      Martin Charlton, Chris Brunsdon           *
*           Stewart Fotheringham                 *
*      University of Newcastle upon Tyne         *
**************************************************

Program limits:
Maximum number of variables.....     35
Maximum number of observations..  78092
Maximum number of fit locations.  78092

Example Run with Georgia Data
Observed data file:        C:\Work\GWRSoftware\data\GeorgiaData.dat
Prediction location file:  GWR estimation at data point locations
Parameter output file:     C:\WORK\GWRSOFTWARE\results\GeorgiaOut.e

Variable names...
 ID       Latitude Longitud TotPop90 PctRural PctBach  PctEld   PctFB
 PctPov   PctBlack

Dependent (y) variable..........PctBach
Easting  (x-coord) variable.....Longitud
Northing (y-coord) variable.....Latitude
No weight variable specified
Dependent variables in this model...
 TotPop90 PctRural PctEld   PctFB    PctPov   PctBlack

View another listing file                    Close
```

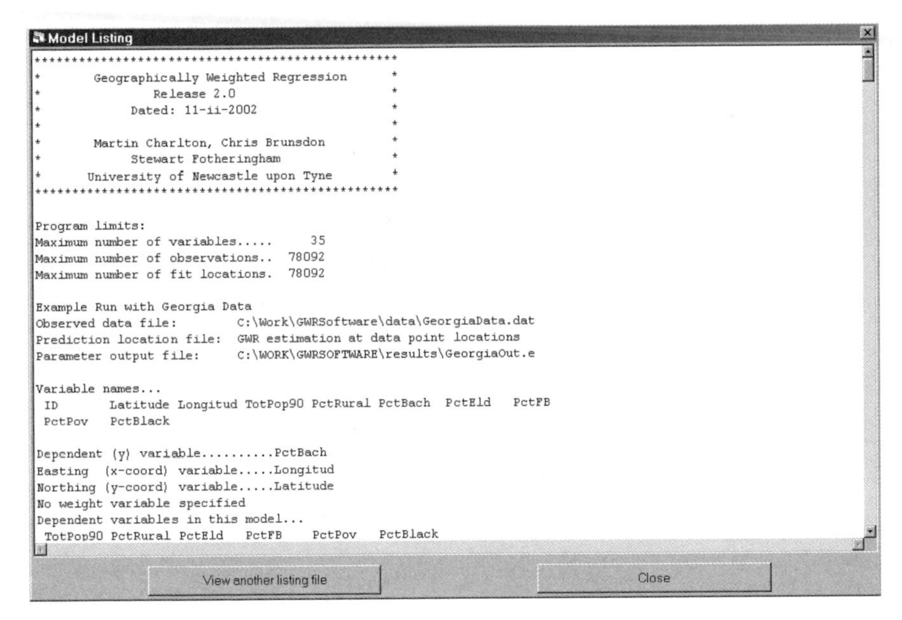

Figure 9.13 *Model listing*

```
*****************************************************************
*                GLOBAL REGRESSION PARAMETERS                   *
*****************************************************************
 Diagnostic information...
 Residual sum of squares.........  1816.21072
 Effective number of parameters..  7.
 Sigma...........................  3.45669708
 Akaike Information Criterion....  855.443391
 Coefficient of Determination....  0.646

Parameter               Estimate                Std Err              T
---------              ---------              ---------          ----
Intercept          14.779297592328        1.705507562188         8.665
 TotPop90           0.000023567534        0.000004746089         4.965
 PctRural          -0.043878182061        0.013715372112        -3.199
 PctEld            -0.061925096691        0.121460075458        -0.509
 PctFB              1.255536084016        0.309690422174         4.054
 PctPov            -0.155421764065        0.070388091758        -2.208
 PctBlack           0.021917908085        0.025251694359         0.867
```

Figure 9.14 *Global regression results*

variance in educational attainment is explained by the model) but there are clearly some factors that are not captured adequately by the global model.

From this point, the output listing contains the results of the GWR. The first section is an optional calibration report which lists the calculated value of the criterion statistic at various bandwidths, as shown in Figure 9.15. The utility of

```
 Bandwidth          AIC
112.183925  841.070913
130.066075  840.288579
141.117851  839.462956
147.948224  839.67504
136.896448  840.209746
143.726821  840.049036
139.505418  839.486703
142.114389  839.722037
140.501956  839.462956
140.121313  839.486703
140.737207  839.462956
140.356563  839.486703
140.591814  839.462956
140.446421  839.486703
140.536279  839.462956
140.480744  839.486703
140.515066  839.462956
140.493854  839.486703
140.506964  839.462956
140.498861  839.486703
140.503869  839.462956
140.500774  839.462956
140.500043  839.462956
140.499592  839.486703
140.500323  839.462956
140.499871  839.486703
140.50015   839.462956
140.499978  839.486703
```

Figure 9.15 *AIC minimisation search*

printing this section is to observe the speed of convergence and also to plot the results to see the shape of the convergence function. If the calibration report is not requested, the program will print only the optimal value of the bandwidth.

The next section of the output (Figure 9.16) presents diagnostics for the GWR estimation. There are two panels in this section. The first panel provides some general information on the model: it includes (a) a count of the number of data points or observations; (b) the number of predictor variables (this is the number of columns in the design matrix); (c) the bandwidth for the type of kernel specified (here it is the number of nearest neighbours to be included in the bi-square kernel); and (d) the number of regression points. The second panel contains similar information to the corresponding panel for the global model. This includes (a) the residual sum of squares; (b) the effective number of parameters; (c) the standard error of the estimate; (d) the Akaike Information Criterion; and (e) the coefficient of determination. The latter is constructed from a comparison of the predicted values from different models at each regression point and the observed values. The coefficient has increased from 0.646 to 0.718 although an increase is to be expected given the difference in degrees of freedom. However, the reduction in the AIC from the global model suggests that the local model is better even accounting for differences in degrees of freedom.

Casewise diagnostics, described in Section 9.8, can be requested as shown in Figure 9.17 for the first 10 observations in the Georgia data set. These include:

```
***********************************************************
*                     GWR ESTIMATION                      *
***********************************************************
 Fitting Geographically Weighted Regression Model...
 Number of observations............ 159
 Number of independent variables... 7
  (Intercept is variable 1)
 Number of nearest neighbours...... 141
 Number of locations to fit model.. 159

 Diagnostic information...
 Residual sum of squares........     1447.30202
 Effective number of parameters..   15.4997717
 Sigma...........................   3.17580105
 Akaike Information Criterion...    839.462956
 Coefficient of Determination...    0.718
```

Figure 9.16 *GWR estimation*

```
***********************************************************
*                 CASEWISE DIAGNOSTICS                    *
***********************************************************
```

Obs	Observed	Predicted	Residual	Std Resid	R-Square	Influence	Cook's D
1	8.20000	9.00600	-0.80600	-0.216137	0.805478	0.028689	0.000089
2	6.40000	6.95755	-0.55755	-0.153571	0.808832	0.079335	0.000131
3	6.60000	8.52393	-1.92393	-0.532626	0.805972	0.088652	0.001780
4	9.40000	8.30820	1.09180	0.300525	0.840237	0.078113	0.000494
5	13.30000	13.83488	-0.53488	-0.148321	0.841259	0.091655	0.000143
6	6.40000	8.90985	-2.50985	-0.684503	0.848977	0.060933	0.001961
7	9.20000	11.75978	-2.55978	-0.686887	0.852006	0.029973	0.000941
8	9.00000	11.44567	-2.44567	-0.656816	0.860664	0.031594	0.000908
9	7.60000	10.23082	-2.63082	-0.714083	0.815341	0.051941	0.001802
10	7.50000	9.10363	-1.60363	-0.437710	0.814244	0.062468	0.000824

... ...

Figure 9.17 *Casewise diagnostics*

(a) the observation sequence number; (b) the observed data; (c) the predicted data; (d) the residual; (e) the standardised residual; (f) the local *r*-square; (g) the influence; and (h) Cook's *D*. While in general it might be helpful to look at a printout of these statistics, it is probably a little more useful to be able to map them: with a large data set you run the risk of being swamped in output. All of these statistics are saved automatically in the output results file so that requesting them in the listing file should be done judiciously. This panel is not available when the regression points are different from the data points.

Another optional set of information that can be printed to the screen concerns the predicted values as shown in Figure 9.18 for the first 10 observations in the Georgia data set. If this option is selected, the following data are printed to the screen: (a) *Obs* – the sequence number of the observation; (b) *Y*(*i*) – the observed value; (c) *Yhat*(*i*) – the predicted value; (d) *Res*(*i*) – the residual; (e) *X*(*i*) – the *x*-coordinate of the regression point; (f) *Y*(*i*) – the *y*-coordinate of the regression point; and (g) an indicator of whether the matrix inverse was computed using

```
Predictions from this model...
   Obs          Y(i)       Yhat(i)        Res(i)        X(i)         Y(i)
     1         8.200         9.006        -0.806     -82.286       31.753   F
     2         6.400         6.958        -0.558     -82.875       31.295   F
     3         6.600         8.524        -1.924     -82.451       31.557   F
     4         9.400         8.308         1.092     -84.454       31.331   F
     5        13.300        13.835        -0.535     -83.251       33.072   F
     6         6.400         8.910        -2.510     -83.501       34.353   F
     7         9.200        11.760        -2.560     -83.712       33.993   F
     8         9.000        11.446        -2.446     -84.839       34.238   F
     9         7.600        10.231        -2.631     -83.220       31.759   F
    10         7.500         9.104        -1.604     -83.232       31.274   F
```

Figure 9.18 *Predicted value information*

```
**********************************************************
*                        ANOVA                           *
**********************************************************
        Source                      SS          DF             MS            F
OLS Residuals                   1816.2        7.00
GWR Improvement                  368.9        8.50        43.4022
GWR Residuals                   1447.3      143.50        10.0857       4.3033
```

Figure 9.19 *ANOVA results*

```
**********************************************************
*             PARAMETER 5-NUMBER SUMMARIES               *
**********************************************************
   Label       Minimum  Lwr Quartile       Median  Upr Quartile       Maximum
Intercept    11.713757     13.673421    15.870931     16.306640     16.794073
TotPop90      0.000011      0.000017     0.000022      0.000027      0.000029
PctRural     -0.065437     -0.052570    -0.039087     -0.030860     -0.021847
PctEld       -0.347454     -0.223961    -0.176703     -0.133954     -0.078959
PctFB         0.442051      0.687714     1.493501      2.230001      2.589042
PctPov       -0.213353     -0.160355    -0.099347     -0.019344      0.032084
PctBlack     -0.047741     -0.024114     0.003978      0.041358      0.085723
```

Figure 9.20 *5-number summary of local parameter estimates*

either the Gauss-Jordan method (F) or a generalised inverse (T). This set of output is not available when the regression points are different from the sample points.

Next in the output listing is a panel of results of an ANOVA in which the global model is compared with the GWR model. The ANOVA tests the null hypothesis that the GWR model represents no improvement over a global model. The results are shown in Figure 9.19 where it can be seen that the F test suggests that the GWR model is a significant improvement on the global model for the Georgia data. More details on this test can be found in Chapter 4.

The main output from GWR is a set of local parameter estimates for each relationship. Because of the volume of output these local parameter estimates and their local standard errors generate, they are not printed in the output listing but are automatically saved to the output file. However, as a convenient indication of the extent of the variability in the local parameter estimates, a 5-number summary of the local parameter estimates is printed. For the Georgia data, this is shown in Figure 9.20. The 5-number summary of a distribution presents the median, upper and lower quartiles, and the minimum and maximum values of the data (Tukey, 1977). This is helpful to get a ‘feel’ for the degree of spatial non-stationarity in a relationship by comparing the range of the local parameter estimates with a confidence interval around the global estimate of the equivalent parameter. Recall that 50% of the local parameter values will be between the upper and lower quartiles and that approximately 68% of values in a normal distribution will be within ± 1 standard deviations of the mean. This gives us a reasonable, although very informal, means of comparison. We can compare the range of values of the local estimates between the lower and upper quartiles with the range of values at ± 1 standard deviations of the global estimate. Given that 68% of the values would be expected to lie within this latter interval, compared to 50% in the inter-quartile range, if the range of local estimates between the inter-quartile range is greater than that of ± 1 standard deviations of the global mean, this suggests the relationship might be non-stationary.

As an example consider the parameter estimates for the two variables PctEld (percentage elderly) and PctFB (percentage foreign born) in the Georgia study. At ± 1 standard deviations of the global estimate, the intervals are:

	−1SD	+1SD
PctEld	−0.1834	0.0596
PctFB	0.9459	1.5653

whereas the inter-quartile ranges of the local parameter estimates are

	LQ	UQ
PctEld	−0.2240	−0.1340
PctFB	0.6877	2.2300

On the basis of this evidence, the local variation in the PctEld parameter estimates does not appear to be very unusual whereas the local variation in the PctFB estimates does (the local range is well outside the range of ± 1 standard deviations of the global mean).

Finally, we can examine the significance of the spatial variability in the local parameter estimates more formally by conducting either a Monte Carlo test or a test developed by Leung *et al.* (2000a). More details on both tests are provided above in Section 9.7. In Figure 9.21 we report the results of a Monte Carlo test on the local estimates which indicates that there is significant spatial variation in the local parameter estimates for the variables PctFB and PctBlack. The spatial variation in the remaining variables is not significant and in each case there is a reasonably high probability that the variation occurred by chance. This is useful information because now in terms of mapping the local estimates, we can concentrate on the two variables, PctFB and PctBlack, for which the local estimates exhibit significant spatial non-stationarity. It is interesting to note that these results reinforce the conclusions reached above with the very informal examination of local parameter variation compared to the distribution of the global mean.

```
**************************************************
*                                                *
*   Test for spatial variability of parameters   *
*                                                *
**************************************************

Tests based on the Monte Carlo significance test
procedure due to Hope [1968,JRSB,30(3),582-598]

Parameter              P-value
----------          -------------
Intercept              0.29000
TotPop90               0.10000
PctRural               0.24000
PctEld                 0.75000
PctFB                  0.00000
PctPov                 0.59000
PctBlack               0.02000
```

Figure 9.21 *Monte Carlo test for spatial non-stationarity*

9.9.3 Running GWR 2.0 under UNIX

The Model Editor is a simple program that produces a control file for the compiled DOS program. You can run the same control file on a UNIX machine (you might need to change the names of the data/output files) and you will get the same results. Running the program on UNIX is more difficult because there is no point-and-click interface to help you construct the control file but it does offer some advantages. For large data sets and complex models, a multiprocessor machine might lead to a considerable saving in time. Additionally, UNIX offers some flexibility which is not always easily found in Windows.

An example of the shell code for running GWR 2.0 under UNIX is given in Figure 9.22. The data lines in the code are marked (a) to (n) inclusive and are used to control the various options available in GWR 2.0. Each of these lines is described in Table 9.1.

The software reads from the standard input and writes to the standard output. A control file can be created using an ordinary text editor. The redirection operators < and > may be used to read from and write to text files. In the above example, GWR is run with a series of different bandwidths; it also shows how to read standard input 'inline'. The '≪**endfile**' means continue reading from what follows as standard input until the control word 'endfile' is reached. We expect that any UNIX users will have sufficient facility with the shell they are using to be able to compile and run the program. In compiling the program we have found the following to be useful:

f 77 -ogwr -w -fast -05 -xtarget=ultra2 -xcache=16/32/1:4096/64/1 gwr.f

Clearly, **xtarget** and **xcache** will be machine dependent. To find the values for **xtarget** and **xcache**, the UNIX utility **fpversion** should be used. On the Newcastle University system the output from fpversion is:

```
(1) #! /bin/csh
(2) foreach i (50 100 150)
(3) gwr <<endfile >/tmp/georgia/gwr$i.lst
(a) /tmp/georgia/georgiadata.dat
(b)
(c) /tmp/georgia/gwr$i.e00
(d) 6
(e) 3,2,0
(f) Predicting Educational Attainment: all variables
(g) FFFTTFTTTT
(h) vaa
(i) $i
(j) 0
(k) 0
(l) 2
(m) 0
(n) 0
(4) endfile
(5) end
```

Figure 9.22 *Shell code for running GWR 2.0 under UNIX*

```
finan [sparc] 19% fpversion
  A SPARC-based CPU is available.
  CPU's clock rate appears to be approximately 250.2 MHz.
  Kernel says CPU's clock rate is 248.0 MHz.
  Kernel says main memory's clock rate is 82.0 MHz.

  Sun-4 floating-point controller version 0 found.
  An UltraSPARC chip is available.
  FPU's frequency appears to be approximately 246.3 MHz.

  Use "-xtarget=ultra2 -xcache=16/32/1:4096/64/1" code-generation option.

  Hostid = 0x8083B6CF.
```

Users may also find it useful to run some large jobs using the batch system. Virtual Memory (VM) may be set by the user's computer centre and so to determine how much VM GWR 2.0 is using, run the job in background and issue the command:

ps -flu *username*

As a guide, a GWR fitted to 78 000 observations and 20 variables run on the machine whose characteristics are listed above took around 100 hours of CPU time.

9.10 Visualising the Output

The main output from GWR is a set of localised parameter estimates and associated diagnostics. Unlike the single global values traditionally obtained in modelling, these

Table 9.1 *Control options for GWR under UNIX*

Option	Description	Notes
a	Location of the input data file	This file should have data separated by commas; the first line should be a list of variables in the data.
b	Location of regression point file	This should be in ArcInfo ungenerated point file format. If the regression points are the same as the data points, leave this line blank.
c	Location of the output parameter estimate and diagnostics file	The filetype depends on the choice for option k.
d	Index of the dependent variable	The variables in the data file are numbered implicitly by the program from 1 to n. This index must be between 1 and n inclusive
e	Indices of data point x-coordinate data point y-coordinate data point weight	The x- and y-coordinate variable indices must be specified. If there is no weight variable, use an index of zero.
f	Title for this run	The first 80 characters of whatever is placed here will be printed on the output.
g	Include string	Place a T in the *i*th position to include the *i*th variable in the set of independent variables. Otherwise place an F.
h	Kernel type:	f = fixed bandwidth kernel v = variable bandwidth kernel
i	Bandwidth	> 0: use this number as the bandwidth = 0: choose a bandwidth by minimising the AIC statistic <0: [decimal $-1.00 <= x < 0$]: sample fraction for calibration with large data sets.
j	Calibration report	1: List the steps in the bandwidth choice process 0: Omit this listing
k	Prediction report	1: List the predictions for the regression points 0: Omit this listing
l	Output file selection	0: Do not produce an output file 1: Produce a comma separated file variable 2: Produce an uncompressed ArcInfo export format file 3: Produce a MapInfo Interchange format file
m	Significance tests	0: No significance tests 1: Use Monte Carlo significance test 2: Use Leung test
n	Diagnostics report	1: List some pointwise diagnostics 0: Omit this listing

local values lend themselves to being mapped. Indeed, in large data sets, mapping, or some other form of visualisation, is the only way to make sense of the large volume of output that will be generated. We now describe ways of visualising the output from GWR. Although we concentrate only on displays of the local parameter estimates, in many instances it might be instructive to plot other local statistics such as the influence and Cook's D statistics. Similarly, it might be useful to plot the local r-square statistic or the local standard deviation. No matter which local statistic is mapped, however, there is a choice of map types that can be employed. We now describe some of these briefly after first discussing mapping the results in a commonly used, PC-based, Geographic Information System (GIS), ArcView.

9.10.1 Viewing the Results in ArcView

We assume that the user has available some software for visualising the results. Most commonly, this will be some mapping package, or preferably, a GIS in which both the results and the data can be manipulated. Saving the output file as either an uncompressed ESRI shapefile or a MapInfo interchange file means the output can be viewed relatively easily within a GIS. For instance, if we convert the .e00 file to a shapefile, it can be viewed in ArcView. This can be done using the Import71 utility which is shipped with ArcView. If the output is saved as a comma separated file, it can be opened in Excel.

An example of using the Import71 utility within ArcView is shown in Figure 9.23. Specify the name of the export file which is located in the GWR **results** folder in the **Export Filename:** box. Then specify the name of the output coverage in the **Output Data Source:** box. Notice that the full pathname to the file and the corresponding coverage must be supplied. However, the coverage need not be placed in

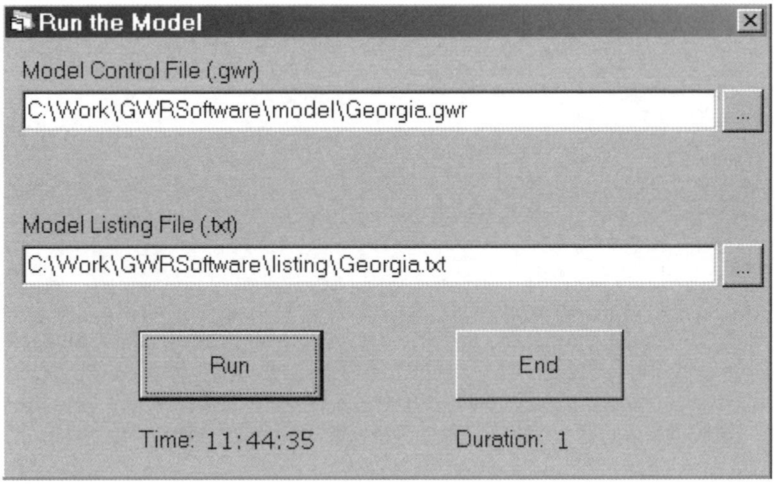

Run the Model

Model Control File (.gwr)

C:\Work\GWRSoftware\model\Georgia.gwr

Model Listing File (.txt)

C:\Work\GWRSoftware\listing\Georgia.txt

Run   End

Time: 11:44:35   Duration: 1

Figure 9.23 *An example of the ArcView Import71 utility*

the same folder as the export file. For those familiar with ArcView, a new 'View' must be opened and the coverage added as a 'Theme' to the 'View'.

9.10.2 Point Symbols

A simple means of data display uses some form of varying size or varying colour point symbolism for the values of each parameter at the regression points. In the ArcInfo and MapInfo versions of the file, the parameter estimate items are numbered PARM_1, PARM_2, PARM_3,... with PARM_1 containing the values for the intercept term. The standard errors are numbered equivalently as SVAL_1, SVAL_2... and the pseudo-*t* values are in TVAL_1, TVAL_2,.... Examples of displaying the GWR results as a point map are shown in Figures 9.24 and 9.25, for

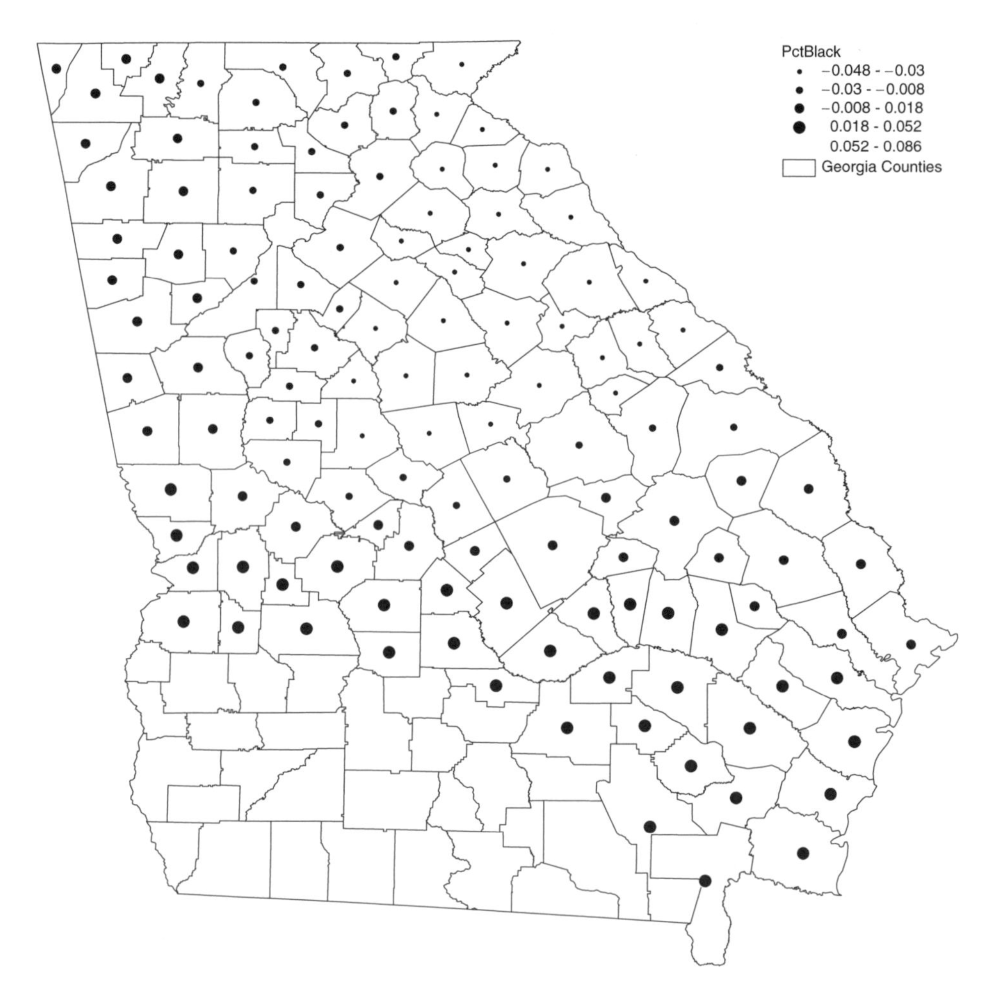

Figure 9.24 Proportional circle map of the local parameter estimates associated with the variable PctBlack

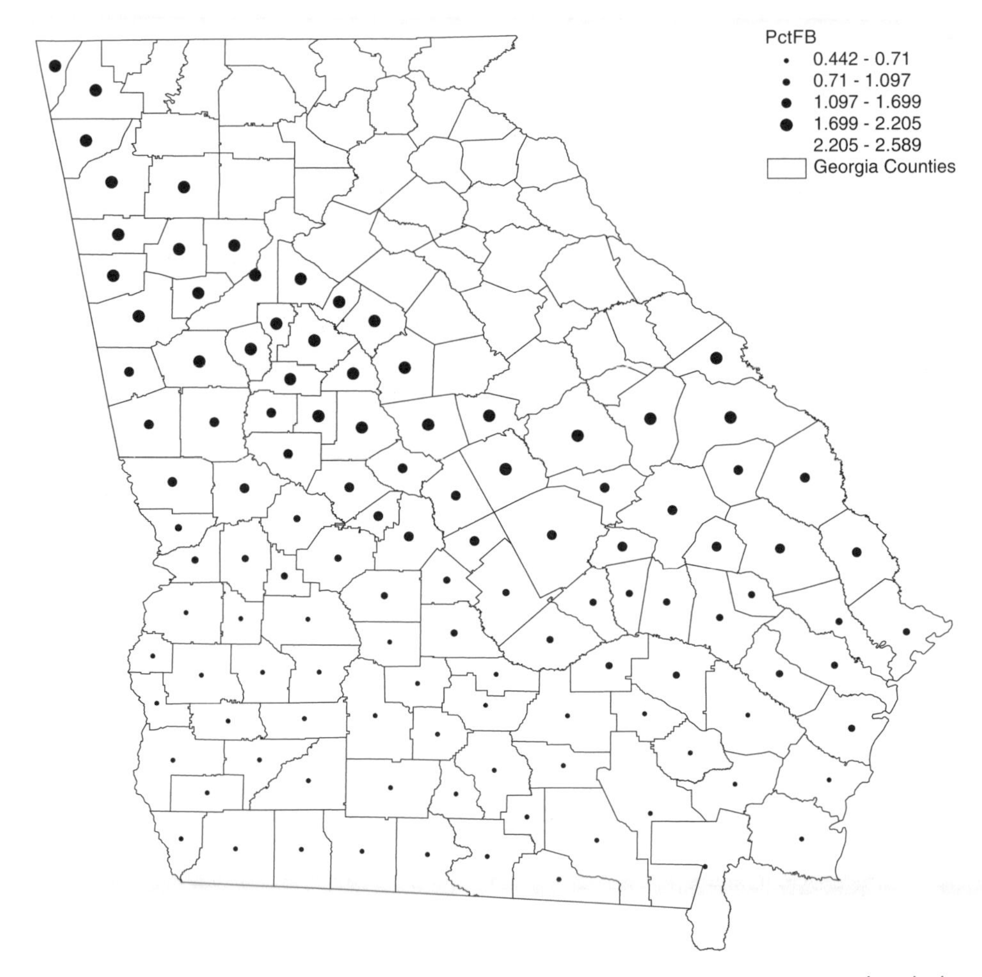

Figure 9.25 *Proportional circle map of the local parameter estimates associated with the variable PctFB*

the local parameter estimates associated with the variables PctBlack and PctFB respectively. In both cases, the point symbols are proportional circles and the points lie at the geographic centroids of the county boundaries. It is clear from both maps that the parameters exhibit considerable spatial variation. In the case of PctBlack, the local parameter estimates change sign over space, being negative in the northeast and positive in the south. In the case of PctFB there is a clear variation in the local parameter estimates with the magnitude of the local estimates increasing towards the south. As noted above, both these sets of local estimates exhibit significant spatial non-stationarity. It remains a matter for more detailed socio-economic investigation to provide possible explanations of the results. Those who know Georgia well, will be aware that there are substantial differences between the counties in the north-west and the southern and western counties bordering north Florida and Alabama.

9.10.3 Area Symbols

A choropleth map is another means of showing the spatial variation in the local statistics and again this can easily be created within a GIS such as ArcView. In this case, the point and polygon theme tables must be merged on the 'Shape' item, the resulting table exported as a .dbf file, added back in to the ArcView 'Project' and then merged with the original polygon theme table. Figures 9.26 and 9.27 show choropleth maps of the spatial variation in the local parameter estimate associated with the PctBlack and PctFB variables, respectively. These maps show the same trends in the local estimates as the proportional circle maps but the patterns are perhaps more clearly seen.

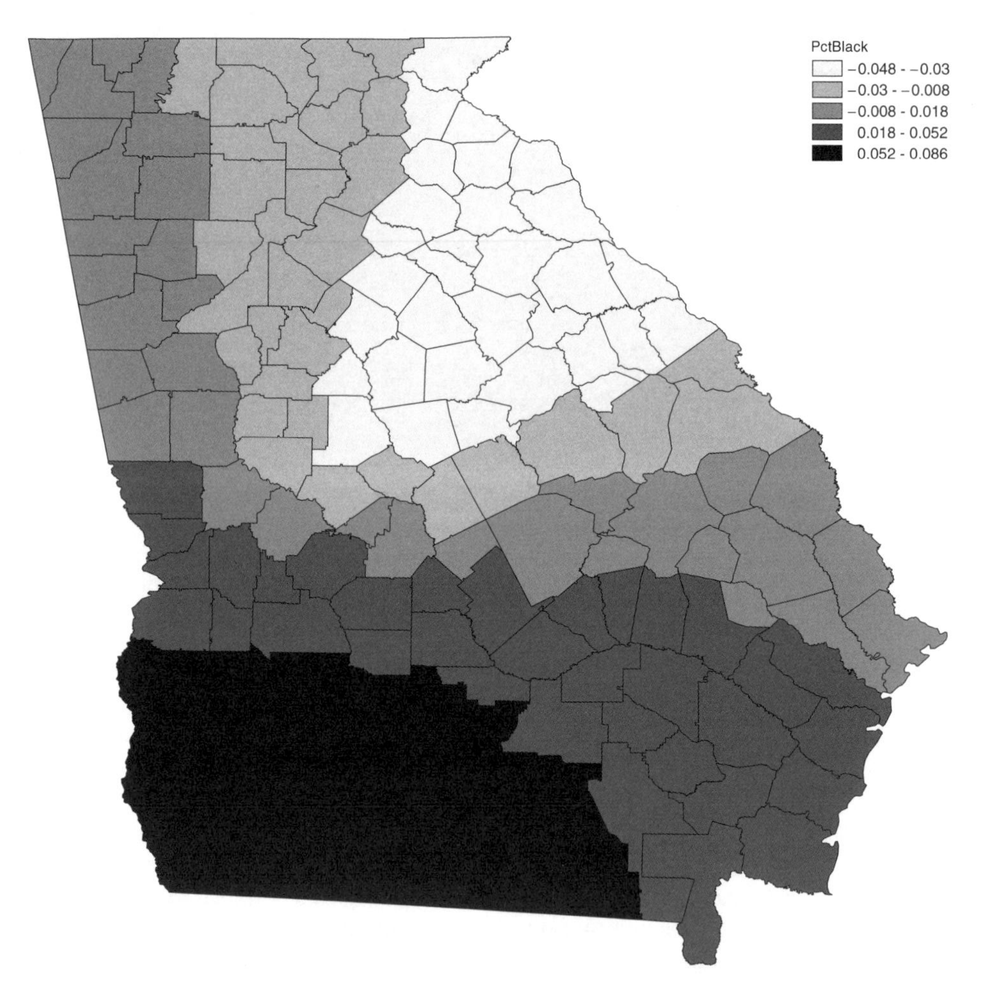

Figure 9.26 *Choropleth map of the local parameter estimates associated with the variable PctBlack*

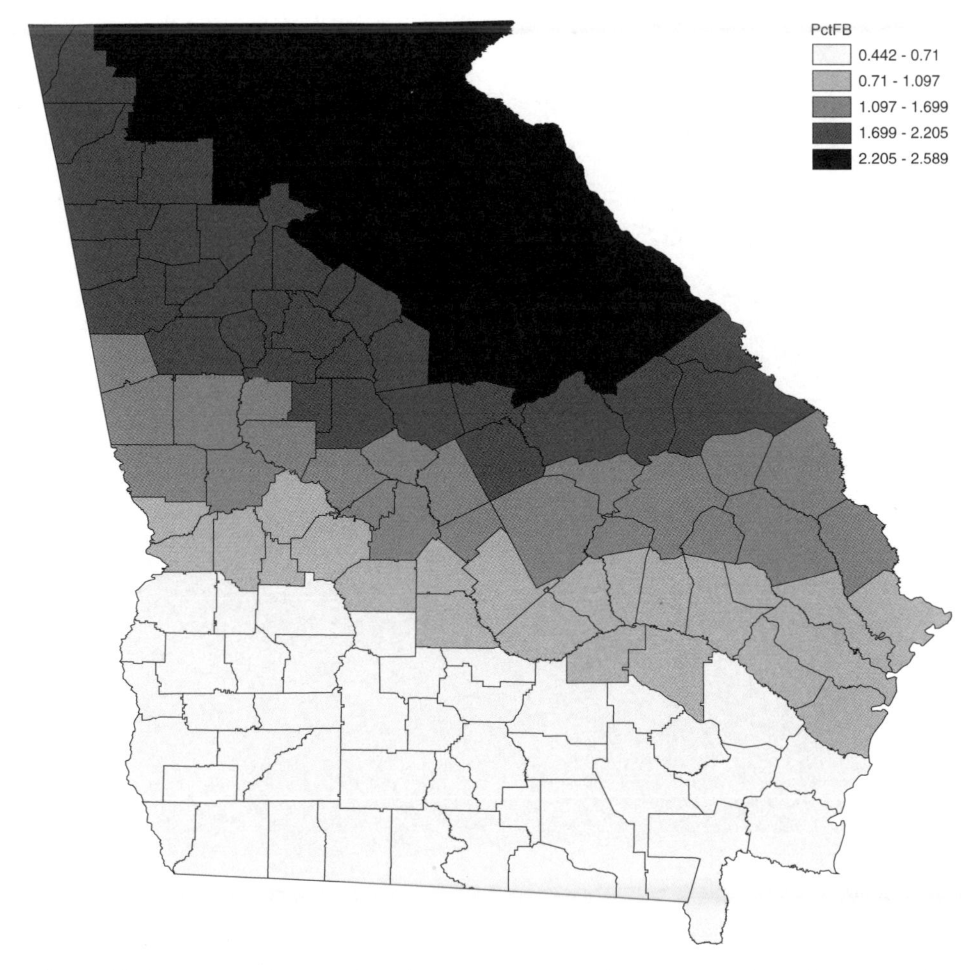

Figure 9.27 *Choropleth map of the local parameter estimates associated with the variable PctFB*

9.10.4 Contour Plots

In many of the examples in this book we have used shaded contours to show the variation in the values of the local parameter estimates and associated statistics. The genesis of these maps is the creation of a gridded surface from the regression point locations and the variable of interest and followed by shading the grid using a suitable lookup table within the GIS. Figure 9.28 presents the local parameter estimates for the variable PctBlack displayed as a shaded contour map. The negative estimates in the north-east are clearly visible as a 'sink' surrounded by positive estimates in the rest of the state.

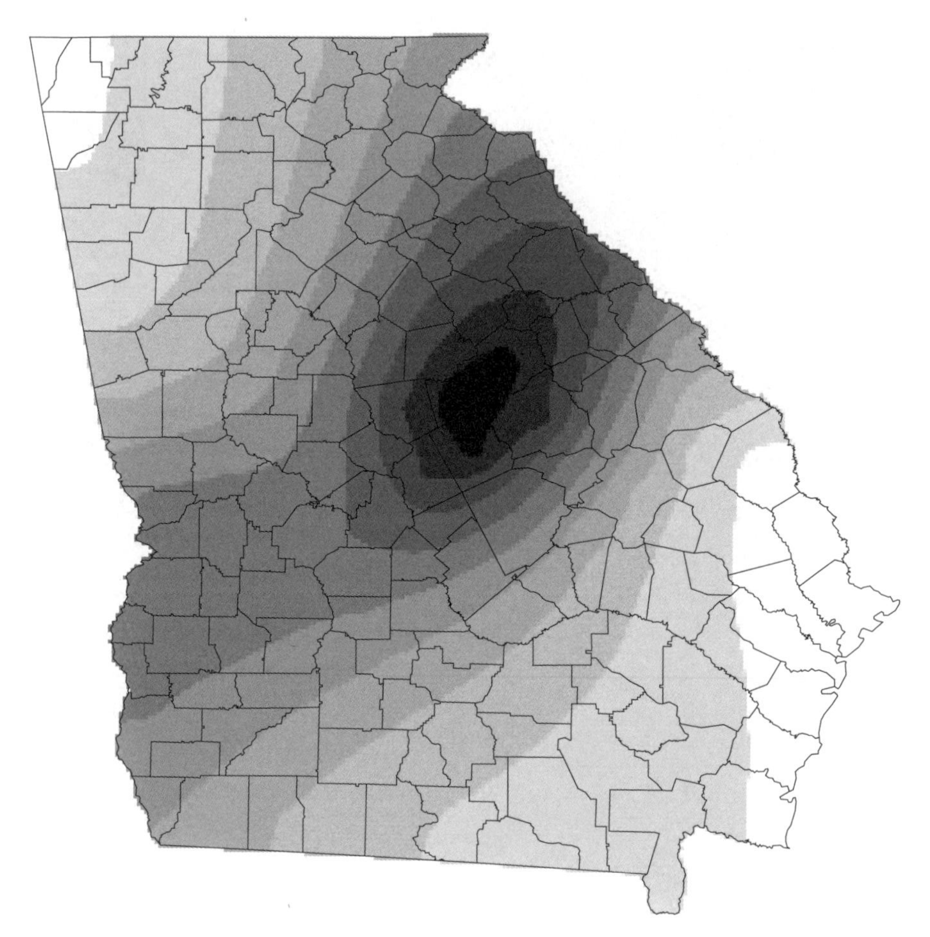

Figure 9.28 Shaded contour map of the local parameter estimates associated with the variable PctBlack

9.10.5 Pseudo-3D Display

A further option is to view the parameter estimates and associated statistics as either wire frame or shaded or coloured surfaces as shown for the variable PctBlack in Figure 9.29. If the requisite grids have been created, then each needs to be loaded into the viewing software and visualised with relevant options. Many software packages give the user control over the following:

1. vertical exaggeration of the surface variation (z-factor);
2. viewing as a wire frame;
3. viewing as a grey-shaded surface;
4. viewing as an illuminated surface (as if the surface were made from a solid material);

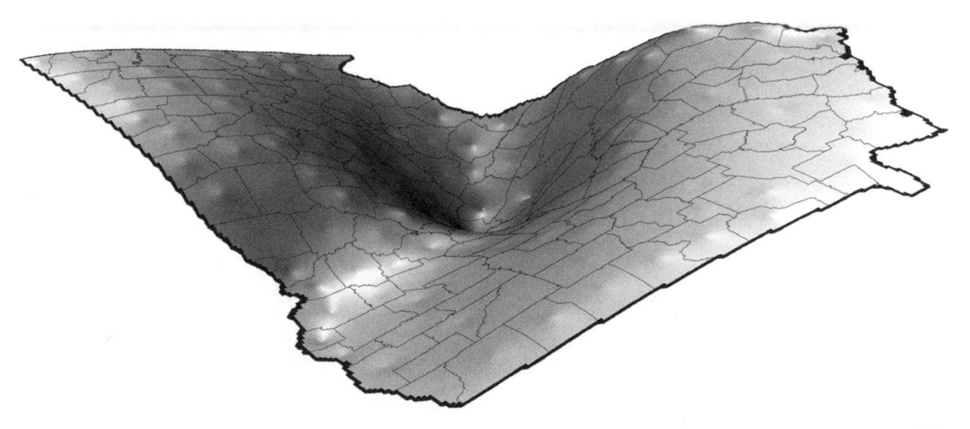

Figure 9.29 *Pseudo-3D map of the local parameter estimates associated with the variable PctBlack*

5. viewing as a coloured surface;
6. position of the observer (distance from the centre of the surface, azimuth, altitude.

It may be that a combination of various forms of output will aid interpretation of the results. Perhaps a salient point is that the graphics should assist rather than hinder and some of the suggestions of Tufte (1983, 1990) with regard to graphic design might be appropriate.

9.11 Summary

In this chapter we have described the operation of a user-friendly program for calibrating GWR models. It is anticipated that updates to the GWR software will be issued as developments take place. The website www.ncl.ac.uk/geography/GWR will have information about the planned updates. Among the general issues we anticipate addressing are:

1. more flexible input facilities;
2. a greater range of output options;
3. a larger range of kernel types.

A more detailed list of topics that might be included in future versions of the software is given in Chapter 10. The authors are sympathetic to suggestions concerning the software and these should be directed to Professor Stewart Fotheringham by email to stewart.fotheringham@ncl.ac.uk.

10
Epilogue

10.1 Overview

This book is based on the premise that relationships between variables measured at different points in space *might* not be constant over space. With a few notable exceptions described in Chapter 1, the prevailing assumption within spatial analysis is that such relationships are constant. However, the assumption of spatial stationarity would appear to be the result of convenience rather than of any serious examination of the issues: it is simply much easier to assume away the problems associated with spatial non-stationarity. If relationships do vary significantly over space, then serious questions are raised about the reliability of global-level analyses. Indeed, as discussed in Chapter 1, the existence of spatial non-stationarity makes the assumption that local relationships can be accurately represented by global parameters equally as untenable as assuming everywhere in the US has a snowfall equal to the country's average.

Why might relationships vary over space? Perhaps it is better to ask two separate questions: 'If measured correctly, would relationships be stationary?' and 'Are our *measurements* of relationships the cause of apparent spatial non-stationarity?' The former question asks whether all relationships can, ultimately, be represented by a global model. It is perhaps more easily answered affirmatively for the natural sciences where laws generally operate in the same way at all locations (at least in this universe), than in the social sciences where such laws are much more difficult to substantiate. Indeed, it remains a debating point within social sciences whether, even if we had the means to measure everything we would like to measure, we would be able to establish globally constant relationships. This leads to the second question posed above about the measurement of relationships. If, as is almost always the case, we are constrained to using models of spatial processes that are in some way a misspecification of reality, do the effects of the misspecification manifest themselves spatially – in most cases through spatial non-stationarity? That is, even if relationships were stationary within social sciences, our ability to record them as stationary is questionable. This is in part because of the paucity of theory in social sciences but also because the subject matter is highly complex. Trying to model the decision-making behaviour of human beings, for example, is a great deal more difficult than modelling the behaviour of atomic particles.

Whatever one's personal beliefs about the predicability of social systems, it seems quite clear that local forms of analysis, such as Geographically Weighted Regression, have their uses. On the one hand, they can be used to demonstrate the degree of local diversity in response to the same stimuli; on the other, they can be used to indicate the degree of misspecification in a global model. In the former case, local analysis leads quite naturally to more detailed, intensive forms of analysis; in the latter case, local analysis leads to improved forms of global models. In both cases, the application of local models generates questions and lines of thought that probably would never otherwise be raised.

10.2 Summarising the Book

Prior to considering, amongst other issues, further research in GWR and related techniques, it is useful to summarise what has been described so far in this book. Following a discussion of the importance of local modelling in Chapter 1, we described in Chapter 2 the mechanics of GWR which was shown to provide a local variation of classic, global regression modelling. We demonstrated, through an extended empirical example of house price determinants in London, how GWR allows spatial variations in relationships to be modelled and mapped. Three extensions to the basic model framework were then described in Chapter 3: (i) the development of mixed GWR estimation techniques in which some parameters in the model are fixed over space while others are allowed to vary locally; (ii) the development of robust methods of GWR calibration in which the influence of local outliers is reduced; and (iii) the construction of spatially heteroskedastic GWR calibration methods which allow for spatial variation in the error variances.

It is quite easy to demonstrate that relationships vary over space in many spatial data sets; it is rather more difficult to demonstrate that such variation is not simply the result of sampling variation. Various types of inferential tests for GWR designed to overcome this issue were described in Chapter 4. These include classical inferential procedures as well as an alternative approach using the Akaike Information Criterion. In Chapter 4 we also derived GWR as a statistical model using the concept of local likelihood.

In Chapters 5 and 6, respectively, we related the problem of spatial non-stationarity to two other major problems in spatial analysis: spatial autocorrelation and spatial scale dependency. In the former we raised the potentially important possibility that a large amount of the observed spatial autocorrelation in error terms from spatial models might result from the application of global models to non-stationary processes. Consequently, a much easier and more intuitive way of dealing with problems of spatially autocorrelated error terms is to apply local modelling techniques such as GWR rather than global spatial regression models. We also demonstrated how information on locally varying levels of spatial autocorrelation can easily be obtained from within a GWR framework. Similarly, in dealing with scale dependency, we examined whether the problems such as the modifiable areal unit problem can be ameliorated by the use of GWR. The logic behind this thinking is that GWR uses a stochastic concept of zone membership rather than the Boolean concept of discrete zone membership.

In Chapters 7 and 8 we turned our attention away from regression and examined the application of geographical weighting to summary statistics and other forms of models. We created a series of local spatial statistics for describing spatial data sets in Chapter 7. We examined in Chapter 8 the applicability of geographical weighting to other established forms of statistical analysis such as Poisson and Binomial modelling, establishing local forms of such procedures. In an important development, we presented a general framework for geographically weighted general linear models that can be used to generate a series of local modelling techniques.

Finally, in Chapter 9 we described software for GWR that is readily available from the authors. The software allows the user to undertake sophisticated and complex forms of GWR within a user-friendly, Windows-based, front-end. An application of the software to data on educational attainment levels in Georgia was presented. This chapter can be used as a users' manual for the software although it does also contain some general statements on the problems of calibrating geographically weighted models.

In the remainder of this chapter we discuss: empirical applications of GWR; other software that is available for GWR calibration; some cautionary notes on GWR; further developments that are planned in this area; and, finally, we speculate on some of the potential impacts of GWR and related techniques within spatial analysis.

10.3 Empirical Applications of GWR

Despite GWR being a relatively new technique, a number of empirical applications have already appeared in the academic literature. For instance, Fotheringham *et al.* (1998) examine spatial variations in the relationships between morbidity and socioeconomic characteristics of areas and find that there are significant spatial variations in the determinants of ill-health across northeastern England. Brunsdon *et al.* (1999) examine the determinants of house prices within the town of Deal in south-eastern England and find that the relationship between house price and size varies significantly across the town. Fotheringham *et al.* (2001) use GWR to investigate spatial non-stationarity in the determinants of school performance across 3 687 schools in northern England, as described partially in Chapter 6.

Other examples of the use of GWR include Páez (2000) and Páez *et al.* (2002a; 2002b) who investigate spatial variations in the determinants of urban temperatures throughout Sendai City in Japan. In a model that relates urban temperature to distance from the city centre and the proportions of various types of land uses, the determinants of urban temperatures are shown to exhibit significant spatial variation across the city. The emphasis of the paper by Páez *et al.* (2002a) is on the development of a method to estimate location-specific bandwidths for GWR directly within the calibration framework for GWR. This procedure contrasts nicely with the way spatially adaptive kernels are derived within the framework of this book through an external *a priori* equation whose parameter optimises a goodness-of-fit function. In Páez *et al.* (2002b), the paper focusses on deriving a GWR version of a general spatial dependency model, in a similar manner to the empirical work described here in Chapter 5.

LeSage (1999a) uses GWR in an examination of changes in gross domestic product (GDP) across Chinese provinces related to various attributes of those provinces. He demonstrates local variations in parameter estimates that lead to a better understanding of regional economic growth processes than would be possible from a simple global analysis of the data. LeSage (1999b; 2001) also describes a Bayesian version of GWR and shows how this can spatially smooth the GWR parameter estimates. He shows that the Bayesian GWR results are very similar to regular GWR results but with the extreme values smoothed. It remains a debating point why one would want to smooth the estimates. LeSage's Bayesian GWR estimates also have smaller standard errors and were therefore more likely to appear as significant.[1] Further discussion of GWR models can be found in Leung *et al.* (2000a, 2000b), Rogerson (2001), Fotheringham *et al.* (2000) and Song (2000).

A different application of GWR is found in Nakaya (2002) who incorporates it into a spatial interaction modelling framework. Spatial interaction models are used to model the flows of people, goods or ideas across space (Fotheringham and O'Kelly 1989). Traditional applications of such models are global although there are numerous examples of the use of origin-specific models being applied (Fotheringham 1981). Origin-specific models are local models in that they are spatial disaggregations of global models and the results of their calibration are specific to each origin. Nakaya takes this one step further and applies GWR to the calibration of each origin-specific model. For each origin, this produces a surface of parameter estimates with any value on this surface describing the relationship being measured *around the destinations close to that location*. Hence, the parameters of the model now become two-dimensional by being both origin-specific and destination-specific. The origin-specific component of the estimates is discrete in that separate surfaces are produced for each origin. The destination component is continuous in that for any origin, a continuous surface of estimates is derived.

Finally, it should be noted that there are parallels between GWR and both kernel regression and Drift Analysis of Regression Parameters (DARP) (Cleveland 1979; Cleveland and Devlin 1988; Casetti 1982). In kernel regression and DARP, $\boldsymbol{y}$ is modelled as a non-linear function of $\boldsymbol{X}$ by weighting data in attribute space rather than geographic space. That is, data points more similar to x_i are weighted more heavily than data points which are less similar and the output is a set of

[1] At the time of writing, LeSage has kindly made available some software for GWR in MATLAB which can be found at http://www.econ.utoledo.edu

localized parameter estimates in x space. However, Casetti and Jones (1983) do provide a limited spatial application of DARP which is very similar in intent to GWR although it lacks a formal calibration mechanism and significance testing framework and so is treated by the authors as a rather limited heuristic method. Nonetheless, this work provides an important landmark in the development of spatially varying parameter models and is very much a forerunner of GWR. An attempt to correct the deficiencies of the Casetti and Jones approach is provided by Casetti and Can (1999) who describe an application of a DARP model to the modelling of per capita state and local government expenditures across the USA. Their approach is to expand the variance of an initial model into a function of distance between regression points and data points. A simple model with two explanatory variables was calibrated by maximum likelihood and the resulting local parameter estimates were mapped. The results indicate seemingly large differences across the USA in the determinants of per capita expenditures between the north-east and the south-west, although still lacking are formal tests of a null hypothesis that the results could have arisen from a stationary process.

In all of these examples, the use of GWR has uncovered strong evidence of spatial non-stationarity which leads to the suspicion that as a geographical phenomenon, spatial non-stationarity is indeed widespread. The use of GWR has also generated insights into the processes being investigated that would have been missed in a traditional global analysis.

10.4 Software Development

For GWR to be used in practical situations, reliable software to carry out the method must be developed. As the basic GWR idea is extended and modified theoretically, new software implementations will need to follow. However, generally available software will inevitably lag behind innovations in the development of methods; there will always be a delay between the development of new techniques and the provision of software that is 'fit for public consumption', in the sense that it can handle large data sets, is reasonably easy to use, and has thorough documentation. Thus, the immediate future of public GWR software is fairly easy to predict – it is likely to consist of more polished versions of current private software being used by those developing and testing extensions to the basic GWR model.

However, in this section an attempt is made to report the more general direction in which GWR software appears to be going, as reported by a number of authors. At the time of writing, there are perhaps two trends worth noting, namely the embedding of GWR into more general, flexible computer packages, and the implementation of more advanced GWR techniques into the software. The second phenomenon is often linked with the first; once GWR is incorporated into a larger package, this package is then used to create new applications and techniques. The following two sections will describe both trends in more detail.

10.4.1 Embedding GWR in Larger Packages

The stand-alone GWR software relies on other software for both pre- and post-processing of data. For example, a typical application may consist of the following steps:

1. Obtain data from a secondary data source. At this stage the raw data may not be in a form understandable by the stand-alone GWR software.
2. Transform and re-format data using a spreadsheet.
3. Feed the output from the previous stage into stand-alone GWR software and produce a set of GWR coefficients.
4. Feed the GWR coefficients into a GIS and produce maps.

Thus, the full analysis requires several software packages in addition to the GWR program. This is a reasonable way to work provided the spreadsheet and GIS are reasonably powerful and user-friendly but it tends to work well mainly when the tasks of data transformation, GWR calibration, and graphical representation are clearly separated. This is not always the case: for example, in an analysis of house prices, one may decide to merge two classes of house type to create a simpler predictor variable and compare maps of GWR surfaces of variables other than those obtained with the original data. One may even decide to map the *differences* between the two surfaces. Furthermore, one may wish to carry out experiments of this sort a number of times, testing different mergers of categories. These operations require a combination of data transformation, GWR computation and graphics routines possibly cycled through a number of times.

An alternative approach is to consider incorporating the GWR function into an existing package capable of the data transformation and graphics operations. Two such packages are R and MATLAB. Both packages offer an ability to process array-based variables in a succinct but easily readable notation. For example in R, if $\mathbf{y}$ is a vector of observations, $y <-y + 1$ adds 1 to each element of $\mathbf{y}$. Similarly, `mean(y)` computes the mean of the elements of $\mathbf{y}$. Brunsdon (2001) and Bivand (2001) both provide examples of incorporating a set of GWR-related functions into R while LeSage (2001) provides a similar facility for MATLAB. The former expands the graphics functionality of R, allowing the drawing of elementary maps of the GWR surfaces. Another major advantage of this approach is that it allows the results of GWR to be linked easily with the wide variety of graphical diagnostics and computational tools already provided in R. Finally, since R is a programming language, it is possible to define new functions automating the use of GWR, for example, iterating through a number of GWR models and systematically producing postscript maps of the results. However, this flexibility does come at a cost: GWR is now called as a command, rather than via a graphical user interface (GUI). This is not unreasonable, since we may now access GWR from R programs, but the GWR function does have a large number of arguments and the user must be reasonably well acquainted with the documentation. Although many of these arguments are set to sensible default values, the GWR-in-R approach is possibly not the best point of departure for the novice – the GUI described in Chapter 9 with its wizard-type series of panels guiding the user through the specification of the procedure is a far better didactic tool.

Another example of embedding is found in recent work by Banos (2001). Here GWR is incorporated into the XLispStat package, another statistical programming language, this one built on the Lisp language and offering powerful linked graphics and interactive exploratory methods such as 'brushing', extending the work of Brunsdon (1998). There are further possibilities for embedding GWR into other packages. One such is to incorporate GWR into a GIS such as ArcView. Although R and MATLAB provide a number of visualisation facilities, they do not have the comprehensive map-making facilities of a GIS, or the tools to manage large and complex spatial datasets. Another alternative is to embed GWR into a spreadsheet. This would provide fewer mapping facilities but may be helpful in scenario modelling where the underlying models exhibit spatial variation.

A final example in which GWR routines have been embedded in a larger package is the development of RIVAM (Regional Information Visualisation and Analysis Machine) which is a suite of programs for the exploratory analysis of spatial data (Song 2000). This package contains a series of routines for local statistics such as the identification of local spatial outliers, the calculation of local Moran's *I* and local Geary's *C* (both measures of spatial autocorrelation) and GWR routines. Although the GWR routines do not involve the calculation of an optimal bandwidth and are rather crude, there is a built-in mapping facility which allows easy viewing of the results.

10.4.2 Software Extending the Basic GWR Idea

As noted above, there is much parallel development between this and the embedding approach described in Section 10.4.1. Since both R and MATLAB are built around programming languages, they provide a good environment for modifying or extending the basic GWR algorithm. For example, by adding a weighting function to the basic GWR routine, the R GWR library provides the function that is at the core of robust GWR and geographically weighted general linear models as described in Chapters 3 and 8, respectively. Similarly, LeSage (2001) recasts GWR into a Bayesian framework in MATLAB and then modifies this model so that the error term has heavier-than-Gaussian tails, yielding a model more robust to outliers.

In both of the above examples, it can be seen that embedding GWR into the packages R and MATLAB facilitates theoretical and practical extensions to the technique. It could be argued that although GWR extensions to MATLAB and R may be released for general public use, the potential audience is somewhat different from that for the basic GUI-based program. The role of the former is to provide a laboratory in which those interested in developing GWR-type methods may experiment. When successful experiments occur, there is another stage where new techniques diffuse into the basic GWR software, to be accessed by a broader public audience.

In the future we envisage several additions to the GWR software described in Chapter 9. These include the following:

1. a suite of programs to calculate a series of geographically weighted descriptive statistics described in Chapter 7;
2. allowing GWR through the origin;
3. the calibration of mixed GWR models;
4. a suite of programs to allow geographically weighted general linear models, as described in Chapter 8, to be calibrated;
5. a full suite of spatial regression models calibrated by GWR;
6. a package allowing various types of local spatial autocorrelation measures to be calculated and local versions of spatial regression models to be calibrated;
7. possibly extending the concept of geographical weighting to other multivariate techniques such as principal components analysis and canonical correlation;
8. possibly extending geographical weighting to handle line and polygon data (currently, polygon data are modelled by collapsing the areal data to a centroid);
9. allowing the user to experiment with a wider choice of kernel functions;
10. allowing the user to experiment with a broader class of bandwidth selection criteria;
11. allowing a greater variety of output format options for mapping and subsequent analysis; and
12. possibly integrating the software directly with mapping functions to avoid having to view the GWR results in a different software package.

10.5 Cautionary Notes

As we have commented in various sections of this book, GWR can be put to at least two very different uses. On the one hand, it can be used as an exploratory tool to pinpoint locations that should be subject to more intensive research. These would be those parts of the study region where interesting and unusual local relationships appear to occur as suggested by the GWR results. On the other hand, GWR can be used as a model-building tool or even as a model diagnostic to check for the presence of spatial non-stationarity. Under this type of usage, the detection of spatial non-stationarity is indicative of a problem with the global model misspecification. The nature of the local variation in measured relationships may hold some clue to the nature of this misspecification. Depending upon one's philosophical viewpoint therefore, it could be argued that it is always useful to subject any spatial model to a GWR-type analysis to check for either real or measured spatial non-stationarity. Clearly what one does if spatial non-stationarity is detected depends on this viewpoint: the non-stationarity could be interpreted as indicative of contextual effects that cannot be modelled or it could be interpreted as a problem with the global model formulation. However, regardless of how one uses GWR (and local statistical methods in general), a few cautionary notes are warranted.

10.5.1 Multiple Hypothesis Testing

If we generate a surface of local t statistics, we should not be surprised if a certain proportion of these exceed some critical value. For example, if we were to generate a surface of local t statistics and use a traditional 95% significance level, we would expect that 5% of the t statistics will generate false positives. To see this more clearly, suppose we were to draw two sets of random numbers and calculate a correlation coefficient between them. Suppose further we repeat this experiment 100 times. Then, by the nature of our significance testing procedure, if we used a 95% significance level, we would expect to find that 5 of the 100 pairs of randomly drawn numbers would yield a significant correlation coefficient. Obviously, such results would not be a product of any real underlying correlation between the two variables; they would merely be an artefact of the significance testing procedure we employed. This is the essence of the multiple hypothesis testing problem.

Sets of local statistics, when subjected to the usual significance testing procedures, suffer from something akin to the multiple hypothesis testing problem because each local statistic is similar to one of the correlation coefficients in the experiment described above.[2] Therefore in any surface of local parameter estimates derived from a large number of model calibrations, we might expect to see some of the local t statistics exceeding the *a priori* defined critical value simply because of the volume of tests we conduct. Caution therefore must be exercised in interpreting such surfaces. In particular, we advise that surfaces of local t statistics be used in a purely exploratory manner as indicating potentially interesting areas for further research. That said, however, a generated surface with a large proportion of locally very diverse parameter estimates would present a more convincing case for spatial non-stationarity than would a surface in which only a small proportion of values exceed some critical value.

There are some partial solutions to the problem of multiple hypothesis testing. One, used in Chapters 6 and 7, is to use Bonferroni adjustments to the critical values to make it more difficult to reject a null hypothesis at any given significance level. The derivation of Bonferroni tests is fairly simple: significance tests of level $(1 - \alpha)$ have a probability of α of not being significant. Assuming n independent tests are carried out, the probability that none is significant is then α^n. The corollary of this is that at least one of the n tests is significant with probability $1 - \alpha^n$. But this is just the significance level for the compound test where we declare a significant result if any one of the individual tests proves positive. If we call the compound significance level α_2 then

$$\alpha_2 = 1 - \alpha^n \tag{10.1}$$

Typically, however, we want to find α from α_2, not the other way round; we would want to know the level at which individual tests should be carried out given a compound level of, say, 5%. We could use

[2] There is one important difference between the two situations, however. In the case of the correlation coefficient experiment, each coefficient is obtained from a random, *independent*, set of observations. The t statistics obtained in a local regression framework are not independent because nearby values will be based on similar sets of data due to the overlapping nature of the spatial kernels used to obtain the estimates.

$$\alpha = (1 - \alpha_2)^{\frac{1}{n}} \tag{10.2}$$

but the Bonferroni approach simplifies equation (10.2) to the first level of the binomial expansion, giving

$$\alpha \cong \left(1 - \frac{\alpha_2}{n}\right) \tag{10.3}$$

which is the usual Bonferroni formula.

Note that the above assumes independent testing. If the tests are dependent (as they will be with GWR results, for example), the derived α level is a conservative test (*inter alia*, Jaccard and Wan, 1996); that is, the significance level of the compound test is *at least* α_2. This raises a number of issues for spatial data:

1. Strictly speaking Bonferroni tests do not provide information on *where* anything is significant in a formal sense; they provide an upper bound for the p-value of the compound test, based on the event that at least one test point is significant. However, we assume, quite reasonably, that the places where the tests are significant are the places of interest.
2. The fact that different α values would be used depending on the value of n is no surprise; in each case we are undertaking a different compound (global) test. The choice of local α is decided in order to fix the conservative α_2 for the global test and as each global test is different, we need different α values in each case.
3. In GWR we have the ability to choose a set of test points (the regression points can be defined independently of the data points). Different choices of test-points will yield global tests with different powers and will make the results somewhat subjective: if we change the number of regression points, how stable are the areas that exceed the critical threshold of the significance test? Two research questions arise from this line of thinking:
   (a) Which arrangement of regression points will give the most powerful test?
   (b) If the ultimate tests are global, then why not just use the Akaike Information Criterion or an F test as tests of global departure from a stationary surface?

The answer to (b) is perhaps that plotting the significant regression points does give a hint of where interesting things occur: AIC and F tests do not allow this. Question (a) is extremely difficult to answer. One observation worth repeating, however, is that the Bonferroni test is conservative for dependent points. The denser the regression points, the more conservative it is.[3] If we were using 10 000 regression points, for example, it is likely that the individual p-values required for significance would be so small that hardly any points would be significant and the power of the test would be nearly zero. Thus, if a Bonferroni adjustment is used, one would

[3] It should be noted that several adjustments to the Bonferroni correction method have been suggested to avoid its excessive conservatism. Some of these are described in Jaccard and Wan (1996); Holm (1979) and Holland and Copenhaver (1988).

probably want to use a relatively small number of points to avoid the conservatism rendering the test virtually useless.

Two possible ways forward might be to attempt to find the 'true' compound significance level rather than the conservative one or, more radically, to move away from the point-based tests altogether and look for an alternative global test that also yields a map, preserving the 'spatial hinting' property. One that springs to mind is to consider the proportion of the total area where the absolute value of the t-surface is greater than some threshold, such as 2, as a test statistic.

One could take this argument further and ask the question, what happens if we select just one location at which to produce local parameter estimates? This would not be equivalent to a global model – the data points would still be geographically weighted around the regression point – but the critical values associated with any statistical test would not need any correction for multiple hypothesis tests (as only one such test is carried out). So, if only one regression point is chosen, we could reach a different conclusion about the underlying process around that location compared to that reached if this one point was part of a set of points that generated a surface of local parameter estimates. Again, the significance test results would depend on the question being asked. Different conclusions might be reached, for example, if the question posed is: 'Is anything interesting happening around this location and this location only?', compared to: 'On this surface of parameter estimates, is there anything interesting happening around this particular location?'

10.5.2 Locally Varying Intercepts

Another note of caution in local regression modelling concerns locally varying intercepts. By allowing the estimated intercept to vary locally, such estimates can absorb some of the explanatory power of the exogenous variables in a model when such variables exhibit spatial clustering. In an extreme case, the locally varying intercept could capture *all* the variation in the independent variable, producing a model with no significant explanatory variables. Such an extreme situation is highly unlikely to exist in reality but vigilance is needed to ensure that local parameter estimates are capturing what is intended. If one suspects that local intercepts are capturing effects that should be incorporated within other locally varying parameter estimates, the solution is to use a mixed GWR model of the sort described in Chapter 3 in which the intercept is not allowed to vary.

10.5.3 Interpretation of Parameter Surfaces

One final cautionary note is a reiteration of a topic raised in Chapter 1 and concerns the interpretation of local models. It is by no means clear whether global statements of human behaviour can be made: there might always be idiosyncratic and contextual effects that prohibit such an action. Consequently, the temptation is great to interpret every set of locally varying parameter estimates in terms of contextual effects. We urge caution on this front and recommend researchers explore the possibility of any obvious model misspecification prior to such an interpretation. The spatial patterns exhibited by local parameter estimates when mapped may provide useful clues as to the nature of the misspecification. Only when such misspecification has been removed should contextual explanations be attempted for any remaining significant spatial variation in local estimates.

10.6 Summary

GWR, as part of a broader research area in local modelling, provides those interested in the analysis of spatial data with new sets of questions. Instead of being restricted to simple global analyses in which interesting local variations in relationships are 'averaged away' and unobservable, GWR allows local relationships to be measured and mapped. In effect, GWR and associated techniques allow the measurement of what Fotheringham *et al.* (1996) refer to as *the geography of parameter space*. In many ways the output from GWR is similar to that presented by a microscope: previously unimagined detail suddenly comes into focus. GWR could be thought of, in fact, as a 'spatial microscope'. The surfaces of parameter estimates from GWR raise new sets of questions about the general issue of parameter variation and the validity of global statements of behaviour and also about the specific nature of the processes being investigated. We suspect that many of the global statements of spatial relationships that have been made in the academic literature await re-examination to see if such statements have validity at the local level.

At another level, local modelling techniques such as GWR raise interesting possibilities about other types of spatial analysis. For example, can we relate issues of spatial dependence and spatial scale to spatial non-stationarity? In Chapter 5 we examine the link between spatial non-stationarity and spatial patterns in residuals. We show how the latter can be reduced considerably by calibrating local models rather than global models. This provides the intriguing possibility that the problems associated with the spatial autocorrelation of error terms in regression models of spatial data might be solvable by local modelling techniques such as GWR. We also show how local estimates of spatial association can be produced within a GWR framework. The latter has the added advantage of being able to produce estimates of the local spatial autocorrelation of one variable conditional on the spatial distribution of other variables; something not possible with univariate measures of local autocorrelation. Similarly, in Chapter 6 we examine the possibility that some problems connected with the sensitivity of spatial analytical results to variations in the spatial scale at which observations are measured might be ameliorated by calibrating local parameter surfaces which appear to be somewhat more robust to scale changes. The investigation of local models thus provides the tantalising possibility of linking the three major problems of spatial analysis: spatial non-stationarity, spatial dependence and spatial scale.

Finally, the issues raised in this book have the potential to provide a bridge between quantitative spatial analysts and qualitative social theorists. Although we do not expect the statistical basis of our arguments to be of great interest to many of our social theory colleagues, the underlying message we are trying to convey should find resonance with their views: *locality is important and measuring local relationships is vital to understanding spatial processes.* Techniques such as GWR should therefore be considered whenever spatial data are analysed, regardless of disciplinary application or philosophical persuasion.

Bibliography

Abramson I S 1982 On bandwidth variation in kernel estimates – a square root law. *Annals of Statistics* **9**: 168–76

Agnew J 1996 Mapping politics: how context counts in electoral geography. *Political Geography* **15**: 129–46

Aitkin M 1997 A general maximum likelihood analysis of overdispersion in generalized linear models. *Statistics and Computing* **6**: 251–62

Akaike H 1973 Information theory and an extension of the maximum likelihood principle. In Petrov B, Csaki F (eds) *2nd Symposium on information theory*. Budapest, Akadémiai Kiadó: 267–81

Akaike H 1981 Likelihood of a model and information criteria. *Journal of Econometrics* **16**: 3–14

Akaike H 1994 Implications of the informational point of view on the development of statistical science. In Bozdogan H (ed.) *Proceedings of the first US/Japan conference on the frontiers of statistical modelling: an informational approach.* Volume 3. Dordrecht, Kluwer Academic

Alker H S 1969 A typology of ecological fallacies. In Dogan M, Rokkan S (eds) *Quantitative ecological analysis*. Boston, MIT Press: 64–86

Altman N S 1990 Kernel smoothing of data with correlated errors. *Journal of the American Statistical Association* **85**: 749–59

Anselin L 1995 Local indicators of spatial association – LISA, *Geographical Analysis* **27**: 93–115

Anselin L 1996 The Moran scatterplot as an ESDA tool to assess local instability in spatial association. In Fischer M M, Scholten H, Unwin D (eds) *Spatial analytical perspectives on GIS*. London, Taylor and Francis

Anselin L 1998 Exploratory spatial data analysis in a geocomputational environment, In Longley P, Brooks A, McDonnell S M, Macmillan B (eds) *Geocomputation: a primer*. Chichester, Wiley

Anselin L 1999 The future of spatial analysis in the social sciences. *Geographic Information Sciences* **5**: 67–76

Appleton D R, French J M, Vanderpump M P 1996 Ignoring a covariate: an example of Simpson's paradox. *American Statistician* **50**: 340–1

Bailey T C, Gatrell A C 1995 *Interactive spatial data analysis.* Harlow, Longman

Banos A 2001 Université de Franche-Comté, personal communication

Bao S, Henry M 1996 Heterogeneity issues in local measurements of spatial association. *Geographical Systems* **3**: 1–13

Barnett W A, Powell J, Tauchen G (eds) 1991 *Nonparametric and semiparametric methods in econometrics and statistics*. New York, Cambridge University Press

Bartlett J 1936 The square root transformation in analysis of variance. *Journal of the Royal Statistical Society* **3**: 68–78

Bavaud F 1998 Models for spatial weights: a systematic look. *Geographical Analysis* **30**: 153–71

Besag J E 1974 Spatial interaction and the statistical analysis of lattice systems. *Journal of the Royal Statistical Society B* **36**: 192–225

Besag J E, 1986 On the statistical analysis of dirty pictures. *Journal of the Royal Statistical Society B* **48**: 259–79

Besag J E, Green P J 1993 Spatial statistics and Bayesian computation. *Journal of the Royal Statistical Society B* **55**: 25–37

Bishop Y M M, Fienberg S E, Holland P W 1975 *Discrete multivariate analysis: theory and practice*. Cambridge, MA, MIT Press

Bivand R 2001 University of Bergen, personal communication

Boots B, Getis A 1988 *Point pattern analysis*. London, Sage

Bowman A W 1984 An alternative method of cross-validation for the smoothing of density estimates. *Biometrika* **71**: 353–60

Bowman A W, Azzalini A 1997 *Applied smoothing techniques for data analysis: the kernel approach with S-plus illustrations*. Oxford, Oxford Science Publications

Brown L A, Goetz A R 1987 Development related contextual effects and individual attributes in Third World migration processes: a Venezuelan example. *Demography* **24**: 497–516

Brown L A, Jones J P III 1985 Spatial variation in migration processes and development: a Costa Rican example of conventional modeling augmented by the expansion method. *Demography* **22**: 327–52

Brown L A, Kodras J E 1987 Migration human resources transfer and development contexts: a logit analysis of Venezuelan data. *Geographical Analysis* **19**: 243–63

Brown P, Marsden J, Batey P, Hirschfield A 1998 Relationships between pupil performance and social conditions: a GIS-based analysis using geodemographics. Paper presented at the conference 'Investigating Locational Data', Lancaster University, 9–10 July

Brunsdon C F 1995 Estimating probability surfaces for geographical points data: an adaptive kernel algorithm. *Computers and Geosciences* **21**: 877–94

Brunsdon C F 1998 Exploratory data analysis and local indicators of spatial association with XLisp-Stat. *The Statistician* **47**: 471–84

Brunsdon C F 2001 *An R library for geographically weighted regression*. Mimeo, Department of Geography, University of Newcastle upon Tyne

Brunsdon C F, Aitkin M, Fotheringham A S, Charlton M E 1999 A comparison of random coefficient modeling and geographically weighted regression for spatially non-stationary regression problems. *Geographical and Environmental Modeling* **3**: 47–62

Brunsdon C F, Charlton M E 1996 Developing an exploratory spatial analysis system in XlispStat. in Parker D (ed.) *Innovations in GIS 3*. London, Taylor and Francis

Brunsdon C F, Fotheringham A S, Charlton M E 1996 Geographically weighted regression: a method for exploring spatial nonstationarity. *Geographical Analysis* **28**: 281–98

Brunsdon C F, Fotheringham A S, Charlton M E 1997 Geographical instability in linear regression modelling – a preliminary investigation. in *New techniques and technologies for statistics II*. Amsterdam, IOS Press: 149–58

Brunsdon C F, Fotheringham A S, Charlton M E 1998 Spatial nonstationarity and autoregressive models. *Environment and Planning A* **30**: 957–73

Brunsdon C F, Fotheringham A S, Charlton M E 1999 Some notes on parametric significance tests for geographically weighted regression. *Journal of Regional Science* **39**: 497–524

Bryk A, Raudenbush S, Seltzer M, *et al.* 1986 *An introduction to HLM: computer program and user's guide*. Chicago: Department of Education, University of Chicago

Burnham K P, Anderson D R 1998 *Model selection and inference: a practical information-theoretic approach*. New York, Springer

Casetti E 1972 Generating models by the expansion method: applications to geographic research. *Geographical Analysis* **4**: 81–91

Casetti E 1982 Drift analysis of regression parameters: an application to the investigation of fertility development relations. *Modeling and Simulation* **13**: 961–6

Casetti E 1997 The expansion method, mathematical modeling, and spatial econometrics. *International Regional Science Review* **20**: 9–32

Casetti E, Can A 1999 The econometric estimation and testing of DARP models. *Geographical Systems* **1**: 91–106

Casetti E, Jones J P III 1983 Regional shifts in the manufacturing productivity response to output growth: sunbelt versus snowbelt. *Urban Geography* **4**: 286–301

Chandler V W, Malek K C 1991 Moving-window Poisson analysis of gravity and magnetic data from the Penokean region east-central Minnesota. *Geophysics* **56**: 123–32

Charlton M E, Fotheringham A S, Brunsdon C 1997 The geography of relationships: an investigation of spatial nonstationarity. in Bocquet-Appel J-P, Courgeau D, Pumain D, (eds) *Spatial analysis of biodemographic data*. Montrouge, John Libbey Eurotext: 23–47

Charnock D 1996 National uniformity and state and local effects on Australian voting: a multilevel approach. *Australian Journal of Political Science* **31**: 51–65

Chatfield C 1995 *Problem solving: a statistician's guide*. London, Chapman & Hall

Cheng Q F, Agterberg P, Bonham-Carter G F 1996 Spatial analysis methods for geochemical anomaly separation. *Journal of Exploration Geochemistry* **56**: 183–95

Cheshire P, Sheppard S 1995 On the price of land and the value of amenities. *Economica* **62**: 247–67

Cleveland W S 1979 Robust locally weighted regression and smoothing scatterplots. *Journal of the American Statistical Association* **74**: 829–36

Cleveland W S, Devlin S J 1988 Locally weighted regression: an approach to regression analysis by local fitting. *Journal of the American Statistical Association* **83**: 596–610

Cliff A D, Ord J K 1972 Testing for spatial autocorrelation among residuals. *Geographical Analysis* **3**: 267–84

Cliff A D, Ord J K 1973 *Spatial autocorrelation*. London, Pion

Cliff A D, Ord J K 1981 *Spatial processes: models and applications*. London, Pion

Conduit E, Brooks R, Bramley G, Fletcher C L 1996 The value of school locations. *British Educational Research Journal* **22**: 199–206

Congdon P 1995 The impact of area context on long-term illness and premature mortality: an illustration of multilevel analysis. *Regional Studies* **29**: 327–44

Coombes M, Raybould S 1997 Modelling the influence of individual and spatial variations in the levels of secondary school examination results. *Environment and Planning A* **29**: 641–58

Coombes M G 1978 *Contiguity: analysis of a concept for spatial studies*. Discussion Paper 15, Newcastle, Centre for Urban and Regional Development Studies, University of Newcastle

Cox K R 1969 The voting decision in a spatial context. *Progress in Geography* **1**: 81–117

Craven P, Wahba G 1979 Smoothing noisy data with spline functions. *Numerische Mathematik* **31**: 377–403

Cressie N 1984 Towards resistant geostatistics. In Verly G M, David A G Journel, Marachel A (eds) *Geostatistics for natural resources characterization* (Part 1) Dordrecht, Reidel: 21–44

Cressman G P 1959 An operational objective analysis system. *Monthly Weather Review* **87**: 367–74

Dacey M F 1960 The spacing of river towns. *Annals of the Association of American Geographers* **50**: 59–61

Dale A, Marsh C 1993 *The 1991 census user's guide*. London, HMSO

Daley R 1991 *Atmospheric data analysis*. New York, Cambridge University Press

Diggle P J 1984 *Statistical analysis of point patterns*. London, Academic Press

Diggle P J, Tawn J A, Moyeed R A 1998 Model-based geostatistics with discussion. *Applied Statistics* **47**: 299–350

Duncan C 1997 Applying mixed multivariate multilevel models in geographical research. In Westert G P, Verhoeff R N (eds) *Places and people: multilevel modeling in geographical research*. Utrecht, Nederlandse Geografische Studies 227 University of Utrecht: 100–17

Duncan C, Jones K 2000 Using multilevel models to model heterogeneity: potential and pitfalls. *Geographical Analysis* **32**: 279–305

Duncan C, Jones K, Moon G 1996 Health-related behaviour in context: a multilevel approach. *Social Science and Medicine* **42**: 817–30

Eldridge J D, Jones J P III 1991 Warped space: a geography of distance decay. *Professional Geographer* **43**: 500–11

Fan J, Gijbels I 1996 *Local polynomial modelling and its applications*. London, Chapman & Hall.

Fleming M M 1999 Growth controls and fragmented suburban development: the effect on land values. *Geographical Information Sciences* **5**: 154–62

Fletcher V 1998 Exam results 'not linked to class size'. *The Times*, 7 December, 4

Flowerdew R 2001 Guest introduction: investigating locational data. *Geographical and Environmental Modelling* **5**: 5–7

Foster S A, Gorr W L 1986 An adaptive filter for estimating spatially varying parameters: application to modeling police hours spent in response to calls for service. *Management Science* **32**: 878–89

Fotheringham A S 1981 Spatial structure and distance-decay parameters. *Annals of the Association of American Geographers* **71**: 425–36

Fotheringham A S 1984 Spatial flows and spatial patterns. *Environment and Planning A* **16**: 529–43

Fotheringham A S 1986 Modelling hierarchical destination choice. *Environment and Planning A* **18**: 401–18

Fotheringham A S 1989 Scale-independent spatial analysis. In Goodchild M, Gopal S (eds) *Accuracy of spatial databases*. London, Taylor & Francis: 221–8

Fotheringham A S 1991 Migration and spatial structure: the development of the competing destinations model. In Stillwell J, Congdon P (eds) *Migration models: macro and micro approaches*, London and New York, Bell-haven: 57–72

Fotheringham A S 1997 Trends in quantitative methods I: stressing the local. *Progress in Human Geography* **21**: 88–96

Fotheringham A S 1999a Trends in quantitative methods III: stressing the visual. *Progress in Human Geography* **23**: 617–26

Fotheringham A S 1999b Guest editorial: local modelling. *Geographical and Environmental Modelling* **3**: 5–7

Fotheringham A S, Brunsdon C 1999 Local forms of spatial analysis. *Geographical Analysis* **31**: 340–58

Fotheringham A S, Brunsdon C, Charlton M E 1998 Geographically weighted regression: a natural evolution of the expansion method for spatial data analysis. *Environment and Planning A* **30**: 1905–27

Fotheringham A S, Brunsdon C, Charlton M 2000 *Quantitative geography: perspectives on spatial analysis*. London, Sage

Fotheringham A S, Charlton M E, Brunsdon C 1996 The geography of parameter space: an investigation into spatial nonstationarity. *International Journal of GIS* **10**: 605–27

Fotheringham A S, Charlton M E, Brunsdon C 1997 Two techniques for exploring non-stationarity in geographical data. *Geographical Systems* **4**: 59–82

Fotheringham A S, Charlton M E, Brunsdon C 2001 Spatial variations in school performance: a local analysis using geographically weighted regression. *Geographical and Environmental Modelling* **5**: 43–66

Fotheringham A S, Curtis A, Densham P J 1995 The zone definition problem and location-allocation modeling. *Geographical Analysis* **27**: 60–77

Fotheringham A S, O'Kelly M E 1989 *Spatial interaction models: formulations and applications*. London, Kluwer

Fotheringham A S, Pitts T C 1995 Directional variation in distance-decay. *Environment and Planning A* **27**: 715–29

Fotheringham A S, Wong D 1991 The modifiable areal unit problem in multivariate statistical analysis. *Environment and Planning A* **23**: 1025–44

Fotheringham A S, Zhan F 1996 A comparison of three exploratory methods for cluster detection in spatial point patterns. *Geographical Analysis* **28**: 200–18

Fox J 2000a *Nonparametric simple regression*. Thousand Oaks, CA, Sage

Fox J 2000b *Multiple and generalized nonparametric regression*. Thousand Oaks, CA, Sage

Friedman J H 1991 Multivariate adaptive regression splines (with discussion). *Annals of Statistics* **19**: 1–141

Gandhi P P, Khassam S A 1991 Design and performance of combination filters for signal restoration. *IEEE Transactions on Signal Processing* **39**: 1524–40

Gangnon R E, Clayton M K 2001 A weighted average likelihood ratio test for spatial clustering of disease. *Statistics in Medicine* **20**: 2977–87

Gehlke C E, Biehl K 1934 Certain effects of grouping upon the size of the correlation coefficient in census tract material. *Journal of the American Statistical Association* Supplement **29**: 169–70

Gelman A J, Stern C. H, Rubin D 1995 *Bayesian data analysis*, London, Chapman & Hall

Getis A, Boots B 1978 *Models of spatial processes: an approach to the study of point line and area patterns*. Cambridge, Cambridge University Press

Getis A, Ord J K 1992 The analysis of spatial association by use of distance statistics. *Geographical Analysis* **24**: 189–206

Goldstein H 1987 *Multilevel models in educational and social research*. London, Oxford University Press

Goldstein H 1994 Multilevel cross-classified models. *Sociological Methods and Research* **22**: 364–75

Goldstein H, Rasbash J, Plewis I, *et al.* 1998 *A user's guide to MlwiN*, London, Institute of Education, University of London

Goodchild M F 1986 *Spatial autocorrelation*. CATMOG 47. Norwich, Geo Books

Goodman A C 1978 Hedonic prices, price indices and housing markets. *Journal of Housing Research* **3**: 25–42

Gorr W L, Olligschlaeger A M 1994 Weighted spatial adaptive filtering: Monte Carlo studies and application to illicit drug market modeling. *Geographical Analysis* **26**: 67–87

Gould P 1975 Acquiring spatial information. *Economic Geography* **51**: 87–99

Green P J, Silverman B W 1994 *Nonparametric regression and generalized linear models: a roughness penalty approach*. London, Chapman & Hall

Greenwood M J, Sweetland D 1972 The determinants of migration between standard metropolitan statistical areas. *Demography* **9**: 665–81

Greig D M 1980 *Optimisation*. London, Longman

Hagerstrand T 1965 A Monte Carlo approach to diffusion. *European Journal of Sociology* **6**: 43–67

Haining R 1979 Statistical tests and process generators for random field models. *Geographical Analysis* **11**: 45–64

Halvorsen R, Pollakowski H O 1981 Choice of functional form for hedonic price equations. *Journal of Urban Economics* **10**: 37–49

Hampel F, Ronchetti E, Rousseeuw P, Stahel W 1986 *Robust statistics: the approach based on influence functions*. New York, Wiley

Hardle W 1990 *Applied nonparametric regression*. New York: Cambridge University Press

Harris R 2001 The diversity of diversity: is there still a place for small area classifications? *Area* **33**: 329–36

Haslett J, Bradley R, Craig P, Unwin A, Wills C 1991 Dynamic graphics for exploring spatial data with applications to locating global and local anomalies. *The American Statistician* **45**: 234–42

Hastie T J, Tibshirani R J 1986 Generalized additive models (with discussion). *Statistical Science* **1**: 297–318

Hastie T J, Tibshirani R J 1990 *Generalized additive models* London, Chapman & Hall

Hastie T J, Tibshirani R J 1993 Varying-coefficient models. *Journal of the Royal Statistical Society (B)* **55**: 757–96

Hauser R M 1970 Context and consex: a cautionary tale. *American Journal of Sociology* **75**: 645–64

Herrmann E, Gasser T, Kneip A 1992 Choice of bandwidth for kernel regression when residuals are correlated. *Biometrika* **79**: 783–95

Hildreth C, Houck J P 1968 Some estimators for a linear model with random coefficients. *Journal of the American Statistical Association* **63**: 584–95

Hjalmars U, Kulldorf M, Gustafsson G, Nagarwalla N 1996 Childhood leukemia in Sweden: using GIS and a spatial scan statistic for cluster detection. *Statistics in Medicine* **15**: 707–15

Hoaglin D C, Welsch R E 1978 The hat matrix in regression and ANOVA. *The American Statistician* **32**: 17–22

Hoerl A E, Kennard R W 1970a Ridge regression: biased estimation for non-orthogonal problems. *Technometrics* **12**: 55–67

Hoerl A E, Kennard R W 1970b Ridge regression: applications to non-orthogonal problems. *Technometrics* **12**: 69–82

Holland B S, Copenhaver M 1988 Improved Bonferroni-type multiple testing procedures. *Psychological Bulletin* **104**: 145–9

Holm S 1979 A simple sequentially rejective multiple test procedure. *Scandinavian Journal of Statistics* **6**: 65–70

Holt D, Steel D G, Tranmer M, Wrigley N 1996 Aggregation and ecological effects in geographically based data. *Geographical Analysis* **28**: 244–61

Hope A C A 1968 A simplified Monte Carlo significance test procedure. *Journal of the Royal Statistical Society Series B* **30**: 582–98

Hordijk L 1974 Spatial correlation in the disturbances of a linear interregional model. *Regional and Urban Economics* **4**: 117–40

Huber P 1981 *Robust statistics*. New York, Wiley

Hurvich C M, Simonoff J S, Tsai C-L 1998 Smoothing parameter selection in nonparametric regression using an improved Akaike information criterion. *Journal of the Royal Statistical Society Series B* **60**: 271–93

Jaccard J, Wan C K 1996 *LISREL approaches to interaction effects in multiple regression*. Thousand Oaks, CA, Sage Publications

Johnson S R, Kau J B 1980 Urban spatial structure: an analysis with a varying coefficient model. *Journal of Urban Economics* **7**: 141–54

Johnston R J 1973 Spatial patterns and influences on voting in multi-candidate elections: the Christchurch City county elections, 1968. *Urban Studies* **10**: 69–81

Jones J P III, Casetti E 1992 *Applications of the expansion method*. London, Routledge

Jones J P III, Hanham R Q 1995 Contingency realism and the expansion method. *Geographical Analysis* **27**: 185–207

Jones K 1991a Specifying and estimating multilevel models for geographical research. *Transactions of The Institute of British Geographers* **16**: 148–59

Jones K 1991b *Multilevel models for geographical research*, Norwich, Environmental Publications

Jones K 1997 Multilevel approaches to modeling contextuality: from nuisance to substance in the analysis of voting behaviour. In Westert G P, Verhoeff R N (eds) *Places and people: multilevel modeling in geographical research*. Utrecht, Nederlandse Geografische Studies 227, University of Utrecht

Jones K, Bullen N J 1993 A multilevel analysis of the variations in domestic property prices: Southern England. *Urban Studies* **30**: 1409–26

Jones K, Gould M I, Watt R 1996 *Multiple contexts as cross-classified models: the Labour vote in the British general election of 1992*. Mimeo, Department of Geography, University of Portsmouth

Kendall S M, Ord J K 1973 *Time series*. Sevenoaks, Edward Arnold

King G 1997 *A solution to the ecological inference problem: reconstructing individual behavior from aggregate data*. Princeton, NJ: Princeton University Press

King L J 1961 A multivariate analysis of the spacing of urban settlements in the United States. *Annals of the Association of American Geographers* **51**: 222–33

Kmenta J 1986 *Elements of econometrics*. 2nd edn. New York, Macmillan

Krige D G 1966 Moving average surfaces for ore evaluation. *Journal of the South African Institute of Mining and Metallurgy* **66**: 13–38

Kullback S, Leibler R 1951 On information and sufficiency. *Annals of Mathematical Statistics* **22**: 79–86

Kulldorf M 1997 A spatial scan statistic. *Communications in Statistics: Theory and Methods* **26**: 1481–96

Kulldorf M, Feuer E, Miller B, Freedman L 1997 Breast cancer in Northeastern United States: a geographical analysis. *American Journal of Epidemiology*, **146**: 161–70

Kulldorf M, Nagarwalla N 1995 Spatial disease clusters: detection and inference. *Statistics in Medicine* **14**: 799–810

Langford I, Leyland A, Rasbash J, Goldstein H 1999 Multilevel modelling of geographical distributions of disease. *Applied Statistics* **48**: 253–68

Leeuw J de, Kreft I 1995 Questioning multilevel models. *Journal of Educational and Behavioral Statistics* **20**: 171–89

LeSage J P 1999a A spatial econometric examination of China's economic growth. *Geographic Information Sciences* **5**: 143–53

LeSage J P 1999b A family of geographically weighted regression models. In Anselin L, Florax J G M (eds) *Recent Developments in Spatial Econometrics*. Berlin, Springer

LeSage J P 2001 *Econometrics toolbox for MATLAB*. URL: http://www.spatial-econometrics.com/

Leung Y, Mei C-L, Zhang W-X 2000a Statistical tests for spatial nonstationarity based on the geographically weighted regression model. *Environment and Planning A* **32**: 9–32

Leung Y, Mei C-L, Zhang W-X 2000b Testing for spatial autocorrelation among the residuals of the geographically weighted regression. *Environment and Planning A* **32**: 871–90

Lilliesand T M, Kiefer R W 1995 *Remote sensing and image interpretation*. New York, Wiley

Lin G, Zeng D 1999 Spatial clusters of diseases: remodeling the concept. *Geographical Information Sciences* **5**: 175–80

Linneman H V 1966 *An econometric study of international trade flows*. Amsterdam, North-Holland

Liu W T, Gauthier C 1990 Thermal forcing on the tropical Pacific from satellite data. *Journal of Geophysical Research* **95**: 13209–17

Loader C 1999 *Local regression and likelihood*. New York, Springer

Loftstgaarden D O, Quesenberry C P 1965 A nonparametric estimate of a multivariate density function. *Annals of Mathematical Statistics* **36**: 1049–51

Maddala G S 1977 *Econometrics*. New York, McGraw-Hill

Majure J, Cressie N 1997 Dynamic graphics for exploring spatial dependence in multivariate spatial data. *Geographical Systems* **4**: 131–58

Mardia K V, Kent J, Bibby J M 1979 *Multivariate analysis*, New York, Academic Press

Martin D 1989 Mapping population data from zone centroid locations. *Transactions of the Institute of British Geographers* **14**: 90–7

McCallum I 1996 The chosen ones? *Education* **187**: 12–13

McLeod A 1985 Remark AS R58: a remark on algorithm AS 183. an efficient and portable pseudo-random number generator. *Applied Statistics* **34**: 198–200

McMillen D P 1996 One hundred fifty years of land values in Chicago: a nonparametric approach. *Journal of Urban Economics* **40**: 100–24

McMillen D P, Thorsnes P 2002 The reaction of house prices to information on superfund sites: a semiparametric analysis of the Tacoma, Washington Market. In Fornby T B, Carter-Hill R (eds) *Advances in econometrics: applying kernel and nonparametric estimation to economic topics*. Stamford, CT, JAI Press

Meen G, Andrew M 1998 *Modelling regional house prices: a review of the literature*. Report prepared for the Department of Environment, Transport and the Regions (now DTLR)

UK, Available from the Centre for Spatial and Real Estate Economics, Department of Economics, University of Reading, UK

Meyer R J, Eagle T C 1982 Context-induced parameter instability in a disaggregate-stochastic model of store choice. *Journal of Marketing Research* **19**: 62–71

Miron J 1984 Spatial autocorrelation. In Gaile G L, Willmott C J (eds) *Regression analysis: a beginner's guide*. Dordrecht, Reidel: 201–22

Mommersteeg H P, Loutre J M, Young M F, Wijmstra R, Hooghiemstra T A, Hooghiemstra H 1995. Orbital forced frequencies in the 975000-year pollen record from Tenagi-Philippon. *Climate Dynamics* **11**: 4–24

Monmonier M 1969 A spatially controlled principal components analysis. *Geographical Analysis* **1**: 192–5

Murray M R, Baker D E 1991 Mwindow – an interactive Fortran-77 program for calculating moving-window statistics. *Computers and Geosciences* **17**: 423–30

Nakaya T 2002 Local spatial interaction modelling based on the geographically weighted regression approach. In Thomas R, Boots B, Okabe A (eds) *Modelling geographical systems: statistical and computational applications*. Dordrecht, Kluwer

Nelder J, Wedderburn R 1972 Generalized linear models. *Journal of the Royal Statistical Society (A)* **135**: 370–84

Nester M 1996 An applied statistician's creed. *Applied Statistics* **45**: 401–10

Oden N 1995 Adjusting Moran's *I* for population density. *Statistics in Medicine* **14**: 17–26

Odland J 1988 *Spatial autocorrelation*, Thousand Oaks, CA, Sage

Openshaw S 1984 *The modifiable areal unit problem*. CATMOG 38, Norwich, Geo-Abstracts

Openshaw S, Charlton M E, Wymer C, Craft A W 1987 A mark I geographical analysis machine for the automated analysis of point data sets. *International Journal of Geographical Information Systems* **1**: 359–77

Openshaw S, Rao L 1995 Algorithms for re-engineering, 1991, census geography, *Environment and Planning A* **27**: 425–46

Openshaw S, Taylor P J 1979 A million or so correlation coefficients: three experiments on the modifiable areal unit problem. In Wrigley N (ed.) *Statistical applications in the spatial sciences*. London, Pion: 127–44

Ord J K 1975 Estimation methods for models of spatial interaction. *Journal of the American Statistical Association* **70**: 120–7

Ord J K, Getis A 1995 Local spatial autocorrelation statistics: distributional issues and an application. *Geographical Analysis* **27**: 286–306

Ord J K, Getis A 2001 Testing for local spatial autocorrelation in the presence of global autocorrelation. *Journal of Regional Science* **41**: 411–32

Pace R K, Barry R 1997 Quick computation of regressions with a spatially autoregressive dependent variable. *Geographical Analysis* **29**: 232–47

Páez A 2000 *Applied statistical analysis of detailed geographical data with emphasis on spatial effects*. PhD thesis, Graduate School of Engineering, Tohoku University, Japan

Páez A, Uchida T, Miyamoto K 2002a *A general framework for estimation and inference of geographically weighted regression models: 1. location-specific kernel bandwidths and a test for spatial nonstationarity*. Laboratory for Regional Planning and Applications, Tohoku University, Japan. Available from authors

Páez A, Uchida T, Miyamoto K 2002b *A general framework for estimation and inference of geographically weighted regression models: 2. spatial association and model specification tests*. Laboratory for Regional Planning and Applications, Tohoku University, Japan. Available from authors

Pattie C, Johnston R 2000 'People who talk together vote together': an exploration of contextual effects in Great Britain. *Annals of the Association of American Geographers* **90**: 41–66

Pavlov A D 2000 Space-varying regression coefficients: a semi-parametric approach applied to real estate markets. *Real Estate Economics* **28**: 249–83

Powe N A, Garrod G D, Willis K G 1995 Valuation of urban amenities using a hedonic price model. *Journal of Property Research* **12**: 137–47

Press W H, Flannery B P, Teukolsky S A, Vetterling W T 1989 *Numerical recipes in Pascal*. Cambridge, Cambridge University Press

Quandt R 1958 The estimation of parameters of a linear regression system obeying two separate regimes. *Journal of the American Statistical Association* **53**: 873–80

Raj B, Ullah A 1981, *Econometrics: a varying coefficients approach*. New York, St Martin's Press

Rao C R 1965 The theory of least squares when the parameters are stochastic and its application to the analysis of growth curves. *Biometrika* **52**: 447–58

Rasbash J, Woodhouse G 1995 *Mln command reference version 1.0*. London, Institute of Education, University of London

Ratdomopurbo A, Poupinet G 1995 Monitoring a temporal change of seismic activity in a volcano. *Geophysical Research Letters* **22**: 775–78

Reijneveld S 1998 The impact of individual and area characteristics on urban socioeconomic differences in health and smoking. *International Journal of Epidemiology* **27**: 33–40

Reynolds R W, Smith T M 1995 A high-resolution global sea surface temperature climatology. *Journal of Climate* **8**: 1571–83

Ripley B 1981 *Spatial statistics*. New York, Wiley

Robinson A H 1956 The necessity of weighting values in correlation analysis of areal data. *Annals of the Association of American Geographers* **46**: 233–6

Robinson W R 1950 Ecological correlation and the behaviour of individuals. *American Sociological Review* **15**: 351–7

Rogerson P A 1999 The detection of clusters using a spatial version of the Chi-square goodness-of-fit test. *Geographical Analysis* **31**: 130–47

Rogerson P A 2001 *Statistical methods for geography*. London, Sage

Rosenberg B 1973 A survey of stochastic parameter regression. *Annals of Economic and Social Measurement* **1**: 381–97

Rosenberg M 2000 The bearing correlogram: a new method of analyzing directional spatial autocorrelation. *Geographical Analysis* **32**: 267–78

Rushton G, Armstrong M P, Lolonis P 1995 Small area student enrollment projections based on a modifiable spatial filter. *Socio-Economic Planning Sciences* **29**: 169–85

Schmid C F, MacCannell E H 1955 Basic problems, techniques and theory of isopleth mapping. *Journal of the American Statistical Association* **50**: 220–39

Schwartz G 1978 Estimating the dimension of a model. *The Annals of Statistics* **6**: 461–4

Sen A 1976 Large sample-size distributions of statistics used in testing for spatial autocorrelation. *Geographical Analysis* **9**: 175–84

Sidak Z 1967 Rectangular confidence regions for the means of multivariate normal distributions. *Journal of the American Statistical Association* **62**: 626–33

Siegmund D O, Worsley K J 1995 Testing for a signal with unknown location and scale in a stationary Gaussian random field. *Annals of Statistics* **23**: 608–39

Silverman B W 1986 *Density estimation for statistics and data analysis*. London: Chapman & Hall

Simpson E H 1951 The interpretation of interaction in contingency tables. *Journal of the Royal Statistical Society B* **13**: 238–41

Smit L 1997 Changing commuter distances in the Netherlands: a macro-micro perspective. In Westert G P, Verhoeff R N (eds) *Places and people: multilevel modeling in geographical research*. Utrecht, Nederlandse Geografische Studies 227, University of Utrecht

Sokal R R, Oden N L, Thomson B A 1998 Local spatial autocorrelation in a biological model. *Geographical Analysis* **30**: 331–54

Song R J 2000 Developing and implementing dynamic geographic visualisation (Gvis) and exploratory spatial data analysis software tools for regional studies: a case study of the People's Republic of China. PhD thesis, The Division of Geography, National Institute of Education, Nanyang Technical University, Singapore

Speckman P 1988 Kernel smoothing in partial linear models. *Journal of the Royal Statistical Society B* **50**: 413–36

Spjotvoll E 1977 Random coefficients regression models: a review. *Mathematische Operationsforschung und Statistik* **8**: 69–93

Staniswalis J G 1987a The kernel estimate of a regression function in likelihood-based models. *Journal of the American Statistical Association* **84**: 276–83

Staniswalis J G 1987b *A weighted likelihood formulation for kernel estimators of a regression function with biomedical applications*. Technical Report 5, Medical College of Virginia, Department of Biostatistics, Virginia Commonwealth University

Steel D G, Holt D 1996 Rules for random aggregation. *Environment and Planning A* **28**: 957–78

Swamy P A V B 1971 *Statistical inference in random coefficient regression models*. Berlin, Springer

Swamy P A V B, Conway R K, LeBlanc M R 1998a The stochastic coefficients approach to econometric modelling, part I: a critique of fixed coefficients models. *Journal of Agricultural Economics Research* **40**: 2–10

Swamy P A V B, Conway R K, LeBlanc M R 1988b The stochastic coefficients approach to econometric modelling, part II: description and motivation. *Journal of Agricultural Economics Research* **40**: 21–30

Swamy P A V B, Conway R K, LeBlanc M R 1989 The stochastic coefficients approach to econometric modelling, part III: estimation, stability testing and prediction. *Journal of Agricultural Economics Research* **41**: 4–20

Tango T 1995 A class of tests for detecting 'general' and 'focused' clustering of rare diseases. *Statistics in Medicine* **14**: 2323–34

Thioulouse J, Chessel D, Champely S 1995 Multivariate analysis of spatial patterns: a unified approach to local and global structures. *Environmental and Ecological Statistics* **2**: 1–14

Thomas E N, Anderson D L 1965 Additional comments on weighting values in correlation analysis of areal data. *Annals of the Association of American Geographers* **55**: 492–505

Thorsnes P, McMillen D 1998 Land value and parcel size: a semiparametric analysis. *Journal of Real Estate Finance and Economics* **17**: 233–44

Thrift N J 1983 On the determination of social action in space and time. *Environment and Planning D: Society and Space* **1**: 23–57

Tibshirani R J, Hastie T J 1987 Local likelihood estimation. *Journal of the American Statistical Association* **82**: 559–67

Tiefelsdorf M 1998 Some practical applications of Moran's *I*'s exact conditional distribution. *Papers of the Regional Science Association* **77**: 101–29

Tiefelsdorf M, Boots B 1997 A note on the extremities of local Moran's I_is and their impact on global Moran's I. *Geographical Analysis* **29**: 248–57

Tierney L 1990 *LISP-STAT: an object oriented environment for statistical computing and dynamic graphics*. Chichester, Wiley

Tinkler K J 1971 Statistical analysis of tectonic patterns in areal volcanism: the Bunyaraguru volcanic field in Western Uganda. *Mathematical Geology* **3**: 335–55

Tobler W R 1989 Frame independent spatial analysis. In Goodchild M, Gopal S (eds) *The accuracy of spatial databases*. London: Taylor and Francis: 115–22

Tomlin C D 1990 *Geographic information systems and cartographic modelling*. Englewood Cliffs, NJ, Prentice-Hall

Tranmer M, Steel D G 1998 Using census data to investigate the causes of the ecological fallacy. *Environment and Planning A* **30**: 817–31

Trigg D W, Leach D H 1968 Exponential smoothing with an adaptive response rate. *Operational Research Quarterly* **18**: 53–9

Tufte E F 1983 *The visual display of quantitative information*. Cheshire, CT, Graphics Press

Tufte E F 1990 *Envisioning information*. Cheshire, CT, Graphics Press

Tukey J W 1977 *Exploratory data analysis*. Reading, MA, Addison-Wesley

Unwin A R, Hawkins G, Hofmann H, Seigl B 1996 Interactive graphics for data sets with missing values – MANET. *Journal of Computational and Graphical Statistics* **5**: 113–22

Unwin A R, Unwin D 1998 Exploratory spatial data analysis with local statistics. *The Statistician* **47**: 415–23

Unwin D J 1981 *Introductory spatial analysis*. London, Methuen
Verheij R A 1997 Physiotherapy utilization: does place matter? In Westert G P, Verhoeff R N (eds), *Places and People: Multilevel Modeling in Geographical Research*, Utrecht, Nederlandse Geografische Studies 227, University of Utrecht: 74–85
Ver Hoef J M, Cressie N 1993 Multivariable spatial prediction. *Mathematical Geology* **25**: 219–40
Visvalingham M 1983 Area-based social indicators: signed Chi-squared as an alternative to ratios. *Social Indicators Research* **13**: 311–29
Voas D, Williamson P 2001 The diversity of diversity: a critique of geodemographic classification. *Area* **33**: 63–76
Wahba G 1990 *Spline models for observational data*. Philadelphia, SIAM
Wand M P, Jones M C 1995 *Kernel smoothing*. London, Chapman & Hall
Wichmann B A, Hill I D 1982 Algorithm AS 183: an efficient and portable pseudo-random number generator. *Applied Statistics* **31**: 188–90
Wichmann B A, Hill I D 1984 Correction: algorithm AS 183: an efficient and portable pseudo-random number generator. *Applied Statistics* **33**: 123
Widrow G, Hoff M E 1960 Adaptive switching circuits. *Institute of Radio Engineers Western Electric Show and Convention Record* part 4: 96–104
Wong D 2001 Location-specific cumulative distribution function (LSCDF): an alternative to spatial correlation analysis. *Geographical Analysis* **33**: 76–93
Yule G U, Kendall M G 1950 *An introduction to statistics*. New York, Hafner Publishing Company

Index